THE INTERNATIONAL SERIES OF MONOGRAPHS ON COMPUTER SCIENCE

General Editors

JOHN E. HOPCROFT   GORDON D. PLOTKIN

JACOB T. SCHWARTZ   DANA S. SCOTT   JEAN VUILLEMIN

THE INTERNATIONAL SERIES OF
MONOGRAPHS ON COMPUTER SCIENCE

1. *The design and analysis of coalesced hashing* Jeffrey S. Vitter
2. *Initial computability, algebraic specifications, and partial algebras* Horst Reichel
3. *Art gallery theorems and algorithms* Joseph O'Rourke
4. *The combinatorics of network reliability* Charles J. Colbourn
5. *Discrete relaxation* Thomas Henderson

Computable Set Theory

Volume 1

DOMENICO CANTONE
Courant Institute
New York University

ALFREDO FERRO
Dipartimento di Matematica
Città Universitaria, Catania, Italy

and

EUGENIO OMODEO
Dipartimento di Matematica e Informatica
Università degli Studi di Udine, Italy

CLARENDON PRESS · OXFORD

1989

Oxford University Press, Walton Street, Oxford OX2 6DP

Oxford New York Toronto
Delhi Bombay Calcutta Madras Karachi
Petaling Jaya Singapore Hong Kong Tokyo
Nairobi Dar es Salaam Cape Town
Melbourne Auckland

and associated companies in
Berlin Ibadan

Oxford is a trademark of Oxford University Press

Published in the United States
by Oxford University Press, New York

British Library Cataloguing in Publication Data
Cantone, A.
Computable set theory.
Vol. 1
1. Set theory. Applications of computer systems
I. Title II. Ferro, A. III. Omodeo, E. IV. Series
511.3'22'0285
ISBN 0-19-853807-3

Library of Congress Cataloging in Publication Data
(applied for)

Printed and bound in
Great Britain by Biddles Ltd,
Guildford and King's Lynn

FOREWORD

Symbolic logic was originally conceived by Leibniz, possibly in analogy with symbolic algebra, as a tool to be used computationally. This same pragmatic aim has been associated with several of the most significant periods of advance of logic during the more than two centuries that have passed since Leibniz's original suggestions: particularly with Boole's systematization of elementary propositional calculus and with the enthusiastic 1920s efforts of Hilbert and his school to demonstrate the full mechanizability of mathematics by solving the predicate decision problem. However, the decisive refutation of Hilbert's attempt by the famous results of Gödel and Church concerning unsolvability and undecidability pushed most later work in logic, up to our own day, in a much less pragmatic direction; in consequence of their brilliant insights, logic took on a largely negative focus, and became principally a tool for demonstrating impossibility results of all kinds.

The rise of the computer in the mid and late 1940s began to revive Leibniz's original emphasis. Propositional calculus, which had been used very successfully for some time as a computational tool for switching circuit design, proved easy to mechanize. It was also clear that the full predicate calculus could also be mechanized (although here a semi-decision procedure was the best that one could hope for). Early expectations that techniques such as resolution would prune predicate proof searches very effectively then focused the energy and enthusiasm of artificial intelligence researchers on the possibility that a wide variety of intelligent actions could be realized by recasting them as predicate-calculus proof searches. Even though this hope has been frustrated by the exponentially large searches that plague the general predicate case even after application of the best available proof-search pruning techniques, the efforts it inspired have left behind a continuing, and steadily growing, stream of more cautiously conceived research, typified by the work on semi-automated proof verifier systems launched by John McCarthy and his collaborators in the 1960s and pursued by various groups up to the present day.

The aim of this verifier-orientated work is to automate simple proofs over a sufficiently wide range of mathematical areas to make computer-checked proof practical, in the sense of no longer requiring manual (steering) inputs which are far more tedious than careful person-to-person (or published) mathematical proof would require. A key to this endeavour is the derivation of meta-theorems to exploit the characteristics of particular mathematical subdomains in ways which mimic the practicing mathematician's intuitive understanding of the most routine aspects of these domains and algorithms which embody these meta-theorems to achieve fully automated treatment of statements belonging strictly to these subdomains. The present book reports on an exceptionally successful continuing series of investigations of one of the most important of these subdomains, namely elementary set theory. The techniques that it describes combine model-theoretic with syntactic considerations in interesting ways, which, as will be seen, grow increasingly challenging and intricate as successively wider sub-languages of set theory are considered.

Work yet to be undertaken will doubtless extend the techniques described in this book to other areas of topology and analysis. These techniques, along with other (entirely different) techniques serving the same general goal, among which the results of Knuth and Bendix, Wu, and Hong are of particular interest, should in time allow mathematicians and computer scientists working together to realize Leibniz's plan of using formulated logic as a practical mechanism for decisive assessment of mathematical discourse.

JACOB T. SCHWARTZ

ACKNOWLEDGEMENTS

Thanks are due to Professor Jacob T. Schwartz (New York University and DARPA), who has given the initial impulse to the research that originated this book and who has conceived the general philosophy of the proof verifier envisaged in Chapter 1. His *eurekas* caused major breakthroughs in many problems (for instance the unionset problem and the maps problem, treated in Chapters 10 and 9).

Professor Martin Davis (New York University) gave big encouragement and inspiration to this research. In particular, he has been the Ph.D. advisor of E. Omodeo.

Doctor Michael Breban (Yeshiva University, New York) contributed to the beginnings of multilevel syllogistic, when he was a Ph.D. student at the New York University.

Professor Franco Parlamento (University of Udine) and a number of Ph.D. students at the New York University (Alberto Policriti, Vincenzo Cutello, Marco Pellegrini) contributed in various ways, including the discovery of new decision techniques, among which the ones described in Sections 5.5 and 5.6.

Susanna Ghelfo, a colleague at ENIDATA, assisted in a SETL implementation of multilevel syllogistic along the lines described in Sections 6.4–6.6, carried out in the framework of the Esprit project no.363. This implementation is currently being brought to perfection by Ron Sigal at the University of Catania.

The Italian companies ENI and ENIDATA have sponsored in many ways, through a project named AXL ('AXiomatic Language') and through a convention with the University of Catania, the preparation of this book—and, of course, much of the research work underlying it. As a matter of fact, E. Omodeo (now at the University of Udine) worked on this book as an employee of ENIDATA, while D. Cantone has enjoyed AXL grants from late 1986 to the present. Some of the researchers mentioned above (namely A. Policriti, M. Pellegrini and R. Sigal) have enjoyed AXL fellowships too.

AXL itself was promoted by Francesco Zambon, responsible of the department of ENIDATA sited in Bologna. We are grateful to him for his involvement with the project, toward which he always demonstrated great enthusiasm.

Expert advise on the use of TeX and LaTeX came from several people, among which Cristina Ruggieri, Daniele Montanari and Gianfranco Valastro. Computer assistance came from the technical staff of the University of Udine too.

Last but not least, we owe much to our wives, whose patience overcomes our merits.

RELATIONSHIPS BETWEEN CHAPTERS

Chapter 1, the Introduction, is addressed to the reader who is already deeply acquainted with set theory. It is, so to say, a firework with sparks thrown in many directions, where formalism and accuracy are slightly sacrificed to the intention of quickly explaining motivations and achievements of our research. Indeed, no proof in this initial chapter has been worked out in detail. Therefore, most readers will find it easier to read the introduction as a sort of appendix to the first part of the book. These readers are nevertheless encouraged to skim through Chapter 1 at the outset.

Generally speaking, Part 1 of the book (formed by Chapters 2 to 7) contains elementary decision methods, while Part 2 (Chapters 8 to 10) contains more complex methods.

Chapters 2, 3, and 4 are introductory to the rest of the book.

Chapter 7 occupies a rather isolated position: its subject matter is somewhat different from what precedes or follows it, but anticipates some of the lines of research that, in our plans, should be deepened in the next volume of this work.

Chapters 5, 6, 8, 9, 10, and 11, are largely independent of one another; however, it is advisable to read them in progression, because they present methods of increasing difficulty. Chapter 7 is probably easier than any of these chapters.

Note: This book results from a strict collaboration among the authors, who therefore share all merits and demerits. Nevertheless, since the book has been largely funded by the Italian companies ENI and ENIDATA, we feel obliged to indicate who, individually, has taken upon himself the preparation of each chapter. Chapters 1, 4, and 6 have been jointly written by Cantone and Omodeo; Chapters 2, 3, and 5, by Omodeo. Chapters 7 and 11 have been written by Ferro; Chapters 8, 9, and 10, by Cantone.

CONTENTS

1. **INTRODUCTION**   **1**
   1.1   Motivation for and general contents of the book   1
   1.2   A proof scenario   3
       1.2.1   Some details on a variant of set theory   4
       1.2.2   Some elementary definitions   8
       1.2.3   The Unique Factorization Theorem   10
   1.3   State of the art of syllogistics   13
       1.3.1   Ground formulae   14
       1.3.2   Satisfiability and validity   15
       1.3.3   Decidability   16
       1.3.4   Known satisfiability decision tests   16
       1.3.5   Examples of syllogisms   21
       1.3.6   An application to program correctness verification   . . .   24
   1.4   A sample satisfiability decision test   25
       1.4.1   Syllogistic diagrams   26
       1.4.2   Computation of the pseudo-ranks   28
       1.4.3   Contradictory diagrams   29
       1.4.4   Dense syllogistic diagrams   29
       1.4.5   The decision method   31
       1.4.6   Syllogistic decomposition   32

Part I   FUNDAMENTALS   **35**

2. **CLASSES AND ORDERINGS**   **37**
   2.1   Introduction   37
   2.2   Basic notation, assumptions, and terminology   37
   2.3   Transitive sets (natural, ordinal, and cardinal numbers). Transfinite induction   42
   2.4   Anti-lexicographic orderings. Concatenation of well-orderings   .   45
   2.5   The von Neumann universe and its ordering   53
   2.6   Well-founded graphs and their realizations. The principle of transfinite recursion   62

3. THE VALIDITY PROBLEM — 77

3.1 Introduction 77
3.2 Language 78
3.3 Semantics 81
3.4 Shorthand notations 83
3.5 Redundancies 88
3.6 Restricted notions of satisfiability and validity 91
3.7 Absolute notions of satisfiability and validity. Decision tests . . 92
3.8 Normalization of formulae 96
3.9 Quantifier elimination 99

4. A PARTIAL SOLUTION TO THE FINITE SATISFIABILITY PROBLEM — 102

4.1 Introduction 102
4.2 Closure properties of superstructures. Safe formulae 104
4.3 König's lemma 107
4.4 n-terms 109
4.5 The characteristic tuple of a superstructure 111
4.6 Normalization of n-terms. Evaluation of safe formulae 117
4.7 First semi-decision procedure 122
4.8 An $\in$-$<$-isomorphism from B_m into $\mathcal{H}$ 123

5. ELEMENTARY SYLLOGISTICS — 131

5.1 Introduction 131
5.2 A technique for realizing well-founded graphs whose nodes are ordered 133
5.3 Application of the reflection lemma to the decision problems of two elementary syllogistics 138
5.4 An improved decision algorithm for an elementary syllogistic involving η 140
5.5 An improved decision algorithm for the elementary syllogistic involving $\subseteq$, $\cap$, and $Finite$ 148
5.6 A collection of safe formulae for which finite satisfiability is decidable 155
5.7 A decision algorithm for an elementary syllogistic involving $\{\bullet, \ldots, \bullet\}$ and η 160

6. MULTILEVEL SYLLOGISTIC — 163

6.1 Introduction 163
6.2 Syllogistic schemes over a family of set variables 165
6.3 Syllogistic schemes as generators of the models of formulae . . 171
6.4 A decision algorithm for MLSF 174

6.5   An improved decision algorithm for MLSF 181
    6.5.1   Determination of $\sim_p$ 181
    6.5.2 p-compatibility test 183
    6.5.3   Generation of a p-$\sim_p$-compatible DAG 183
    6.5.4   Algorithm 186
6.6   Proof of correctness for the improved decision algorithm 188
6.7   How to embed two-level syllogistic into MLS 192
6.8   Venn diagrams and their affinity with syllogistic schemes 194

7. RESTRICTED QUANTIFIERS, ORDINALS, AND ω 197
7.1   Introduction 197
7.2   The elementary syllogistic including only $=, \in, \emptyset$ 198
7.3   A satisfiability algorithm for a class of quantified formulae . . . 200
7.4   A quantified formula expressing infinite sets 204
7.5   An elementary syllogistic including integers and ordinals 205
7.6   A quantified syllogistic including integers and ordinals 208

Part II   EXTENDED MULTILEVEL SYLLOGIS-
TICS   213

8. THE POWERSET OPERATOR   215
8.1   Introduction 215
8.2   Preliminaries 216
8.3   The main result 219
8.4   Proof of the Main Inductive Lemma 236
8.5   Termination proof 247

9. MAP CONSTRUCTS   254
9.1   Introduction and preliminaries 254
9.2   The decision algorithm 259
9.3   Completeness of the decision algorithm 266

10. THE UNIONSET OPERATOR   286
10.1 Preliminaries 286
10.2 The decision algorithm in the absence of trapped places 290
10.3 The decision algorithm when trapped places are present 300

11. THE CHOICE OPERATOR   317
11.1 Introduction 317
    11.1.1 Scenario of a proof exploiting the properties of η 318
11.2 Syllogistic diagrams 319

11.2.1 Dense syllogistic diagrams 322
11.2.2 Sets inducing syllogistic diagrams 324
11.3 Decidability of the theory MLSSFR 326
11.4 Ordered syllogistic diagrams 328
11.4.1 The completeness assumption 329
11.4.2 Sets inducing ordered syllogistic diagrams 330
11.4.3 Consistency of the completeness assumption 331
11.5 Decidability of the theory MLSSFRC 331

REFERENCES   **333**

INDEX OF REFERENCED STATEMENTS   **343**

1. INTRODUCTION

1.1 Motivation for and general contents of the book

It is well-known that any mathematical theory can be embedded in set theory. It hence follows that an automated theorem-prover for set theory can be used as an *all-purpose* theorem prover, or as a useful component of a program that checks the correctness of mathematical proofs. Also, various specialized dialects of set theory—often dealing with finite sets only—have been designed and successfully exploited for high-level specifications of algorithms, and, more generally, for expressing in an unambiguous form problems that one intends to eventually submit to computers. It is therefore legitimate to expect that the automation of deduction methods for set theory will be very helpful in formal verifications of program correctness.

Much of the research that has been carried out in the field of automated deduction beginning in the years around 1960 was devoted to first-order predicate calculus; and since set theory can be regarded as an extension of predicate calculus by means of suitable axioms, it is possible to obtain a theorem prover for set theory by simply adapting a theorem prover originally designed for predicate calculus. It must be noted, however, that even if one resorts to a formulation of set theory which is finitely axiomatizable—e.g. the one due to Bernays and Gödel (see [BLM*86]), with proper classes—any approach to theorem-proving based on predicate calculus will presumably have an utterly unmanageable computational complexity. It follows that it is important to develop set theory as an autonomous calculus, no longer designed within the framework of predicate calculus. As long as one's research on set theory is motivated by foundational interests, the approach that grounds set theory on predicate calculus seems reasonable; it reveals itself as too limiting, however, when one moves towards a prospect of 'deduction engineering'.

Another even more compelling motivation for developing computational set logic independently of predicate calculus comes from the observation that computer-verified proofs based on today's technology still require, even for very elementary mathematical theorems, entry of an excessive mass of tedious detail. Unless this fundamental difficulty can be alleviated very substantially, proof verifiers will remain intriguing toys rather than tools useful to the working mathematician. The same remark applies even more forcefully to the much more pragmatic problem of program verification. Any hope of making such verification practical must presume that existing proof verifiers can be greatly improved, since in attempting to prove anything but very tiny programs correct one will convert them into heavy masses of mathematical statements, all of which must be verified formally if the correctness of the original program is to be established in any rigorously checked sense.

Thanks to its greater expressive power, set theory promises to be much more easily viable than 'naked' predicate calculus in the automation of deductive reasoning. Among others, the power of set theory is shown by its strong attitude for self-reference. Indeed, it has been proved in recent years (see [PP88a]) that a minimal set theory comprising only the extensionality axiom and axioms stating the existence of the empty set and the closure of the class of all sets with respect to the operations 'with' and 'less' (addition and removal of an element), suffices for the representability of syntactic notions and for the diagonalizability of the theory. The ensuing Gödel's incompleteness, as well as the unsolvability of the provability problem, both apply, even for such a modest theory, to formulae of an extremely low syntactic complexity.

Decidability results regarding various portions of set theory have been obtained in the last ten years. The research that has led to such results aims at laying the foundation stones upon which a formal set theory, both neat and orientated towards the needs of automated deduction, can develop. In the envisaged deductive system, the decision algorithms for particular fragments of set theory will act as broad axiom schemes; powerful inference rules will be admitted, in addition to the minimal traditional endowment. It is easy to predict that the development of such a set theory will soon bifurcate in two directions:

- The 'submerged' part of the theory, serving as a support for the deductions that the computer will carry out autonomously, 'behind the scene'.

- The part of the theory that is user-orientated, i.e. some kind of natural deduction system which:

  - on the one hand, will ease the exposition of the proofs produced by the users to the computer programs devoted to checking their correctness;

  - on the other hand, can be used as an explanatory medium for the proofs generated by the computer.

Most of the work that has been carried out so far has been orientated towards singling out a convenient 'submerged calculus'. The book in your hands exclusively reports about this direction of research. For the time being, the problem of proving theorems in first-order set theory is not addressed in full generality. More modestly, a number of situations are classified where an algorithm can detect whether or not a formula is valid. The criteria adopted in this classification are purely syntactical, and rooted on several constructs of set theory, most of which (e.g. intersection, set difference, powerset) will be familiar to everybody, while a few others (e.g. the rank and choice operators) are somewhat more esoteric.

Proving that the validity problem is unsolvable for certain classes of set-theoretic formulae is outside the scope of this treatise. This enables us to work out our proofs in a rather naive set theory, which is formalized only to the little necessary extent. We therefore believe that an advanced background in mathematical logic is not strictly needed for undertaking the reading. On the other hand, some familiarity with programming languages or, at least, with computability theory is welcome.

1.2   A proof scenario

The task of a proof verifier is not to discover proofs (except proofs of the elementary individual steps within larger, more challenging proofs). Indeed, the mathematician or computer scientist using such a verifier will always know the proof of any theorem that they wish the verifier to accept. The problem is that, as known to the system user, this proof ordinarily has far too many missing details, intuitive leaps justified by geometric or other relatively informal insight, steps to be handled by analogy with other arguments given earlier or 'well-known in the literature', etc., for a computerized verifier to follow what is intended. Thus, in designing a proof verifier, one must provide some means for the formalization of this layer of missing detail. The result must be convenient enough for the mathematician to accept it as posing a reasonable rather than an unbearable burden.

The formalism upon which we propose to base a verifier is set theory, which will provide all basic set-theoretic constructs (see Sections 1.2.1 and 1.2.2 and Chapter 3). This formalism is extremely powerful. For example, it makes it possible, starting from the barest set-theoretic rudiments, to define cardinals, integers, rationals, reals, and complex numbers in entirely precise and formal fashion, and to culminate in a full formal *statement* of, say, the Cauchy Integral Theorem, all within the space of no more than 150 lines. This makes it plain that if it were possible to advance through set-theoretic *proofs* at anything close to 1/10th this rate, such a verifier would be a very powerful tool for the mathematician. This, however, is the very core of the problem that must be faced.

That set theory has a great pragmatic advantage as a basic language for mathematical discourse should require little argument beyond the simple remark that it is in fact the common language of mathematical reasoning, used everywhere throughout the vast corpus of number theory, algebra, analysis, geometry, and topology. It is nevertheless worth commenting somewhat more explicitly on the technical advantages of set theory as the basis for a proof verifier—the example discussed later (cf. Section 1.2.3) hints at these advantages too. The ordinary set-former construct of set theory provides a very powerful notation for explicit instantiation of the objects crucial to predicate arguments. Without this mechanism, i.e. in a pure predicate-calculus environment like that provided in the early versions of Stanford FOL [Wey77], such instantiation (well-known to be the crux of predicate reasoning, since by Herbrand's theorem this reduces to purely propositional reasoning once one has pinpointed all necessary elements in the Herbrand universe defined by the operative set of axioms) can be very clumsy. That is, the objects to be instantiated cannot be written in the relatively explicit, 'algebraic' fashion which set theory allows, but instead appear as relatively anonymous variable names whose significance needs to be defined by carrying along extensive sets of characterizing propositions. This makes many useful, relatively direct manipulations of set-former expressions much more roundabout. This is just the opposite of what we think a verifier should do, namely to facilitate set-former-based arguments by supplying an extensive suite of built-in manipulations and decision algorithms which make many equalities, inclusions, disjointness relations, map-related relationships, etc. between set formers easy to deduce.

The main goal of this section is to illustrate the style of man-machine interaction that we envisage for a set-based theorem-prover. This is done by developing the scenario of a proof of an important mathematical proposition. Necessarily, an illustration of this kind must be presented in somewhat formalized terms. This motivates us to produce a dialect of set theory tailored to our current purposes. The latter is outlined in Section 1.2.1, and will not be used in the rest of this book. Indeed, since the major aim of the book is to contribute to the design of a 'heavy-duty' set theory adequate to the needs of automated theorem-proving, we cannot take one such theory for granted now. Improvements to the theory sketched below can arise, among others, from the analysis of a long series of significant proof scenarios such as, for example:

- the proof of the fundamental theorem of Galois theory [Art44];

- the proof of the Cauchy Integral Theorem [Ahl66];

- the proof of Sylow's theorems in group theory [BM67].

In the context of this introductory book, we content ourselves with the scenario of the fundamental theorem of arithmetic, or 'unique factorization' theorem.

1.2.1 Some details on a variant of set theory

Set theory is based on an amazingly small number of primitive constructions, supplemented by a very few mechanisms for definition. Its fundamental components are:

(a) *Its predicate logic foundation.* The operatois $\&, \vee, \leftrightarrow, \rightarrow$, and $\neg$ of propositional calculus are available, also the quantifiers $(\forall x)$ and $(\exists x)$ of predicate calculus, together with all the standard rules for manipulating them.

(b) *Equality, inclusion and membership.* We have the equality relation $x = y$, the inclusion relation $\subseteq$ and the membership relation $x \in y$. Equality has all its standard properties and in addition we have the

axiom: $u = v \leftrightarrow (\forall x)(x \in u \leftrightarrow x \in v)$.

Inclusion is defined by the

axiom: $u \subseteq v \leftrightarrow (\forall x)(x \in u \rightarrow x \in v)$.

(c) *Set-formers.* One is allowed to write set-formers

$$\{e : x_1, \ldots, x_n \ : \ C\},$$

where e is any expression, C any predicate expression, and $x_1, \ldots, x_n$ any list of variables. This syntactic construct designates the set of all values that e can take on as $x_1, \ldots, x_n$ vary over all values satisfying the condition C. Notice therefore that the variables $x_1, \ldots, x_n$ are supposed to be existentially quantified. In such a set-former, the expression e and condition C are essentially arbitrary, except that certain technical restrictions are imposed that prevent formation of 'paradoxical' sets, for example, the set of all sets that are not members of themselves. The technical restrictions that we impose on set-formers to eliminate paradoxical sets are as follows. Let a set-former have the form

$$\{e : x_1, \ldots, x_n \ : \ C\}, \tag{1.1}$$

so that $x_1, \ldots, x_n$ are its bound variables. Then C must be *safe* with respect to the bound variables $x_1, \ldots, x_n$ according to the definition given below.

Definition 1.1 *(I) A predicate expression P is* sufficiently restrictive *with respect to the collection $V = \{x_1, x_2, \ldots, x_n\}$ of variables, if, for some variable x_i in V, P has one of the following forms:*

$$x_i \in e^*, \ x_i \subseteq e^*, x_i \in e^* \ \& \ P', \ x_i \subseteq e^* \ \& \ P',$$

where e^* is any expression not involving any of the bound variables $x_1, x_2, \ldots,$ x_n, and where P' is sufficiently restrictive with respect to the collection of variables $V \setminus \{x_i\}$.

(II) A predicate expression C is said to be safe *with respect to the variables $x_1, x_2, \ldots, x_n$ if*

- *C has the form C_0 or the form $C_0 \ \& \ C_1$, with C_0 sufficiently restrictive with respect to the variables $x_1, x_2, \ldots, x_n$, and with C_1 any predicate expression.*

- C *is any disjunction of predicate expressions that are safe with respect to the set of variables* $x_1, x_2, \ldots, x_n$.

- C *is logically equivalent to a predicate expression which can be shown to be safe with respect to* $x_1, x_2, \ldots, x_n$ *on the basis of the preceding two clauses.*   □

If C_0 is a simple conjunction of clauses having the form

$$x_1 \otimes_1 e_1 \,\&\, x_2 \otimes_2 e_2(x_1) \,\&\, \ldots \,\&\, x_n \otimes_n e_n(x_1, \ldots, x_{n-1}),$$

where each $\otimes_j$ is either $\in$ or $\subseteq$, e_j depends only on $x_1, \ldots, x_{j-1}$ (along with variables that are not bound in (1.1)), then we may abbreviate (1.1) into

$$\begin{aligned} \{e \ : \ \ & x_1 \otimes_1 e_1, \ x_2 \otimes_2 e_2(x_1), \ldots, \ x_n \otimes_n e_n(x_1, \ldots, x_{n-1}) \\ : \ \ & C_1(x_1, \ldots, x_n)\} \end{aligned} \qquad (1.2)$$

(or just

$$\{e : x_1 \otimes_1 e_1, \ x_2 \otimes_2 e_2(x_1), \ldots, \ x_n \otimes_n e_n(x_1, \ldots, x_{n-1})\},$$

if the component C_1 of C is missing).

If e is simply x_1, then 'e :' can be omitted in (1.2).

Some examples illustrating the use of set-formers are:

- the singleton $\{x\}$ can be written as $\{y : y : y \subseteq x \,\&\, y = x\}$, and then as $\{y : y : y = x\}$;

- the union $Un(x)$ as $\{z : y, z : y \in x \,\&\, z \in y\}$;

- the powerset $Pow(x)$ as $\{y : y : y \subseteq x\}$;

- the unordered pair $\{x, y\}$ as

$$\{x, y\} = \{z : z : z \in \{x\} \ \lor \ z \in \{y\}\};$$

- and the union $x \cup y$ as

$$x \cup y = \{z : z : z \in x \lor z \in y\}.$$

(d) *Definition of new functions and predicates.* If f is any function (respectively, predicate) symbol that has never been used before, and e is any expression (respectively, predicate) whose only free variables are $x_1, \ldots, x_n$, the rules of logic allow us to introduce a new equality

$$f(x_1, \ldots, x_n) = e,$$

as a definition. For emphasis, we will generally write such definitions as

$$f(x_1, \ldots, x_n) =_{\text{Def}} e.$$

It is also convenient to allow functions introduced in this way to be written as infix or prefix operators, or to be indicated by use of other convenient syntactic forms, for example, special brackets; this is strictly a syntactic matter.

(e) *Definition of a function that selects some element with a given defining property.* For any predicate $P(x_1, \ldots, x_n, y)$ whose only free variables are $x_1, \ldots, x_n$, y the rules of logic allow introduction of a new function symbol f of n variables and a defining statement

$$(\forall x_1, \ldots, x_n)(P(x_1, \ldots, x_n, f(x_1, \ldots, x_n)) \ \leftrightarrow \ (\exists y)P(x_1, \ldots, x_n, y)).$$

(f) *Choice operator and foundation.* A special 'choice' operator η which selects a (fixed) element from each nonnull set (or class) is available, together with the

axiom: $(s \cap \eta s = \emptyset \ \& \ \eta s \in s) \lor (s = \emptyset \ \& \ \eta s = \emptyset).$

(Here $\emptyset$ denotes the null set.)

(g) *Recursive definitions ('transfinite' recursion).* Under appropriate technical restrictions, definitions are allowed to be recursive, that is, the function symbol appearing on the left of a definition can also appear on the right. Specifically, let a definition have the form

$$f(x_1, \ldots, x_n) =_{\text{Def}} r. \tag{1.3}$$

Then there must exist some other, already defined, function $a(x_1, \ldots, x_n)$, called the *auxiliary function* of the definition (1.3), such that every occurrence of f within r is part of a sub-expression

$$\text{if } a(e_1, \ldots, e_n) \notin a(x_1, \ldots, x_n)$$
$$\text{then } \emptyset$$
$$\text{else } f(e_1, \ldots, e_n).$$

Similarly, for recursive predicate definitions

$$P(x_1, \ldots, x_n) =_{\text{Def}} r,$$

one must insist that every occurrence of P within r be part of a predicate subterm

$$a(e_1, \ldots, e_n) \in a(x_1, \ldots, x_n) \to P(e_1, \ldots, e_n).$$

It should be understood that only the initial form of a recursive definition is subject to these rules. This initial form can then be rewritten in any equivalent form.

(h) *Infinite set.* The final necessary assumption that there exists at least one infinite set that can be formulated (for an explanation, see end of Section 10.1) as the

axiom: $(\exists u)(u \subseteq Un(u) \ \& \ u \neq \emptyset)$ (axiom of infinity).

(Here *Un* designates the general set union operation—see above).

1.2.2   Some elementary definitions

In this subsection we emphasize the very great definitional power of set theory by giving all the definitions which will be subsequently used to prove that each natural number greater than 1 has a unique prime factorization, up to reorderings.

As von Neumann showed in the 1920s (cf. [vN77b, vN77a]), one can leap very directly from basic set theory to the foundations of arithmetic (and then, along standard lines, to rationals, reals, and real analysis too). One might, for instance, proceed according to the following path:

$\{u\} =_{\text{Def}} \{y : y = u\}$	(singleton)
$\{u, v\} =_{\text{Def}} \{y : y = u \lor y = v\}$	(pair)
$\emptyset =_{\text{Def}} \{y : y \neq y\}$	(null set)
$u \cup v =_{\text{Def}} \{y : y \in u \lor y \in v\}$	(union)
$u \cap v =_{\text{Def}} \{y : y \in u \ \& \ y \in v\}$	(intersection)
$u \setminus v =_{\text{Def}} \{y : y \in u \ \& \ y \notin v\}$	(difference)
$u \subseteq v =_{\text{Def}} u \setminus v = \emptyset$	(inclusion)
$0 =_{\text{Def}} \emptyset, \ 1 =_{\text{Def}} \{0\}, \ 2 =_{\text{Def}} \{0, 1\}$	(zero, one, and two)
$[x, y] =_{\text{Def}} \{\{0, \{x\}\}, \{2, \{y\}\}\}$	(ordered pair)
$\underline{hd} \, u =_{\text{Def}} \eta\eta\{x \setminus \{0\} : x : x \in u \ \& \ 0 \in x\}$	(first component of ordered pair)
$\underline{t\ell} \, u =_{\text{Def}} \eta\eta\{x \setminus \{2\} : x : x \in u \ \& \ 2 \in x\}$	(second component of ordered pair)
$Pow(u) =_{\text{Def}} \{y : y \subseteq u\}$	(powerset)
$f\{u\} =_{\text{Def}} \{y : [u, y] \in f\}$	(multivalued map application)
$f(x) =_{\text{Def}} \eta f\{x\}$	(image element)
$range \ f =_{\text{Def}} \{y : x, y : [x, y] \in f\}$	(range)
$domain \ f =_{\text{Def}} \{x : x, y : [x, y] \in f\}$	(domain)
$f_{|s} =_{\text{Def}} \{[x, y] : [x, y] \in f \ \& \ x \in s\}$	(restriction of map to set)
$f[s] =_{\text{Def}} range(f_{|s})$	(multi-image of map on set)
$f^{-1} =_{\text{Def}} \{[y, x] : x, y : [x, y] \in f\}$	(inverse mapping)
$single\text{-}valued(f) =_{\text{Def}} (\forall x \in domain \ f)$	
$\quad (f\{x\} = \{f(x)\})$	(single-valued map)
$one\_one(f) =_{\text{Def}} single\text{-}valued(f)$	
$\quad \& \ single\text{-}valued(f^{-1})$	(one-to-one map)
$s \times t =_{\text{Def}} \{[x, y] : x \in s \ \& \ y \in t\}$	(cartesian product)

$smaps(s,t) =_{\text{Def}} \{f : f \subseteq s{\times}t$
  $\&\ domain\ f = s\ \&\ single\text{-}valued(f)\}$ (set of all functions from s to t)
$injections(s,t) =_{\text{Def}} \{f \in smaps(s,t)\ :$
  $one\_one(f)\}$ (set of all injections from s to t)
$bijections(s,t) =_{\text{Def}} \{f \in injections(s,t) :$
  $f^{-1} \in injections(t,s)\}$ (set of all bijections from s onto t)
$Un(s) =_{\text{Def}} \{x : x,y : x \in y\ \&\ y \in s\}$ (union set of s, i.e. union of all the individual elements of s).

Von Neumann's trick for mapping arithmetic into set theory is to encode the integers recursively by sets, putting $0 = \emptyset$, $1 = \{0\}$, $2 = \{0,1\}$, $3 = \{0,1,2\}$, etc.. This is the intuitive basis for his famous one-line definition of the class of all finite and infinite ordinal numbers:

$$is\_an\_ordinal(s) =_{\text{Def}} s \supseteq Un(s)\ \&\ (\forall x \in s)(\forall y \in s)(x \in y \vee y \in x \vee x = y).$$

One more line defines the cardinality number of any set s, that is, the first ordinal number that is in one-to-one correspondence with s:

$$|s|\ =_{\text{Def}}\ \eta\{x : is\_an\_ordinal(x)\ \&\ (\exists f \in smaps(x,s))(range\ f = s)\}$$
$$\text{(cardinality).}$$

The sum and product of numbers are then defined in just two more lines by

$$s * t\ =_{\text{Def}}\ |s \times t| \qquad \text{(cardinal product)}$$
$$s + t\ =_{\text{Def}}\ |(\{0\} \times s) \cup (\{1\} \times t)| \qquad \text{(cardinal sum)}.$$

Moreover, integer division can be defined by

$$s/t =_{\text{Def}} \eta\{u \in s+1 : t * u = s\}.$$

Finally, if u_0 is any set having the property stated in the axiom of infinity, then we can put

$$\omega =_{\text{Def}} \eta\{x : x \in |Pow(u_0)|\ \&\ x = Un(x)\ \&\ x \neq \emptyset\},$$

thus defining the set of all nonnegative integers. Also we put

$$\omega^+ =_{\text{Def}} \omega \setminus \{0\}$$

to denote the set of all positive integers.

Next we introduce signed integers with addition and multiplication and define the embedding of nonnegative integers into signed integers.

$eq\_class(m,n) =_{\text{Def}} \{[m_0,n_0] : m_0 \in \omega, n_0 \in \omega : m + n_0 = n + m_0\}$
$\mathbf{Z} =_{\text{Def}} \{eq\_class(m,n) : m \in \omega, n \in \omega\}$ (signed integers)
$p_1 +_Z p_2 =_{\text{Def}} eq\_class(\underline{hd}\,\eta p_1 + \underline{hd}\,\eta p_2, \underline{t\ell}\,\eta p_1 + \underline{t\ell}\,\eta p_2)$
  (addition of signed integers)
$p_1 *_Z p_2 =_{\text{Def}} eq\_class(\underline{hd}\,\eta p_1 * \underline{hd}\,\eta p_2 + \underline{t\ell}\,\eta p_1 * \underline{t\ell}\,\eta p_2,$
  $\underline{hd}\,\eta p_1 * \underline{t\ell}\,\eta p_2 + \underline{t\ell}\,\eta p_1 * \underline{hd}\,\eta p_2)$ (multiplication of signed integers)
$Imbed(n) =_{\text{Def}} eq\_class([n,0])$ (imbedding of ω in $\mathbf{Z}$).

Having thus introduced the very basic definitions of elementary arithmetic, we next show how the unique factorization theorem could be proved in a set-theoretic environment.

1.2.3   The Unique Factorization Theorem

We will proceed roughly along lines which we would expect to be suitable for a verifier based on the set-theoretic formalism. Though an entire level of detail is missing from this sketch, it should give some idea (and flavour) of the type of proof we expect such a verifier to accept. The line of argument is of course quite standard.

We begin with some definitions.

$$factors(n) =_{\text{Def}} \{m : m \in n, k \in n : m * k = n\},$$
$$factors_1(n) =_{\text{Def}} factors(n) \cup \{n\},$$
$$primes =_{\text{Def}} \{m : m \in \omega \setminus \{0, 1\} : factors(m) = \emptyset\},$$
$$largest\_factor(n) =_{\text{Def}} Un(factors(n)),$$
$$smallest\_factor(n) =_{\text{Def}} n/largest\_factor(n).$$

Then the following lemma holds.

Lemma 1.1 $(n \in \omega \ \& \ 1 \in n \ \& \ factors(n) \neq \emptyset) \rightarrow$
$(n = largest\_factor(n) * smallest\_factor(n) \ \& \ largest\_factor(n) \in n)$
$\& \ smallest\_factor(n) \cap factors(n) = \emptyset.$ $\qquad\qquad \square$

Next, we make the following recursive definitions.

$$factorization(n) =_{\text{Def}} \text{if } factors(n) = \emptyset$$
$$\text{then } \{[0, n]\}$$
$$\text{else } factorization(largest\_factor(n))$$
$$\cup\{[|factorization(largest\_factor(n))|,$$
$$smallest\_factor(n)]\},$$

$$last\_component(f) =_{\text{Def}} \eta\{x \in f : (\underline{hd}\,x) + 1 = |f|\},$$

$$\textstyle\prod f =_{\text{Def}} \text{if } f = \emptyset \text{ then } 1 \text{ else}$$
$$(\underline{tl}\ last\_component(f)) * \textstyle\prod(f \setminus \{last\_component(f)\}),$$

where we recall that $\underline{hd}\,p$ and $\underline{tl}\,p$ designate the first and second component of the pair p, and η is a selection operator. One easily recognizes that $\prod f \neq 0$ if and only if $domain\ f = |f|$, so that $\prod f \neq 0$ implies $f \in smaps(|f|, \omega^+)$, for all $f \subseteq \omega \times \omega^+$ with $|f| \in \omega$. Also, assuming $\prod f \neq 0$, $|f| \in \omega \setminus \{0, 1\}$, one has $range\ f \subseteq factors(\prod f)$.

We have:

Theorem 1.1 $n \in \omega \ \& \ 1 \in n \rightarrow$

$$\left(n = \prod factorization(n) \ \& \ range \ factorization(n) \subseteq primes\right).$$

Proof: If not, there is a least falsifying n_0. (This is a general inductive principle, to be built into the verifier.) Hence $n_0 \in \omega \ \& \ 1 \in n_0 \ \& \ (n_0 \neq \prod factorization(n_0) \lor range \ factorization(n_0) \not\subseteq primes)$. Then two cases are considered:

Case $factors(n_0) = \emptyset$. Then $factorization(n_0) = \{[0, n_0]\}$—where $n_0 \in primes$—by definition, and therefore

$$\prod factorization(n_0) = n_0, \quad range \ factorization(n_0) = \{n_0\} \subseteq primes,$$

contradicting our initial assumption.

Case $factors(n_0) \neq \emptyset$. Putting

$$\ell = largest\_factor(n_0), \quad p = smallest\_factor(n_0),$$

we obtain from Lemma 1.1 that $p \in primes$, since otherwise

$$p = largest\_factor(p) * smallest\_factor(p),$$

so $largest\_factor(p) \in factors(n_0)$, contradicting $largest\_factor(p) \in p$. Therefore $\ell \in n_0 \setminus \{0, 1\} \subseteq \omega \setminus \{0, 1\}$, which, by the minimality of n_0, implies

$$\ell = \prod factorization(\ell), \quad range \ factorization(\ell) \subseteq primes.$$

Since

$$n_0 = p * \ell = p * \prod factorization(\ell),$$

it follows from the definition of $factorization(n_0)$ that $n_0 = \prod factorization(n_0)$ and

$$range \ factorization(n_0) = range \ factorization(\ell) \cup \{p\} \subseteq primes,$$

contradicting our initial assumption. ∎

Another definition which is needed at this point is that of the greatest common divisor

$$gcd(p, q) \quad =_{\text{Def}} \quad Un(\{m : m \in \omega, n \in \omega, \ell \in \omega : p = m * n \ \& \ q = m * \ell\})$$
$$\text{(greatest common divisor)}.$$

Lemma 1.2 $p \in \omega \ \& \ q \in \omega \ \& \ 0 \in p \ \& \ 0 \in q \rightarrow$
$(\exists m \in \mathbb{Z})(\exists n \in \mathbb{Z})(Imbed(gcd(p, q)) = m *_Z Imbed(p) +_Z n *_Z Imbed(q)).$ □

This lemma, due to Euclid, is proved inductively, i.e. by considering the smallest falsifying p, q and deriving a contradiction. (Details omitted.) We use it to introduce two auxiliary functions $a_1(p, q)$ and $a_2(p, q)$ satisfying

$$Imbed(gcd(p, q)) = a_1(p, q) *_Z Imbed(p) +_Z a_2(p, q) *_Z Imbed(q).$$

The integers modulo p are introduced by

$$\mathbf{Z}_p =_{\text{Def}} \{\{n +_Z p *_Z k : k \in \mathbf{Z}\} : n \in \mathbf{Z}\},$$

and the natural homomorphism of $\mathbf{Z}$ into $\mathbf{Z}_p$ by

$$\varphi_p =_{\text{Def}} \{[n, \{n +_Z p *_Z k : k \in \mathbf{Z}\}] : n \in \mathbf{Z}\}.$$

The natural multiplication in $\mathbf{Z}_p$ is given by

$$a *_p b =_{\text{Def}} \varphi_p(\eta a *_Z \eta b),$$

and easily proved to satisfy

$$a *_p b = b *_p a, (a *_p b) *_p c = a *_p (b *_p c), \varphi_p(m *_Z n) = \varphi_p(m) *_p \varphi_p(n).$$

We also have the

Lemma 1.3 *(a)* $p \in \omega^+$ & $q \in \omega^+$ & $factors_1(p) \cap factors_1(q) \subseteq \{1\} \rightarrow gcd(p, q) = 1$, *so*
  (b) $p \in \omega^+$ & $q \in \omega^+$ & $factors(p) = \emptyset$ & $p \notin factors_1(q) \rightarrow \varphi_p(a_2(p, q)) *_p \varphi_p(Imbed(q)) = \varphi_p(Imbed(1))$.   □

It follows by induction that

Lemma 1.4 *(a)* $p \in primes$ & $f \in smaps(n, \omega^+)$ & $\{x : x \in range\ f : p \in factors_1(x)\} = \emptyset \rightarrow \varphi_p(Imbed(\prod f)) *_p \varphi_p(Imbed(\prod\{[m, \varphi_p(a_2(p, f(m)))] : m \in n\})) = \varphi_p(Imbed(1))$;
  (b) $p \in primes$ & $f \in smaps(n, \omega^+)$ & $p \in factors_1(\prod f) \rightarrow \{x : x \in range\ f : p \in factors_1(x)\} \neq \emptyset$.   □

Now we can let $n \in \omega$ & $1 \in n$, and let f be any non-increasing monotone sequence of natural numbers such that $range\ f \subseteq primes$ and $\prod f = n$. We need to prove the

Theorem 1.2 $f = factorization(n)$.

Proof: If not, there is a smallest falsifying n_0. If $factors(n_0) = \emptyset$ then f must be a singleton, hence $f = \{[0, f(0)]\}$, so $n_0 = \prod f = f(0)$, and thus $f = \{[0, n_0]\} = factorization(n_0)$, where $n_0 \in primes$, contradicting our initial assumption. So, suppose $factors(n_0) \neq \emptyset$. Let $\ell = largest\_factor(n_0)$ and $p = smallest\_factor(n_0)$, so that $p \in primes$, as already observed in the proof of Theorem 1.1. Therefore by the preceding lemma $p \in \textit{range } f$, and since f is monotone $f(|f \setminus \{\eta f\}|) = p$. Thus

$$\prod f = p * \prod (f \setminus \{ \textit{last\_component}(f) \}) \,,$$

whence, as $\prod f = n_0 = p * \ell$ by hypothesis, and $p \neq 0$, we obtain

$$\prod (f \setminus \{ \textit{last\_component}(f) \}) = \ell \in n_0 \setminus \{ 01 \} \,.$$

It follows by the minimality of n_0 that

$$f \setminus \{ \textit{last\_component}(f) \} = \textit{factorization}(\ell) \,,$$

and therefore—recalling that $\textit{last\_component}(f) = [|f \setminus \{\eta f\}|, p]$—

$$f = \textit{factorization}(n_0) \,,$$

again contradicting our initial assumption. ∎

1.3 State of the art of syllogistics

This book exposes a number of computational techniques for deciding formulae in set theory; this is to say, methods for accepting or rejecting, on a sound basis, any given alleged theorem. We are interested in methods that could be automated, at least in principle; it is hence obvious that none of our methods will be applicable to full set theory. It is in fact known that no variant of set theory that deserves this name is decidable (see [PP88a]). We do not treat the problem of semi-deciding collections of formulae, except *en passant* in Chapter 4. We will not analyse the complexity of any of our decision problems either; as a matter of fact, most of our methods have an exponential or, even worse, hyper-exponential computational behaviour, and very little is known today about the intrinsic difficulty of the problems that we address (an exploratory paper in this direction is [COP89], which shows that multilevel syllogistic—see Section 1.3.4—is NP-complete [AHU76]).

This section presents a detailed taxonomy of the formulae whose validity can be established or refuted by state-of-the-art procedures. Many—but not all—of the procedures that are mentioned below will be described later in the book. Most of the decidable collections of formulae are unquantified, or *ground*, as we call them. It is our expectation that some of these classes of formulae, or similar ones, will demonstrate, in a set calculus of the future, an importance at least equal to that of propositional formulae in predicate calculus. In this perspective, we widen the meaning of a word that occurs in classical systems of logic somewhat simpler than ours, namely the word *syllogism*, to refer to the 'tautologies' of this future calculus to be.

1.3.1   Ground formulae

We outline here the language of ground set theory (see Chapter 3 for more details), and the intended interpretation of its formulae.

Variables, assumed to range over the class of all sets, are available. As is customary in set theory, we assume that sets can be nested into an arbitrary, even transfinite, degree of complexity; moreover a set can have finitely or infinitely many elements. Membership is assumed to be well-founded, i.e. it is assumed that no chain $\cdots \in x_n \in \cdots \in x_1 \in x_0$ extends indefinitely to the left. Exploiting the well-known von Neumann hierarchy (see Section 2.5), one can associate with each set x an ordinal number $rk\,x$ which is called the *rank* of the set. A very compact recursive definition of rank is:

$$rk\,x = \bigcup_{y \in x} \left((rk\,y) \cup \{rk\,y\} \right),$$

which is assumed to hold for every set x.

Let us now briefly introduce *terms*: these are formal expressions, usually involving set variables. Every term T denotes a set, when the variables appearing in T are replaced by sets. The *simple terms* are: $\emptyset$ (denoting the empty set), and all set variables. A *compound term* is obtained by combining together two terms by means of one of the binary operators $\cap$, $\setminus$, $\cup$ (denoting intersection, difference and union of two sets, respectively). Moreover, the expression $\{T_0, \ldots, T_n\}$ is a compound term whenever $T_0, \ldots, T_n$ are terms. Finally, $Un\,T$, $Pow\,T$ and $\eta\,T$ are compound terms whenever T is a term.

Un denotes the unionset operation, i.e. $Un\,T$ stands for the set

$$\{\, x \,:\, x \text{ belongs to some member of the set denoted by } T \,\}.$$

Pow denotes the powerset operation, i.e. $Pow\,T$ stands for the set

$$\{\, x \,:\, x \text{ is a subset of the set denoted by } T \,\}.$$

The semantics of η, the *choice operator* (see Sections 2.5 and 3.3), is as follows. We define a set x to be q *times infinite* if q is the first natural number such that there exists a chain $x_q \in x_{q-1} \in \cdots \in x_1 \in x_0$ with $x_0 = x$ and x_q endowed with finitely many members. Then we assume that an ordering $<$, extending the usual order of natural and ordinal numbers, has been imposed to the universe of all sets, so that:

- every non-empty family X of sets has a member x such that $x < y$ holds for every y in X, except for $y = x$;

- $x < y$ for any two sets x, y with $rk\,x \in rk\,y$;

- $x < y$ for any two sets x, y such that x is n times infinite and y is m times infinite, with $rk\,x = rk\,y$ and $n < m$;

- $<$ is anti-lexicographic (see Section 2.4) between any two finite sets.

Finally, we require that ηX is the least member of X with respect to $<$, for every non-empty set X. For definiteness, we also put $\eta \emptyset = \emptyset$.

Similarly to terms, one introduces *formulae*: these too are formal expressions, and each of them denotes either truth or falsehood when its variables are replaced by sets. *Simple formulae* are obtained by inserting one of the relators $\in$, $\subseteq$, $=$ (which denote membership, inclusion and equality) between two terms. A *compound formula* is obtained by prefixing the negation connective, $\neg$, to a (simple or compound) formula, or by inserting one of the binary connectives $\&$, $\vee$, $\rightarrow$, $\leftrightarrow$ (which denote conjunction, disjunction, implication, and bi-implication) between two formulae. Note that we are *not* making use here of the existential or universal quantifier.

1.3.2   Satisfiability and validity

Let P be a formula not involving the choice operator η. If P is true for some replacement of set values for the variables appearing in P, then P is said to be *satisfiable*. Any interpretation of the variables that so renders P true is called a *model of P*. If P has no models, P is said to be unsatisfiable. If the negation $\neg P$ of P is unsatisfiable, then P is said to be *valid*. In other words, P is valid if and only if every interpretation of the variables of P is a model of P.

Matters become more intricate when η appears in P because the well-ordering of all sets that defines η is far from being uniquely characterized by its aforementioned properties. This in turn implies that there are myriads of legal interpretations of η, called *versions of η*. Satisfiability is now affected by an annoying dependence on the version of η: indeed, for a given interpretation of the variables the formula P may turn out to be true in some version of η but false in another.

Nevertheless, at least for formulae P that do not involve the operators *Un* and *Pow*, the following theorem holds. There is a version of η that makes satisfiable every P that is satisfiable in any version of η. Whether this remains true for formulae that involve *Un* or *Pow* is not known.

Let us now introduce a few variants of the satisfiability notion: in the definition of *finite* satisfiability, only the substitutions of finite sets for the variables will be taken into account, i.e. P is finitely satisfiable if and only if there exist finite sets which make P true when substituted for the variables appearing in P. *Hereditarily finite* satisfiability is defined similarly, notably taking into account only the substitutions of hereditarily finite sets for the variables. We anticipate (see Section 2.5 and Chapter 4) that a set is hereditarily finite if and only if it is finite and each of its members is hereditarily finite. The hereditarily finite sets are denoted by variable-free terms involving only $\emptyset$ and the construct $\{T_0, \ldots, T_n\}$, with $n = 0, 1, 2, \ldots$.

A formula P is said to be *easily satisfiable* if there exist finite sets which, when substituted for the variables appearing in P, not only make P true, but also allow one to interpret *every term* appearing in P as a finite set. Trivially, every easily satisfiable formula is also finitely satisfiable. However, easy satisfiability and finite satisfiability become conceptually distinct when one is dealing with formulae that involve η or Un (see Section 3.6).

1.3.3   Decidability

A collection $\mathcal{C}$ of formulae is said to be *decidable* (or, better to say, to have a decidable satisfiability problem) if there exists an algorithm that can determine, for every given P in $\mathcal{C}$, whether or not P is satisfiable. Such an algorithm $\mathcal{A}$ is called a *decision algorithm* for $\mathcal{C}$. It is desirable that $\mathcal{A}$ does not simply provide a yes/no answer, but is also able to produce a model of P whenever a model exists. Ideally, $\mathcal{A}$ ought to produce the entire collection of models of P, but doing this explicitly is seldom possible; for example, in order to exhibit all models of the formula $x = x$, $\mathcal{A}$ would have to produce the assignments $\{(x, X)\}$ with X ranging over the class of all sets!

It may be the case, however, that every P in $\mathcal{C}$ can be transformed into the disjunction of finitely many satisfiable formulae P_i, such that every model of P is a model of some P_i and vice versa (see Chapter 6). If each P_i is simple and expressive enough as to convey all relevant information about the full collection of its models, then we may regard the collection of the P_is, which is equivalent to the original P, as a convenient substitute for the collection of all models of P. A decision algorithm $\mathcal{A}$ which produces such a disjunction for every input formula P taken from $\mathcal{C}$, is called a *simplifier for $\mathcal{C}$* [LS80a].

Note that in this definition of a simplifier we do not require that the disjuncts P_i, produced when some input P is fed into $\mathcal{A}$, belong to $\mathcal{C}$. In practice, it may even be convenient to let the P_is belong to a formalism that is richer than the language outlined at the beginning of this section.

1.3.4   Known satisfiability decision tests

We now present various *syllogistics*, i.e. decidable collections of formulae, and make a few remarks concerning each of them. Each syllogistic is closed with respect to some of the constructs of our unquantified language. In particular, $\emptyset$, variables, equality, membership and the propositional connectives are always allowed constructs: they form our *initial language endowment*.

- By adding the choice operator to these constructs, a first syllogistic is obtained. Indeed, it has been proved in [Omo84] and in [FO87] (see also Chapter 5) that if P is satisfiable then P has a model M in which, for every variable x,

$$rk\, Mx \leq 2 \cdot (k-1)\,,$$

where rk denotes the *rank* function and k is the total number of terms appearing (possibly inside other terms) in the formula $p\&\emptyset = \emptyset$.

- By adjoining the construct $\{\bullet, \ldots, \bullet\}$ too, decidability is not lost (see Section 5.7). This was first proved in [PP88b].

- By adding $\cap$ and $\subseteq$, instead of η, to the initial constructs, one obtains a quickly decidable syllogistic, that might be called *Horn syllogistic* (see Section 5.5). By adding $\setminus$ and $\cup$ too, one obtains the so-called *multilevel syllogistic* [FOS80a], for which a simplifier can also be designed [CGO88].

  Even if well-foundedness of membership is not assumed, multilevel syllogistic remains decidable [Pol87]. Notice however that in this varied framework the notions of satisfiability and validity cannot be defined as naively as we have done so far, because one cannot think of them in terms of a standard universe of sets.

- The theory obtained by adding the constructs $\cup$, $\{\bullet, \ldots, \bullet\}$ to the initial endowment has been proved decidable in [Vil71].

- Any one of the constructs

$$\{\bullet, \ldots, \bullet\},\, Un,\, Pow,\, rk$$

  can be safely added to the constructs of multilevel syllogistic, without undermining decidability [FOS80a, CFS85, CFS87, CFMS87] (see Chapters 8, 10). Moreover, decidability is preserved if $\{\bullet, \ldots, \bullet\}$ and Pow are added together [Can88, Can87] (see Chapter 8). Note that when the operator rk is present, one can express the fact that x is an ordinal number by the equality $x = rk\, x$. It is conjectured that all of the above four constructs can be simultaneously dealt with. A positive partial result in this direction is: one can treat formulae in which terms of the form $\{\bullet, \ldots, \bullet\}$ occur freely and only one term of the form $Un\, t$ (but no term of the form $Pow\, t$) occurs [Bre82].

  It has been proved that any formula P, possibly involving Pow and $\{\bullet, \ldots, \bullet\}$ in addition to the constructs of multilevel syllogistic, is satisfiable if and only if it is hereditarily finitely satisfiable [Can88, Can87]. This is no longer true if Un, instead of Pow, is allowed to appear in P. In this case it may be necessary, in order to satisfy P, to interpret some variables as infinite sets (see Chapter 10).

- Let us add η to the constructs of multilevel (respectively, of the Horn) syllogistic. In either case, *easy* satisfiability is decidable [Omo84, FO87].

When an easy model exists, then there is an easy model in which, for every variable x,

$$rk\,Mx < 2^{k+1} - 1$$

(respectively, $rk\,Mx \leq 4 \cdot (k-1)$),

where k is as before. If $T_1, \ldots, T_k$ are the terms in the formula, then, in the proposed model, each set $\widehat{W}$ of the form

$$\widehat{W} = \left(\bigcap_{i \in W} M\,T_i \right) \setminus \bigcup_{j \notin W} M T_j$$

has exactly one member not of the form MT, and this member is the maximum of $\widehat{W}$.

There are several known decidable fragments of set theory that cannot be embedded in our language. In order to mention them the language must be enriched.

- One obtains decidable extensions of multilevel syllogistic [CFMS87] by adding either the binary operator Σ, where ΣT denotes $\{\{S\} : S \in T\}$, or the unary relator *Finite*, whose intended meaning is

  Finite(T) holds iff T has finitely many members.

No more than one term of the form ΣT can be treated, at the current state of the art.

The operator Σ is not very interesting in its own right. However, proving its decidability seems to be related to proving the (conjectured) decidability of the cartesian product operator.

A decision algorithm also exists that can treat multilevel syllogistic extended with both the construct $\{\bullet, \ldots, \bullet\}$ and the relator *Finite*. This theory even admits a simplifier [CGO88, COP89] (see Chapter 6).

If one adds to the latter language the constructs η, $<$, and a rank comparison operator $\mathcal{R}$ with the meaning

$$x\,\mathcal{R}\,y \text{ if and only if } rk\,x \leq rk\,y,$$

then it can be proved (see [Fer88b, Fer88a, CFOP88]) that a version of η exists that renders all simultaneously satisfiable those formulae that are satisfiable in at least one version of η (see Chapter 11). If one adopts one such version of η, then a simplifier exists for the language (for the outline of a proof with a similar flavour, which, however, does not take η or $<$ into account, see Section 1.4).

Alternatively, instead of η, $<$, and $\mathcal{R}$ (but retaining $\{\bullet,\ldots,\bullet\}$ and *Finite*, as before), one can introduce a new unary operator $pred_<$, whose meaning is

$$pred_<(x) = \{y : rk\,y < rk\,x\}.$$

The resulting fragment of set theory is decidable too [CC89c].

- A decidable extension of multilevel syllogistic is obtained [CC89b] by adding the constructs $\{\bullet,\ldots,\bullet\}$ and $\mathcal{R}$, together with the unary intersection operator $\bigcap$.

- Yet another decidable extension of multilevel syllogistic results [FOS80a] from adding the construct $\{\bullet,\ldots,\bullet\}$, a unary cardinality operator $|\bullet|$, and operators denoting addition, subtraction and comparison of cardinals. Together with these, one can also add the rank comparison operator $\mathcal{R}$ [CC89a].

- The decidable fragment of set theory last mentioned can be further extended—leaving out $\mathcal{R}$, however—, by introducing maps as a new variable type and the operators D (domain), R (range) and $\bullet[\bullet]$ (restricted range, or 'multi-image'), which associate maps with sets [FOS80b], [BFOS81]. In such an extension of decidable syllogistic the language is enriched by an infinity of map variables. The definition of set terms is also extended by postulating that Df, Rf and $f[s]$ are set terms whenever f is a map variable and s is a set term. Note that map variables are simply sets of pairs, not necessarily functions; it is therefore useful to introduce atoms of the form

$$single\text{-}valued(f), \; one\text{-}to\text{-}one(f),$$

where f is a map variable. None of these extensions disrupts decidability. Moreover, various new constructs can be defined easily in terms of the primitive ones, e.g.

$$y = f^{-1}[x] \leftrightarrow y \subseteq Df \,\&\, f[y] = x \cap Rf \,\&\, f[Df \setminus y] \cap x = \emptyset.$$

- This result about maps has been refined (see [CS88] and Chapter 9), by treating maps as 'first class citizens' in the universe of sets. In fact, even if the syntactic distinction between set variables and map variables is abolished (so that, e.g. the union of a map with an arbitrary set becomes expressible, and a sense is attributed to the notation $D\,x$ for any set x), the constructs of multilevel syllogistic, plus the map constructs just mentioned, remain decidable. Moreover, other constructs such as $f = g_{|s}$, $f = g^{-1}$ can be safely added.

Other available decidability results concern classes of quantified formulae as shown below.

- [BFOS81] solves the satisfiability problem for the class of propositional combinations of simple formulae over the language consisting of set variables, the constants $\emptyset$, ω (set of the positive integers), *Ord* (class of all ordinals), and the predicates = and $\in$. We report about this result in Chapter 7. By a *simple* formula over an unquantified language $\mathcal{L}$ it is meant a prenex formula $Q_1 Q_2 \ldots Q_n p$, with p in $\mathcal{L}$, such that for $i = 1, 2, \ldots, n$ either every Q_i is $(\exists y_i \in z_i)$ or every Q_i is $(\forall y_i \in z_i)$, and no z_j is also a y_i for any $i, j = 1, 2, \ldots, n$.

- A partial result concerning classes of quantified formulae is contained in [CCF88]. Specifically, it is proved that the class of propositional combinations of simple formulae over the above unquantified language extended with the rank operator rk has a solvable *finite* satisfiability problem.

- Another class of formulae which has been proved decidable, with respect to the ordinary satisfiability problem, is the class of simple formulae over the language $\emptyset$, =, $\in$ and the predicate *Finite* with the restriction that no occurrences of *Finite* and of equalities between bound variables are allowed within the scope of quantifiers (cf. [CC88]).

- Two classes of sentences have been shown to be complete—hence decidable—in [Gog78] and [Gog79]. These are

  - the class of closed formulae of the type

$$(\forall x_1) \cdots (\forall x_n)(\exists x_{n+1})\, p,$$

    where p is an unquantified formula in the language containing only the predicates = and $\in$, and

  - the class of closed prenex formulae with three quantifiers in the predicates = and $\in$.

  It is interesting to note that the satisfiability problem of multilevel syllogistic extended with the construct $\{\bullet, \ldots, \bullet\}$ easily translates into the validity problem of the first of these two Gogol's classes. A drawback of this translation is that it does not yield any practical decision algorithm (akin to those to be described in Chapter 6 of this book) for multilevel syllogistic.

We end by mentioning two decidability results concerning multisorted set languages. These are:

- Propositional combinations of literals of type $x = y$, $x \in X$, $X = Y \cup Z$, $X = Y \setminus Z$, where x, y stand for *individual variables* and X, Y, Z stand for *set variables*. This language, called *2-level syllogistic* (see [FO78]), is briefly treated in Section 6.7, where it is extended by means of the constructs $Finite(X)$ and $X = \{x_1, \ldots, x_n\}$, and its satisfiability problem is reduced to the problem of multilevel syllogistic.

- The extension of 2-level syllogistic by various topological constructs, namely the Kuratowski closure operator (cf. [Kel55]), variables that stand for continuous and closed function, and a few elementary map constructs (direct multi-image, inverse image, and point evaluation), is proved decidable in [CO88] (see also [CO89b, CO89a]).

Syllogistics like these two, that involve individuals and do not provide means for expressing membership *among sets*, are not considered in this book except very seldom and quite incidentally. It is in fact felt by the authors that these theories are more naturally studied in the framework of first or second order predicate calculus, or within restricted subsystems of these, such as monadic calculus (cf. [Ack62]).

1.3.5   Examples of syllogisms

In the design of a set calculus, a decision method $\mathcal{M}$ can be exploited as a very general axiom scheme. To show this, we call *ground $\mathcal{M}$-syllogism* (or just 'ground syllogism') any formula that is recognized valid by the method $\mathcal{M}$. By arbitrarily replacing each free able x in a ground syllogism by a term t_x (where t_x may involve such high-level constructs as the set-formers of Section 1.2), one obtains an *instance of* the ground syllogism. By prefixing a string

$$\forall x_0 \ldots \forall x_n$$

of universal quantifiers to an instance of a ground syllogism, one obtains a *universal syllogism*.

To obtain *existential* syllogisms, one will proceed dually. Let P be a formula that is recognized satisfiable by $\mathcal{M}$, and let P' result from the substitution in P of variables to terms (distinct terms being replaced by distinct variables). By prefixing a string

$$\exists x_0 \ldots \exists x_n$$

of existential quantifiers to a formula P' obtained in this manner, so that every variable of P' is one of the x_is, one gets an *existential* syllogism.

We now begin to show examples of syllogism that fall under the jurisdiction of one or another of the decision methods known today.

Our first group of examples consists of trivial ground syllogisms, that are nevertheless important for historical reasons. These are

$$\neg\, z \in \emptyset,$$
$$z \in x \cap y \leftrightarrow (z \in x \,\&\, z \in y),$$
$$z \in x \setminus y \leftrightarrow (z \in x \,\&\, \neg\, z \in y),$$
$$z \in x \cup y \leftrightarrow (z \in x \,\vee\, z \in y),$$
$$z \in \{t_0, \ldots, t_n\} \leftrightarrow (z = t_0 \,\vee\, \cdots \,\vee\, z = t_n),$$

and (see [BFOS81])

$$x \subseteq y \leftrightarrow \forall z \in x\ z \in y,$$
$$x = y \leftrightarrow ((\forall z \in x\ z \in y) \,\&\, (\forall z \in y\ z \in x)),$$
$$x \in Un\, y \leftrightarrow \exists z \in y\ x \in z;$$
$$x \in Pow\, y \leftrightarrow \forall z \in x\ z \in y.$$

The reader will easily identify, among these, variants of axioms that are traditionally known as axiom of extensionality, axiom of the empty set, axiom of pairing, axiom of the union, and axiom of the power set (cf., e.g., [Man77]).

Our second group of syllogisms embody much of the semantics of η. Before we list them, we must observe (see Chapter 4) that when t is an unquantified term without variables—so that t denotes a hereditarily finite set X—then it is possible to determine a finite collection $\mathcal{S}_t$ of terms, also unquantified and not involving variables, such that the sets denoted by the terms s in $\mathcal{S}_t$ are precisely those sets that precede X in the well-ordering $<$ of all sets (see Section 1.3.1):

$$\eta\, \emptyset = \emptyset,$$
$$y \in x \rightarrow \eta\, x \in x,$$
$$y \in x \rightarrow \eta\, x \leq y,$$
$$y \in x \rightarrow y < x,$$
$$(x = \{x_0, x_1, \ldots, x_n\} \,\&\, x \setminus y < y \setminus x \rightarrow x < y,$$
$$x_0, x_1, \ldots, x_n < y \rightarrow \{x_0, x_1, \ldots, x_n\} < \{y\},$$
$$(\&_{i=0}^{n}\, \{x_i\} < z \,\&\, y \subseteq z \,\&\, x\, Swells\, y) \rightarrow \{x_0, x_1, \ldots, x_n\} < z,$$
$$(x = \{x_0, x_1, \ldots, x_n\} \,\&\, x < y) \rightarrow x < y \cup z,$$
$$x < t \rightarrow \bigvee_{s \in \mathcal{S}_t} x = s,$$

where t is a term containing no variables, $n = 0, 1, 2\ldots$, and $x\, Swells\, y$ is defined as follows:

$$x\, Swells\, y =_{\text{Def}} y \subseteq \eta\, x \,\&\, \neg\, y = \eta\, x \,\&\, y \in x$$

(and implies—as will be explained in Section 3.6—that y is infinite).

The existence of infinite sets can be expressed by means of existential syllogisms in various manners. The most obvious way is by the syllogism

$\exists x \neg \, Finite\, x$; a rather esoteric way is by the syllogism $\exists x \, \exists y \, x \; Swells \, y$; ways of an intermediate difficulty (cf. (10.4) and (9.8) in Sections 10.1, 9.2) are:

$$\exists u \, (\neg u = \emptyset \, \& \, u = Un\, u),$$
$$\exists u \, (\neg u = \emptyset \, \& \, u = D\, u),$$

where $D\, u$ denotes the domain of u, i.e. the set formed by the first components of ordered pairs belonging to u.

Examples of very useful syllogisms are the De Morgan laws:

$$a \setminus (b \cup c) = (a \setminus b) \cap (a \setminus c),$$
$$a \setminus (b \cap c) = (a \setminus b) \cup (a \setminus c),$$

and the following:

$$a \subseteq b \;\rightarrow\; Pow(a) \subseteq Pow(b),$$
$$Un(a \cup b) = Un(a) \cup Un(b).$$

We like to conclude with syllogisms that are perhaps more recreational than useful:

$x \in y \cap z \;\rightarrow\; \neg z \in x$
        (note that $\cap$ could not be replaced by $\cup$ here),

$x \subseteq z \, \& \, y \subseteq z \;\rightarrow\; y \setminus (z \setminus x) = y \cap x,$

$\neg \, ((\&_{i=0}^{n} x_i \in x_{i+1}) \, \& \, x_{n+1} \in x_0),$

plato = socrates & socrates $\in$ men & men $\subseteq$ mortals $\rightarrow$
        plato $\in$ mortals,

socrates $\in$ philosophers $\cap$ immortals & men $\subseteq$ mortals $\rightarrow$
        socrates $\not\subseteq$ men $\vee$ mortals $\cap$ immortals $\neq \emptyset$,

amanita $\in$ purple_things $\cap$ mushrooms &
    purple_things = violet_things &
    mushrooms $\subseteq$ plants &
    violet_things $\cap$ plants $\subseteq$ poisonous_things $\rightarrow$ amanita $\in$
        poisonous_things $\cap$ purple_things $\cap$ violet_things $\cap$ plants
        $\cap$mushrooms,

$y \not\subseteq x \, \& \, \eta(y \setminus x) \subseteq y \;\rightarrow\; \eta(y \setminus x) \subseteq x,$

$|y \setminus x| < |y \cap x| \, \& \, |y \setminus z| < |y \cap z| \;\rightarrow\; x \cap z \neq \emptyset,$

$s \cap q = \emptyset \, \& \, p \subseteq q \cup r \, \& \, (p = \emptyset \rightarrow q \neq \emptyset) \, \& \, q \cup r \subseteq s \;\rightarrow$
    $p \cap r \neq \emptyset,$

$p \neq \emptyset \, \& \, f \cap g \cap r = \emptyset \, \& \, p \subseteq g \cap f \, \& \, (p \subseteq q \vee p \cap r \neq \emptyset) \;\rightarrow$
    $q \cap p \neq \emptyset,$

$(p \neq \emptyset \leftrightarrow q \neq \emptyset) \, \& \, (\forall x \in p)(\forall y \in q)(x \in r \leftrightarrow y \in s) \;\rightarrow$
    $(((\forall x \in p)x \in r) \leftrightarrow ((\forall x \in q)x \in s)).$

(The last three of these are translations of the problems 24, 25, and 26 of [Pel86]. The fourth-to-last translates De Morgan's numerical syllogism 'If most ys are xs, and most ys are zs, then some xs are zs'—cf. [Gar68].)

1.3.6 An application to program correctness verification

A potential area of application of syllogistic algorithms is the proof of partial correctness of imperative SETL-like programs (for a description of SETL, see [SDDS86]). Consider for instance the problem of finding the smallest superset S of S_0 which enjoys the following closure property:

'S contains as a member any y such that $x \, F \, y$, where x belongs to S and F is a given binary relation.'

Here is an annotated algorithm that solves the problem at hand:

hypothesis: $S_1 \supseteq S_0 \,\&\, F[S_1] \subseteq S_1$;
$S \leftarrow S_0$;
(while $F[S] \not\subseteq S$)
invariant: $S_1 \supseteq S \,\&\, S \supseteq S_0 \,\&\, F[S_1] \subseteq S_1$;
    $S \leftarrow S \cup F[S]$;
end while;
thesis: $S_1 \supseteq S \,\&\, S \supseteq S_0 \,\&\, F[S] \subseteq S$;

This algorithm can be translated automatically into the following four statements that can then be submitted directly to a suitable syllogistic decider (see [FOS80b]), which will demonstrate the correctness of the procedure by proving them valid.

(a) $S_1 \supseteq S_0 \,\&\, F[S_1] \subseteq S_1 \,\&\, S = S_0 \,\&\, \neg(F[S] \not\subseteq S) \rightarrow$
$S_1 \supseteq S \,\&\, S \supseteq S_0 \,\&\, F[S] \subseteq S$

(b) $S_1 \supseteq S_0 \,\&\, F[S_1] \subseteq S_1 \,\&\, S = S_0 \,\&\, F[S] \not\subseteq S \rightarrow$
$S_1 \supseteq S \,\&\, S \supseteq S_0 \,\&\, F[S_1] \subseteq S_1$

(c) $S_1 \supseteq S \,\&\, S \supseteq S_0 \,\&\, F[S_1] \subseteq S_1 \,\&\, \overline{S} = S \cup F[S] \,\&\, F[\overline{S}] \not\subseteq \overline{S} \rightarrow$
$S_1 \supseteq \overline{S} \,\&\, \overline{S} \supseteq S_0 \,\&\, F[S_1] \subseteq S_1$

(d) $S_1 \supseteq S \,\&\, S \supseteq S_0 \,\&\, F[S_1] \subseteq S_1 \,\&\, \overline{S} = S \cup F[S] \,\&\, \neg(F[\overline{S}] \not\subseteq \overline{S}) \rightarrow$
$S_1 \supseteq \overline{S} \,\&\, \overline{S} \supseteq S_0 \,\&\, F[\overline{S}] \subseteq \overline{S}$.

Abstracting from this example, we can say that to each decidable fragment F of set theory there corresponds a collection Φ of decidable annotated programs, each of which can be translated into a collection $f_1, f_2, \ldots, f_N$ of formulae in F, which are all valid simultaneously if and only if the given annotated program is partially correct (i.e. produces correct answers when its execution terminates). If a program φ is not correct, then the decision algorithm for F will produce counter-examples that can be used as sample data for testing the program after it has been corrected. Only if these tests are successful, it will make sense to again submit φ to the validation procedure, which, in general, is extremely time-consuming.

We may conjecture that the formulae $f_1, f_2, \ldots, f_N$ into which φ is translated are, in a way, a declarative specification of the algorithm encoded by φ. In order to make this idea explicit one must find a convenient operational semantics for the collection F of formulae. The effort of defining operational semantics for set theoretic formulae would be pointless, however, at the rather low level exemplified by (a), (b), (c), (d) above. In the RAPTS system (see [PH85] and [Pai86]) the algorithm described above admits the declarative specification

$$\text{the } S \supseteq S_0 \text{ such that } F[S] \subseteq S, \text{ minimizing } |S|\,.$$

Even though this specification does not fully correspond to the 'natural' statement of the problem (in fact it minimizes with respect to cardinality instead of with respect to inclusion, and a simple lemma is needed to show the equivalence between the two minimizations), it is clear that this is the appropriate specification level. We recall that RAPTS will perform, *automatically*:

- a preliminary check that the S characterized by the above description is indeed computable in a finite amount of time (under the implicit assumption that all sets are hereditarily finite);

- transformations that translate the above code into imperative code with optimal asymptotic complexity.

1.4   A sample satisfiability decision test

The correctness of a satisfiability decision test of the kind described in Section 1.3 usually rests on two propositions, called *soundness* theorem and *completeness* theorem. Taken together, these two propositions form a *reflection theorem*, which—unlike the other two—has a mathematical significance totally independent of the satisfiability problem. Sometimes in this book (see, for example, Chapter 5) a reflection theorem is presented first and then from it a decision test for one or more collections of set-theoretic formulae is derived. With this approach, the soundness and completeness theorems descend as easy corollaries of the reflection theorem.

In the soundness theorem, a finite list $X_1, \ldots, X_m$ of sets is given, and an algorithmic data-structure $\mathcal{D}$ (typically a graph with various superimposed relations) is associated with it, so as to retain in the most essential form information about particular properties (for instance finiteness) enjoyed by the X_is, or mutual relations (for instance membership relations) that hold among them. It is then proved that $\mathcal{D}$ fulfils a number of combinatorial conditions (that is, conditions whose truth or falsity can be detected by means of an algorithm) that force $\mathcal{D}$ to be inside a finite range $\mathcal{D}_1, \ldots, \mathcal{D}_{F(m)}$ of possibilities.

In the completeness theorem, $\mathcal{D}$ and m are given, and $\mathcal{D}$ is assumed to fulfil the same combinatorial conditions that appear in the soundness theorem. Then it is proved that sets $Y_1, \ldots, Y_m$ exist subject to certain properties and mutual relations, more or less explicitly encoded by $\mathcal{D}$. Hence, if $\mathcal{D}$ originates from sets $X_1, \ldots, X_m$ as above, certain relations among the Y_is faithfully correspond to, or—so to say—reflect, relations among the X_is. It must be noted, however, that the Y_is generally enjoy various other properties and mutual relations, about which $\mathcal{D}$ does not say anything, that make them more interesting than the X_is. For instance, their ranks may be smaller than some (finite or hereditarily denumerable) ordinal $\mathcal{O}(m)$.

In this section we illustrate the typical pattern of a proof of existence of a satisfiability decision test. We take into account a rather rich class $\mathcal{L}$ of formulae, and outline a proof that $\mathcal{L}$ has a solvable satisfiability problem. The details of the proof are omitted, because our aim here is to telegraphically convey the flavour of many parts of this book. The reader will find it easy to fill in such details after reading Chapter 6. At any rate, a similar proof will be presented in Chapter 11.

$\mathcal{L}$ consists of all formulae that can be construed, starting from set variables and the constant $\emptyset$, by proper use of the following operators, relators, and propositional connectives: $\cap$, $\setminus$, $\cup$, $\{\bullet, \ldots, \bullet\}$, $\in$, $=$, $\subseteq$, $Finite$, $=_\mathcal{R}$, $=_\mathcal{R}^+$, $>_\mathcal{R}^+$, $\neg$, and $\&$. We are indicating here by $Finite\,X$ that X is finite, and by $X =_\mathcal{R} Y$, $X =_\mathcal{R}^+ Y$, $X >_\mathcal{R}^+ Y$ that $rk\,X = rk\,Y$, $rk\,X = rk\,Y + 1$, and $rk\,X > rk\,Y + 1$, respectively. Thus, e.g., $X >_\mathcal{R}^+ Y$ indicates that the rank of X exceeds the rank of Y by more than one. In what follows we are also denoting by the symbol ω the set of all natural numbers, and by V_α, where α is an ordinal, the family of all sets having rank smaller than α (see Sections 2.3 and 2.5).

1.4.1 Syllogistic diagrams

We begin by introducing a data-structure that may seem, at first, unrelated to $\mathcal{L}$ (it will turn out, however, that this is not the case).

Definition 1.2 *Consider a tuple $\beta_1, \ldots, \beta_n$ with $n \in \omega$, $\beta_i \subseteq Pow\{1, \ldots, i - 1\}$ for $i = 1, \ldots, n$, and $\beta_i \cap \beta_j = \emptyset$ when $i \neq j$.*

Consider also $\delta_1, \ldots, \delta_n \in \{0, 1, \omega\}$ with $\delta_i \neq 0$ when $\beta_i = \emptyset$.

Finally, suppose that $\varrho_1, \ldots, \varrho_n \in \{0, 1\}$, with $\varrho_1 = 1$.

Such a triple β, δ, ϱ is called a syllogistic diagram, *and n is said to be the* size *of the diagram.* □

The most typical example of a syllogistic diagram (indeed, the only example of interest) is obtained as follows. Take sets $\sigma_1, \ldots, \sigma_n$ with $\emptyset < \sigma_1 < \cdots <$

σ_n in the ordering introduced in Section 1.3.1, and $\sigma_i \cap \sigma_j = \emptyset$ when $i \neq j$. Put $\Sigma_J = \bigcup_{j \in J} \sigma_j$ for every $J \subseteq \{1, \ldots, n\}$, and $\partial_i = |\sigma_i \setminus \{\Sigma_J : J \subseteq \{1, \ldots, n\}\}|$ for every $i \in \{1, \ldots, n\}$. Moreover put $\sigma_0 = \emptyset$ and $\sigma_{n+1} = V_\alpha \setminus \Sigma_{\{1, \ldots, n\}}$, where α is the first limit ordinal exceeding $rk \, \sigma_n$. Finally, put

(1) $$\beta_i = \{J \subseteq \{1, \ldots, n\} : \Sigma_J \in \sigma_i\};$$

(2) $$\delta_i = \begin{cases} 0 \text{ if } \partial_i = 0, \\ \omega \text{ if } \partial_i \geq \omega, \\ 1 \text{ otherwise}; \end{cases}$$

(3) $$\varrho_i = \begin{cases} 0 \text{ if } rk \, \sigma_i = rk \, \sigma_{i-1}, \\ 1 \text{ otherwise}; \end{cases}$$

for $i = 1, \ldots, n$. This is—as one easily ascertains—a syllogistic diagram.

Definition 1.3 *The syllogistic diagram satisfying (1)–(3) above is said to be induced by the σ_is.* □

The essential aspects of the combinatorics of the Σ_Js are as follows:

Lemma 1.5

$$\begin{array}{ll} \Sigma_I \in \Sigma_J & \textit{iff } I \in \bigcup_{j \in J} \beta_j; \\ \Sigma_I \subseteq \Sigma_J & \textit{iff } I \subseteq J; \\ \Sigma_I = \Sigma_J \cap \Sigma_K & \textit{iff } I = J \cap K; \\ \Sigma_I = \Sigma_J \setminus \Sigma_K & \textit{iff } I = J \setminus K; \\ \Sigma_I = \Sigma_J \cup \Sigma_K & \textit{iff } I = J \cup K; \\ \textit{Finite } \Sigma_I & \textit{iff } (\forall i \in I)\delta_i < \omega; \\ rk \, \Sigma_I \leq rk \, \Sigma_J & \textit{iff } \exists j \in J \forall i \in I(j < i \rightarrow (\forall k \in \{j+1, \ldots, i\})\varrho_k = 0). \end{array}$$

Moreover, for every q in ω,

$$\Sigma_J = \{\Sigma_{I_1}, \ldots, \Sigma_{I_q}\} \quad \textit{iff} \quad \{I_1, \ldots, I_q\} = \bigcup_{j \in J} \beta_j \, \& \forall j \in J \delta_j = 0.$$

□

Not every syllogistic diagram is induced by suitable sets σ_i in the manner explained above. In fact, each one of the conditions (1), (2), (3) imposes some constraint on the ranks of the σ_is, and it may turn out that the different requirements are incompatible.

In order to bring the difficulties to light, we need to introduce the concept of *pseudo-rank*.

1.4.2   Computation of the pseudo-ranks

Given a syllogistic diagram β, δ, ϱ of size n, we convene, in harmony with the preceding conventions regarding the values σ_0 e σ_{n+1}, that:

$$\beta_0 = \emptyset, \ \delta_0 = 0, \ \varrho_0 = 0,$$

$$\beta_{n+1} = Pow\,\{1, \dots, n\} \setminus \bigcup_{i=1}^{n} \beta_i, \ \delta_{n+1} = \omega, \ \varrho_{n+1} = 1.$$

We now wish to characterize the n-tuples $\psi = (\psi_1, \dots, \psi_n)$ that fulfil the identity

$$(\star) \qquad \psi_i = \begin{cases} \psi_{i-1} & \text{if } rk\,\sigma_i = rk\,\sigma_{i-1}, \\ \psi_{i-1} + 1 & \text{if } rk\,\sigma_i = rk\,\sigma_{i-1} + 1, \\ \psi_{i-1} + 2 & \text{if } rk\,\sigma_i > rk\,\sigma_{i-1} + 1, \end{cases}$$

for $i = 1, \dots, n$, where $\sigma_1, \dots, \sigma_n$ are sets that induce β, δ, ϱ according to Definition 1.3, and where ψ_0 is assumed to be 0. Putting $\psi_{n+1} = \psi_n + 2$, we have

- $\psi_1 \neq 0$ and moreover, for $i = 1, \dots, n(+1)$,

- $\psi_{i-1} \leq \psi_i \leq \psi_{i-1} + 2$;

- $\psi_j < \psi_i$ holds whenever $j \in J \in \beta_i$;

- if $\psi_{i-1} < \psi_i$ and $j \in \{i, \dots, n\}$ is such that $\delta_j = 0$ and $\beta_j \subseteq Pow\{1, \dots, i-1\}$, then $\psi_j = \psi_{i-1} + 1$.

Definition 1.4 *A pseudo-rank of β, δ is an n-tuple $\psi = (\psi_1, \dots, \psi_n)$ such that (putting, as before, $\psi_0 = 0$ and $\psi_{n+1} = \psi_n + 2$) the above four conditions are fulfilled. The threshold of such a pseudo-rank is the first s in $\{0, 1, \dots, n\}$ for which $\psi_{s+1} = \psi_s + 2$.*

A pseudo-rank ψ is said to be compatible with ϱ *iff*
$$\varrho_i = 0 \text{ if and only if } \psi_{i-1} = \psi_i, \text{ for } i = 1, \dots, n. \qquad \square$$

When ψ is compatible with ϱ, the triple β, δ, ψ (from which one immediately derives ϱ) is called a *syllogistic diagram endowed with a pseudo-rank*. In particular ψ is compatible with ϱ if $(\star)$ holds, where the σ_is induce β, δ, ϱ. In the latter case, β, δ, ψ is said to be *induced* by the σ_is. After defining $\Sigma_J = \bigcup_{j \in J} \sigma_j$ as usual, and putting also

$$prk(J) = \bigcup_{j \in J} \psi_j,$$

for all $J \subseteq \{1, \dots, n\}$, we can enrich Lemma 1.5 as follows:

Lemma 1.5 (continued)

$$rk\,\Sigma_I = rk\,\Sigma_J \qquad \textit{iff} \quad prk(I) = prk(J)\,,$$
$$rk\,\Sigma_I = (rk\,\Sigma_J) + 1 \quad \textit{iff} \quad prk(I) = prk(J) + 1\,,$$
$$rk\,\Sigma_I > (rk\,\Sigma_J) + 1 \quad \textit{iff} \quad prk(I) > prk(J) + 1\,.$$
$\qquad\qquad\square$

One easily finds an algorithm which, given β, δ, ϱ, produces in a finite number of steps the set of all pseudo-ranks that are compatible with ϱ. Unless it is empty, this set has a member π such that any ψ compatible with ϱ satisfies $\psi_i \leq \pi_i$ for $i = 1, \ldots, n$. This π is called the *main pseudo-rank* of β, δ, ϱ. The threshold of π is said to be the *threshold* of β, δ, ϱ too.

1.4.3 Contradictory diagrams

There are circumstances under which a syllogistic diagram β, δ, ϱ (respectively, a diagram β, δ, ψ endowed with a pseudo-rank) clearly *does not admit* sets σ_i inducing it. We will now make a provisional list (to be completed in Section 1.4.4) of such manifestly contradictory situations.

Definition 1.5 *We will say in the first place that β, δ, ϱ is* contradictory *if β, δ has no pseudo-ranks compatible with ϱ, or—equivalently—if the main pseudo-rank π of β, δ, ϱ is undefined. If π exists, then β, δ, ϱ is* contradictory *if the diagram β, δ, π is contradictory, in the sense explained below.* $\qquad\square$

Definition 1.6 *Let β, δ, ϱ be a syllogistic diagram endowed with a pseudo-rank ψ having threshold s. We will say that this diagram is* contradictory *if $\delta_i = \omega$ for some $i \leq s$, or there are i, j with $s < i < j \leq n$, $\psi_i = \psi_j$, $\delta_i = \omega$, $\delta_j < \omega$.*

Finally, let $\beta_*, \delta_*, \varrho_*$ be the three tuples

$$\beta_1, \ldots, \beta_s, \qquad \delta_1, \ldots, \delta_s, \qquad \varrho_1, \ldots, \varrho_s.$$

If there exist no sets $\sigma_1, \ldots, \sigma_s$ with $rk\,\sigma_i = \psi_i (\leq \psi_s \leq s)$ for $i = 1, \ldots, s$ that induce the syllogistic diagram $\beta_*, \delta_*, \varrho_*$, a fortiori no sets will exist that induce β, δ, ϱ or β, δ, ψ. Even in this situation (which is algorithmically verifiable, as will result from the study of the hereditarily finite sets in Chapter 4), the diagram β, δ, ψ is said to be contradictory. $\qquad\square$

1.4.4 Dense syllogistic diagrams

A syllogistic diagram β, δ, ϱ is said to be *dense* if it is not contradictory by Definition 1.5 or by Definition 1.6 and if, furthermore, the diagram $\beta_*, \delta_*, \varrho_*$ obtained by 'restricting'—as explained in the preceding subsection—β, δ, ϱ to the threshold s coincides with the syllogistic diagram induced by the singleton members of $V_{\pi_s + 1}$.

Plainly, for a dense diagram, $\beta_*, \delta_*, \varrho_*$ are induced *solely* by the family $\{\{x\} : x \in V_{\pi_s}\}$: in fact, $|\beta_i| = 1$, $\delta_i = 0$, $\sigma_i = \{\Sigma_J : \beta_i = \{J\}\}$, for $i = 1, \ldots, s$. When a dense diagram is endowed with a non-contradictory pseudo-rank ψ, it turns out that $\pi_i = \psi_i$ for $i = 1, \ldots, s+1$; therefore, the threshold of π and ψ is the same.

Given arbitrary sets $X_1, \ldots, X_m$, we can speak of the *dense* syllogistic diagram $\mathcal{D}$ induced by $X_1, \ldots, X_m$, in the following sense. Let $\tau_1, \ldots, \tau_g$, with $\tau_1 < \cdots < \tau_g$, be all sets different from $\emptyset$ that can be written in the form

$$\bigcap W \setminus \bigcup (\{X_1, \ldots, X_m\} \setminus W) ,$$

with $\emptyset \neq W \subseteq \{X_1, \ldots, X_m\}$. Put, moreover, $\tau_0 = \emptyset$, $\tau_{g+1} = V_\alpha \setminus (\tau_1 \cup \cdots \cup \tau_g)$, where α is the first limit ordinal that exceeds $rk\, \tau_g$, and indicate by q the first subscript in $\{0, \ldots, g\}$ for which $rk\, \tau_q + 1 < rk\, \tau_{q+1}$. Let

$$\{\sigma_1, \ldots, \sigma_s\} = \{\{x\} : x \in V_{rk\, \tau_q}\},$$
$$\{\sigma_{s+1}, \ldots, \sigma_{s+r}\} = \{\tau_\ell \setminus V_{rk\, \tau_q + 1} : \ell \in \{q+1, \ldots, g\}\},$$

with $\sigma_1 < \cdots < \sigma_{s+r}$. The diagram induced by the σ_ks is dense and has threshold s. Moreover, for every $j \in \{1, \ldots, m\}$ there exists a $K \subseteq \{1, \ldots, s+r\}$ such that $X_j = \bigcup_{k \in K} \sigma_k$. Finally, one has $s + r < C(m)$, where C is the computable function sending k to $|V_{2^k}| + 2^k$ for every k. In order to fix once and for all the size of a dense diagram induced by an m-tuple of sets, we prolongate the list of σ_ks, putting

$$\sigma_{k+1} = V_{\alpha_k} \setminus (\sigma_1 \cup \cdots \cup \sigma_k), \text{ for } k = s + r, \ldots, C(m) - 1,$$

where α_k is the first limit ordinal after $rk\, \sigma_k$, and define $\mathcal{D} = \mathcal{D}(X_1, \ldots, X_m)$ to be the diagram induced by $\sigma_1, \ldots, \sigma_{C(m)-1}$.

In conclusion, we have the following:

Lemma 1.6 (Soundness) *Given arbitrary sets $X_1, \ldots, X_m$, there is a dense syllogistic diagram $\mathcal{D}$ induced by the X_is, whose size is $C(m)$, for a specific computable function C.* □

Note that all dense syllogistic diagrams of a given size can be generated by a computation of a finite duration.

Suppose now that a dense syllogistic diagram β, δ, ϱ of size n and threshold s is given, endowed with a non-contradictory pseudo-rank ψ (cf. Definitions 1.5, 1.6). If β, δ, ψ is induced by sets $\sigma_1, \ldots, \sigma_n$, then the linear ordering $\lhd$ defined by

$$J \lhd I \text{ if and only if } \Sigma_J < \Sigma_I ,$$

for $I, J \subseteq \{1, \ldots, n\}$, enjoys the following properties:

(1) $J \lhd K$ when $prk(J) < prk(K)$;

(2) $\{j\} \lhd \{k\}$ when $1 \leq j < k \leq n$;

(3) $J \lhd K$ when the following conditions hold together: $\delta_j < \omega$ for all j in $J \cup K$, and either $J \subsetneq K$ or $J \not\subseteq K$, $K \not\subseteq J$, $max(J \setminus K) < max(K \setminus J)$;

(4) $J \lhd K$ when the following conditions hold together: $\delta_j < \omega$ for all j in J, $\delta_k = \omega$ for some k in K, and $prk(J) = prk(K)$;

(5) when $\delta_j = \delta_k = 0$, $prk(\{j\}) = prk(\{k\})$, and $j < k$, then β_j anti-lexicographically precedes β_k with respect to $\lhd$.

Definition 1.7 *If no ordering $\lhd$ meeting the conditions (1)–(5) exists, then β, δ, ψ is said to be* contradictory.   □

On the other hand, it is possible to prove the following proposition:

Lemma 1.7 (Completeness) *For every number N of the form $C(m) - 1$, any dense syllogistic diagram $\mathcal{D}$ of size N, endowed with a pseudo-rank, is induced by suitable sets $\sigma_1, \ldots, \sigma_n$, unless contradictory by Definition 1.6 or by Definition 1.7.*   □

1.4.5   The decision method

We are finally able to establish, given any formula p of $\mathcal{L}$, whether or not p is satisfiable. Let us explain how.

Consider the distinct terms $t_1, \ldots, t_m$ that appear (possibly within other terms) inside p. If $t \mapsto t^\varphi$ is any function sending $\{t_1, \ldots, t_m\}$ to sets, then the dense syllogistic diagram β, δ, ϱ with pseudo-rank ψ, induced by $t_1^\varphi, \ldots, t_m^\varphi$, belongs to a finite family of syllogistic diagrams, that can be generated algorithmically.

Indicating by n the size of β, δ, ψ, we are in the following situation. There is a tuple $\sigma_1, \ldots, \sigma_n$ of non-empty disjoint sets (see the preceding subsection) that induces β, δ, ψ; and as usual we put $\Sigma_J = \bigcup_{j \in J} \sigma_j$ for all $J \subseteq \{1, \ldots, n\}$. There is also a function $\gamma : \{1, \ldots, m\} \to Pow\{1, \ldots, n\}$ such that $t_j^\varphi = \Sigma_{\gamma(j)}$ for $j = 1, \ldots, m$.

The relations of type

$$\Sigma_I \in \Sigma_J, \Sigma_I \subseteq \Sigma_J, \Sigma_I = \Sigma_J \cap \Sigma_K, \Sigma_I = \Sigma_J \setminus \Sigma_K, \Sigma_I = \Sigma_J \cup \Sigma_K,$$
$$Finite\, \Sigma_I, \Sigma_J = \{\Sigma_{I_1}, \ldots, \Sigma_{I_q}\}, \Sigma_I =_{\mathcal{R}}^{+} \Sigma_J, \Sigma_I =_{\mathcal{R}}^{+} \Sigma_J, \Sigma_I >_{\mathcal{R}}^{+} \Sigma_J$$

that hold among sets of the form Σ_Q reflect certain combinatorial situations (see Lemma 1.5) which, given β, δ, ψ, can be ascertained even without knowing the σ_is. In order that φ can simply be an *interpretation* of p, a number of relations of this kind must be fulfilled (for instance, if t_i has the form $t_j \cap t_k$, then t_i^φ must equal $t_j^\varphi \cap t_k^\varphi$). In order that φ can be a *model* of p, it is necessary that the *evaluation* of p—whose result only depends from the truth values of a number of relations of this kind—gives the result *TRUE*.

In order to check whether p is satisfiable, it will suffice to 'guess' the diagram β, δ, ψ and the function γ, and then verify that the function $t_j \mapsto \Sigma_{\gamma(j)}$ (where the Σ_Js are associated with any tuple of sets σ_i inducing β, δ, ψ— for example the one constructed in the proof of Lemma 1.7) is an interpretation satisfying p. After a limited number of fruitless attempts, we will at last obtain a model of p, or will be authorized to conclude that p has no models at all.

1.4.6   Syllogistic decomposition

The decision method outlined in the preceding subsection is perhaps better understood in terms of the decomposition method that will be described now. By this new method, every formula p of $\mathcal{L}$ can be rewritten in the form

$$(*) \qquad \exists x_1 \exists x_2 \cdots \exists x_N (p_1 \vee p_2 \cdots \vee p_Q),$$

where each p_j involves the same variables that occur in p, plus the auxiliary variables x_i. Each disjunct p_j is satisfiable, whereas no combination $p_j \& p_k$ with $j \neq k$ is satisfiable; thus, in a sense, the decomposition $(*)$ is irredundant. Since p is logically equivalent to $(*)$, establishing that p is satisfiable simply amounts to checking that $Q \neq 0$. Moreover—as will be clear at the end of this section—if q is a formula of $\mathcal{L}$ that involves the same variables as p, then checking each implication $p_j \rightarrow q$ amounts to a very straightforward evaluation of q. Hence, rewriting p in the form $(*)$ may be advantageous when one has to prove many implications of the form $p \rightarrow q$.

Referring to a pre-established enumeration $\xi_0, \xi_1, \xi_2, \ldots$ of all set variables, we define:

- h to be the first i for which ξ_{i+1} does not occur in p;

- $t_1, \ldots, t_m$ to be the distinct terms that appear (possibly within other terms) inside p;

- $M = \{ i \in \{1, \ldots, m\} : t_i \text{ is a variable} \}$;

- $N = C(m) - 1$, the size of a dense syllogistic diagram induced by an m-tuple of sets (see Section 1.4.4);

- Γ_m to be the set of all quadruples $\beta, \delta, \psi, \gamma$ such that

  - β, δ, ψ is a dense syllogistic diagram of size N, endowed with a non-contradictory pseudo-rank;

  - $\gamma : \{1, \ldots, m\} \rightarrow Pow\{1, \ldots, N\}$, induces an interpretation (see preceding subsection) which makes p true.

Next, we define the terms σ_j, Σ_J, $\mathcal{S}_{\mathcal{J}}$ for $j \in \{0, \ldots, n\}$, $J \subseteq \{1, \ldots, n\}$, $\mathcal{J} \subseteq Pow\{1, \ldots, n\}$. These must not be confused with the *sets* σ_j and Σ_J used so far, but are meant to *designate* precisely those sets.

$\sigma_0 =_{\text{Def}} \emptyset$, $\sigma_j =_{\text{Def}} \xi_{h+j}$ for $j \neq 0$
(this is the only place where h intervenes);

$\Sigma_\emptyset =_{\text{Def}} \emptyset$, $\Sigma_J =_{\text{Def}} \sigma_{j_0} \cup \cdots \cup \sigma_{j_h}$
(in particular $\Sigma_J =_{\text{Def}} \sigma_{j_0}$ if $h = 0$) if
$J = \{j_0, \ldots, j_h\}$ with $j_0 < j_1 < \ldots < j_h$;

$\mathcal{S}_\emptyset =_{\text{Def}} \emptyset$, $\mathcal{S}_{\mathcal{J}} =_{\text{Def}} \{\Sigma_{J_0}, \ldots, \Sigma_{J_h}\}$
(in particular $\mathcal{S}_{\mathcal{J}} =_{\text{Def}} \{\Sigma_{J_0}\}$ if $h = 0$) if
$\mathcal{J} = \{J_0, \ldots, J_h\}$ with $J_0 < J_1 < \cdots < J_h$.

Furthermore, we put $\mathcal{S} =_{\text{Def}} \mathcal{S}_{Pow\{1,\ldots,n\}}$.
Then, for $j = 1, \ldots, n$, we define

$$R_j \quad =_{\text{Def}} \quad \begin{cases} \sigma_j =_{\mathcal{R}} \sigma_{j-1} & \text{if } \psi_j = \psi_{j-1}, \\[2mm] \sigma_j =_{\mathcal{R}}^+ \sigma_{j-1} & \text{if } \psi_j = \psi_{j-1} + 1, \\[2mm] \sigma_j >_{\mathcal{R}}^+ \sigma_{j-1} & \text{if } \psi_j = \psi_{j-1} + 2; \end{cases}$$

$$D_j \quad =_{\text{Def}} \quad \begin{cases} \sigma_j \subseteq S & \text{if } \delta_j = 0 \\ \neg Finite(\sigma_j \setminus S) & \text{if } \delta_j = \omega \\ Finite(\sigma_j \setminus S) \& \sigma_j \not\subseteq S & \text{if } \delta_j = 1. \end{cases}$$

Finally,

$$\Phi_{\beta\delta\psi} =_{\text{Def}} \left(\underset{0 < i < j \leq n}{\&} \sigma_i \cap \sigma_j = \emptyset \right) \& $$
$$\underset{0 < j \leq n}{\&} \left(\sigma_{j-1} < \sigma_j \& R_j \& D_j \& \sigma_j \cap S = S_{\beta_j} \right).$$

The syllogistic decomposition of p is

$$\exists \xi_{h+1} \cdots \exists \xi_{h+N} \bigvee_{(\beta,\delta,\psi,\gamma) \in \Gamma_m} \left(\Phi_{\beta\delta\psi} \& \underset{j \in M}{\&} t_j = \bigcup_{i \in \gamma(j)} \xi_{h+i} \right).$$

Part I

FUNDAMENTALS

2. CLASSES AND ORDERINGS

2.1 Introduction

In this chapter we informally sketch a theory of classes which will serve as a basis for investigating, in later chapters, the notions of validity and satisfiability, referred to a formalized language of sets. We particularly focus on certain classes, namely:

- the family ω of all *natural numbers*;

- *well-founded graphs, well-orderings, anti-lexicographic orderings*, the class *Ord* of all *ordinal numbers*;

- the class *Card* of all *cardinal numbers*; and

- the class $\mathcal{V}$ of all sets, called the *von Neumann universe*.

After reviewing the most fundamental closure properties of *Ord*, we introduce the *rank* function and recall its main properties.

A theorem in this chapter (Theorem 2.3) states that an anti-lexicographic well-ordering $<$ of the von Neumann universe exists which agrees with rank comparison, in the sense that $x < y$ holds whenever x has smaller rank than y. This ordering will play an important role in this book, because the semantics of the choice operators, to be presented in the next chapter, will be based on it.

Another result in this chapter (Lemma 2.11) shows how to establish a mapping from any given well-founded graph into a suitable family of sets, which is a 'homomorphism' in the sense that two arbitrary nodes x, y are connected by an edge (x, y) of the graph if and only if the set X that corresponds to x belongs to the set Y that corresponds to y.

2.2 Basic notation, assumptions, and terminology

The background of our discussion is an extensional theory of *classes* similar to Morse's theory (cf. [Kel55, Mor65]). Some classes qualify as *sets*, according to the criterion:

x is a set if and only if x belongs to some class y .

'*Family*' and '*collection*' are synonymous with 'set' and 'class' respectively. As is customary nowadays, a class f whose members are pairs (see definition below) will be called a *function*, provided simply that it fulfils the *uniqueness condition*:

> for each first component x of a pair in f,
> there is only one pair (x, y) in f;

the y that corresponds to x in this manner, is written $f x$ or $f(x)$.

We postulate the existence of an empty set, to be denoted $\emptyset$, and of a 'universal' class, $\mathcal{V}$, to which every set belongs. The internal structure of $\mathcal{V}$ will be discussed in later sections.

Enumerations of the two kinds

$$\{ x_1, \ldots, x_n \} \text{ and } (x_1, \ldots, x_n)$$

can always be used to aggregate a finite number of sets into a single set. Enumerations of the first kind are insensitive to any rearrangement or repetition of their components; for instance,

$$\{ x_1, \ldots, x_n \}, \{ x, x_1, \ldots, x_n \}, \text{ and } \{ x_{\sigma_1}, \ldots, x_{\sigma_n} \}$$

all three denote the same set if x equals some x_i with i from among $1, \ldots, n$ and σ is any permutation of $\{1, \ldots, n\}$. Enumerations of the form $(x_1, \ldots, x_n)$, sometimes written $[x_1, \ldots, x_n]$, are called *tuples*. The identity

$$(x_1, \ldots, x_n) = (y_1, \ldots, y_m)$$

holds between two tuples if and only if both $n = m$ and $x_i = y_i$ for $i = 1, \ldots, n$.

Membership, denoted by $\in$, is assumed to be *well-founded*: that is, no infinite descending chain

$$\ldots \in x_2 \in x_1 \in x_0$$

exists, nor, as a consequence, does any cycle

$$x_0 \in x_1 \in \cdots \in x_n \in x_0.$$

Inclusion, written $x \subseteq y$, holds whenever all members of x also belong to y, in which case x is said to be a *subclass of y*. Every subclass of a set y is itself a set, and is said to be a *subset of y*. The familiar set-theoretic operations

$$\cap, \setminus, \cup, \times \quad \text{(binary),}$$
$$Pow, \; Un \quad \text{(unary)}$$

are permitted to operate not only on sets, but also on classes.

Note that every non-empty class x has a member y with $x \cap y = \emptyset$. In fact, we can arbitrarily select a member y_0 in x and next, for $i = 0, 1, 2, \ldots$, we can choose some y_{i+1} in $x \cap y_i$ until the latter set is empty. Since the chain $\cdots \in y_2 \in y_1 \in y_0$ cannot be infinitely long, it eventually terminates with the y sought.

The class consisting of all first (respectively, second) components of pairs belonging to a function f, is called the *domain* (respectively, the *range*) of f; the fact that $fv = w$ is sometimes written

$$v \overset{f}{\mapsto} w \quad (\text{read: } f \text{ sends } v \text{ into } w).$$

If x is the domain of f, and the range of f is a subclass of y, we say that f is a *function from x into y*, and write

$$f : x \rightarrow y.$$

Functions from x *onto* y, *one-to-one correspondences between x and y*, and other related concepts are defined in agreement with the tradition; we therefore omit the definitions. Note that one can characterize *finite* sets in terms of one-to-one correspondences: *a set x is finite if and only if there is no subset y of x such that $y \neq x$ holds and a one-to-one correspondence exists between x and y.*

Let f and g be functions such that the domain D of g is a subclass of the domain of f and

$$fx = gx \text{ for all } x \text{ in } D.$$

Then g is called the *restriction of f to D* and is denoted as $g = f|_D$. The reader is warned that the word 'restriction' will often be used with a slightly different meaning: if

$$R \subseteq T \times T, \ S \subseteq T, \text{ and } R^* = R \cap (S \times S),$$

then we say that R^* is the *restriction of R to S*. Which of the two senses is meant should always be clear from the context.

The class consisting of all sets x that satisfy a certain condition $p(x)$ will be denoted by the *abstraction term*

$$\{ x : p(x) \}.$$

For instance, we have

- $\emptyset = \{x : x \in x\} = \{x : x \neq x\}$,

- $\mathcal{V} = \{x : x \notin x\} = \{x : x = x\}$,

- $\{t_1, \ldots, t_n\} = \{x : x = t_1 \text{ or } \cdots \text{ or } x = t_n\}$,

- $y \cap z = \{x : x \in y \text{ and } x \in z\}$,

- $y \setminus z = \{x : x \in y \text{ and } x \notin z\}$,

- $y \cup z = \{x : x \in y \text{ or } x \in z\}$,

- $Pow(y) = \{x : x \text{ is a set and } x \subseteq y\}$,

- $Un(y) = \{x : x \in z \text{ for some } z \text{ in } y\}$,

- $Dom(f) = \{x : (x, y) \in f \text{ for some } y\}$,

- $range(f) = \{y : (x, y) \in f \text{ for some } x\}$,

- $y \times z = \{x : x = (v, w) \text{ for some } v \text{ in } y \text{ and some } w \text{ in } z\}$.

The following notations will be frequently used:

- $\bigcup_{p(x)} x$ to denote the class $Un(\{x : p(x)\})$, and even

- $\bigcup_{p(x)} t(x)$, where $t(x)$ is any term, to denote the class

$$Un(\{z : z = t(x) \text{ for some } x \text{ such that } p(x)\}) \, ;$$

- $\bigcap_{p(x)} x$ to denote the class

$$\{y : y \in x \text{ for all } x \text{ such that } p(x)\} \, ,$$

  and even

- $\bigcap_{p(x)} t(x)$, where $t(x)$ is any term, to denote the class

$$\{y : y \in t(x) \text{ for all } x \text{ such that } p(x)\} \, ;$$

- $\{x \in y : p(x)\}$ to denote the class

$$\{x : x \in y \text{ and } p(x)\} \, ,$$

  and even

- $\{t(x) \in y : p(x)\}$ to denote the class

$$\{z : z = t(x) \text{ for some } x \text{ such that both } p(x) \text{ and } t(x) \in y\} \, ;$$

- $f[z]$, where f is a function from a class x into a class y, and $z \subseteq x$, to denote the class

$$\{v : v = fw \text{ for some } w \text{ in } z\} \, ;$$

- $\prod_{x \in y} f\,x$, where f is a function from the class y into a class z, to denote the class

$$\{g\ :\ g \text{ is a function from } y \text{ into } Un(z)$$
$$\text{such that } g\,x\ \in\ f\,x\ \text{ for all } x \text{ in } y\}\,;$$

- $Pow^*(x)$, to denote the class

$$\left\{\, Un(\,f[x]\,)\ :\ f \text{ in } \prod_{y \in x}\,(\,Pow(y)\ \setminus\ \{\emptyset\}\,)\,\right\},$$

formed by the subsets of $Un(\,x\,)$ that have a non-empty intersection with every one of the members of x.

Note that if $p(x)$ holds for no x, then

$$\bigcap_{p(x)} x\ =\ \mathcal{V}, \text{ and } \bigcup_{p(x)} x\ =\ \emptyset\,.$$

Also note that $Pow^*(\emptyset)\ =\ \{\,\emptyset\,\}$.

The theory of classes that we are now sketching will serve as a basis for investigating a formalized theory of sets in later chapters; for this reason, we shall refer to it as to our *meta-theory*. We avoid the cumbersome process of entirely formalizing the meta-theory; in particular, we have been vague about the meaning of the words 'condition' and 'term', and are leaving implicit the rules for proving that an abstraction term $\{\,x\ :\ p(x)\,\}$ denotes a set. The reader who wants to see these things in more detail is again referred to [Kel55] and [Mor65].

Most of this chapter centres around the notion of well-ordering. Let us briefly recall the definition:

Definition 2.1 *Suppose that C is a class and that $<$ is a subclass of $C \times C$. We indicate by $x \not< y$ the fact that $(\,x\,,\,y\,) \notin\,<$, and by $x_0 < x_1 < \cdots < x_n$ the fact that*

$$(x_0\,,\ x_1\,),\ (x_1\,,\ x_2\,),\dots,\ (\,x_{n-1}\,,\ x_n\,)\ \in\,<\,.$$

If the conditions

(i) $x \not< x$, and

(ii) $x\ <\ y\ <\ z$ implies $x\ <\ z$

are both satisfied for all $x\,,\,y\,,\,z$ in C, then $<$ is said to be a (partial) ordering of C. If, in addition, either $x\ <\ y$ or $x\ =\ y$ or $y\ <\ x$ holds for every pair $(\,x\,,\,y\,)$ in $C\ \times\ C$, then $<$ is said to be a total (or 'linear') ordering. If, for every non-empty subclass K of C, there is a 'minimum' k in K such that $k\ <\ c$ for all c in $K\ \setminus\ \{k\}$, the ordering is said to be a well-ordering. Equivalently, $<$ is a well-ordering iff $<$ is total and well-founded, in this sense: every 'descending chain' $\dots\ <\ c_2\ <\ c_1\ <\ c_0$ has finite length. □

2.3   Transitive sets (natural, ordinal, and cardinal numbers). Transfinite induction

A class x is said to be *transitive* iff:

$$x \subseteq Pow(x)\,;$$

that is: if and only if every member of x is a subset of x. Equivalently, x is transitive iff $Un(x) \subseteq x$; that is: if and only if every member of a member of x itself belongs to x. Observe that $\emptyset$ belongs to every non-empty transitive class x. Indeed, since x is non-empty, there is a y in x such that $x \cap y = \emptyset$. Since x is transitive, it follows that $y \subseteq x$, which shows that $\emptyset = x \cap y = y$ belongs to x.

A class x is said to be *finitely transitive* iff every member of a finite member of x itself belongs to x, i.e. for $n = 0,\, 1,\, 2,\, \ldots$:

$$y \in x\,,\ \text{together with}\ y = \{\, x_1,\ldots, x_n \,\}\,,\ \text{implies that}\ x_1,\ldots,\ x_n \in x\,.$$

It has become customary (after von Neumann), in presentations of class theory, to identify each natural number i with the set $\{\, 0,\, 1,\ldots,\, i-1 \,\}$ of its predecessors. In particular

$$0 = \emptyset\,,\quad 1 = \{\emptyset\}\,,\quad 2 = \{\, \emptyset, \{\emptyset\} \,\}\,.$$

Having this set-theoretic encoding of numbers in mind, and noticing that $x \in y$ always implies that neither $x = y$ nor $y \in x$, one is led to the following definition.

A *natural number* is a finite transitive set x which is totally ordered by the membership relation $\in$; the *successor* of a set x is the set $x \cup \{x\}$, which is also denoted $x + 1$. We also inductively define:

$$\begin{aligned}
&\text{if } k = 0, &&\text{then } x + k = x;\\
&\text{if } k = h + 1, &&\text{then } x + k = (\,x + h\,) + 1.
\end{aligned}$$

One can prove that when x and k are natural numbers —so that both notations $x + k$ and $k + x$ make sense—, then $x + k = k + x$.

One generalizes the notion of a natural number to that of an ordinal, by simply dropping the finiteness requirement. Formally: *an* ordinal number *(or just an 'ordinal') is a transitive set x which is totally ordered by the membership relation $\in$.*

We will denote by ω the set of all natural numbers. Thus ω and $\omega \cup \{\omega\}$ provide examples of infinite ordinals. Note that ω cannot be written in the form, $x \cup \{\, x \,\}$, of a successor; it is therefore called a *limit ordinal*. Formally: *a* limit ordinal *is an ordinal number, other than 0, which cannot be written as $x + 1$, for any set x.*

Note that each ordinal x is actually well-ordered by $\in$. In fact, if y is a non-empty subset of x, then y has a member z with $y \cap z = \emptyset$. For every w

in y, since $w \notin z$, we have either $w = z$ or $z \in w$. We conclude that z is the minimum member of y with respect to $\in$.

There seems to be no difficulty involved in the assumption that there exists a class of all ordinals:

$$Ord = \{ x \ : \ x \text{ is an ordinal} \} .$$

One can prove that this class is transitive and well-ordered by $\in$; however, Ord cannot be an ordinal, else we would derive

$$Ord \in Ord ,$$

which leads to a contradiction. Even more counter-intuitively, the successor $Ord \cup \{ Ord \}$ of Ord (see below) would be 'smaller' than Ord, because $Ord \cup \{ Ord \} \in Ord$. In order to escape this Burali-Forti paradox, one will agree that Ord cannot be a set: Ord represents, indeed, our first example of a *proper class*, i.e. a class which is not a set. As a consequence, the universe $\mathcal{V}$ of all sets cannot be a set either, because $Ord \subseteq \mathcal{V}$.

Ord has a lot of nice closure properties. To mention just a few of them (in addition to transitivity):

- If $x, y \in Ord$, then
  $$x \in y \text{ if and only if } x \subsetneq y, \text{ and}$$
  $$x \notin y \text{ if and only if } y \subseteq x.$$

- For each x in Ord, also $x + 1 = x \cup \{x\} \in Ord$.

- If $Y \subseteq Ord$ and Y is a set, then $Un(Y) \in Ord$; moreover, if Y is a non-successor ordinal, then $Y = Un(Y)$.

- Every class Y of ordinals has an upper bound m, which is
  $$m = \bigcup_{y \in Y} (y + 1).$$

  This m is either an ordinal or the class Ord; in particular m is the least member of $Ord \setminus Y$ when Y is a set.

- For every x in Ord, the set $\bigcup_{k \in \omega} (x + k)$, to be denoted $x + \omega$, is the least limit ordinal that exceeds x.

By far the most useful closure property of Ord, however, is the well-known

Principle 2.1 (Transfinite induction) *If a given class Y includes among its members every ordinal x such that $z \in Y$ for all z in x, then $Ord \subseteq Y$.*$\Box$

A more convenient statement of this principle is the following:

Principle 2.1 (Transfinite induction) *Let Y be a class. If*

(i) $0 \in Y$;

(ii) $y + 1 \in Y$ whenever $y \in Y \cap Ord$;

(iii) $x \in Y$ for every limit ordinal x such that $z \in Y$ for all z in x;

then $Ord \subseteq Y$.   □

Even more powerful inductive principles will be examined in Section 2.6.

Without proof, we recall two very fundamental propositions:

Theorem 2.1 *Suppose that S is a set included in a well-ordered class $(T, <)$. Then the restriction of $<$ to S is a well-ordering of S. Moreover, there is exactly one ordinal number α for which a one-to-one correspondence*

$$f : S \to \alpha$$

exists between S and α that satisfies, for all r, s in S, the equivalence

$$r < s \text{ if and only if } fr \in fs.$$

This ordinal α will be written

$$ord(S, <).$$

The correspondence f is also uniquely determined by $(S, <)$; as a matter of fact, $fs = ord(\{r \in S : r < s\}, <)$, for all s in S.   □

Theorem 2.2 *For every set S, the set*

$$Q = \{ ord(S, <) : < \text{ is a well-ordering of } S \}$$

has a member C such that

$$C \in \sigma \text{ for all } \sigma \text{ in } Q \setminus \{C\}.$$

Such an ordinal is called the cardinality *of S, written $|S|$.*

Hint of the Proof. Q is non-empty, according to the well-known Zermelo principle, and is indeed a set, because every well-ordering of S belongs to the set $Pow\,(S \times S)$.

Since $Q \subseteq Ord$ and $(Ord, \in)$ is a well-ordered class, Q has a minimum C with respect to membership.   ■

The class

$$\{|s| : s \in \mathcal{V}\}$$

consisting of the cardinalities of all sets is called *Card*; its members are called *cardinal numbers* or just *cardinals*. For each cardinal C, one has $C = |C|$.

Let $S \subseteq Card$. Choose for each C in S a set s_C, so that $C = |s_C|$ and the sets s_C are pairwise disjoint. For instance it would suffice to let

$$s_C = C \times \{C\} = \{(x, C) : x \in C\}.$$

A summation operation is defined by putting

$$\Sigma S = \Sigma_{C \in S} C = |\bigcup_{C \in S} s_C|.$$

For a finite ordinal x, one has $x = |x|$; hence natural numbers are indistinguishable from their cardinalities. Moreover, when restricted to finite sets S of natural numbers, the summation operation just introduced coincides with the ordinary arithmetic summation.

2.4   Anti-lexicographic orderings. Concatenation of well-orderings

Lemma 2.1 (Anti-lexicographic orderings) *Suppose $(S, <)$ is a totally ordered class. Let F be a class consisting of finite subsets of S. Define the binary relation $\lhd$ on F as follows:*

if $\{r_0, \ldots, r_n\}$, $\{s_0, \ldots, s_m\} \in F$ and $r_n < r_{n-1} < \cdots < r_0$, $s_m < s_{m-1} < \cdots < s_0$, then $\{r_0, \ldots, r_n\} \lhd \{s_0, \ldots, s_m\}$ holds if and only if either

   (i) $n \in m$ and $r_i = s_i$, for $i = 0, \ldots, n$; or

   (ii) there is some j such that $n, m \notin j$, and
   $r_0 = s_0, \ldots, r_{j-1} = s_{j-1}, r_j < s_j.$

Then $\lhd$ is a total ordering of F which is said to be anti-lexicographically *induced by $<$ on F. If S is a set and $<$ is a well-ordering, then $\lhd$ is a well-ordering too; $r \lhd s$ is read: r anti-lexicographically precedes s, with respect to $<$.*

Proof. Let us prove first that either $x_0 \lhd x_1$ or $x_0 = x_1$ or $x_1 \lhd x_0$ when $x_0, x_1 \in F$. In fact, if we scan the sets x_0, x_1 in parallel, proceeding from 'larger' to 'smaller' elements, sooner or later one of the following will happen:

   (a) The element which is being scanned in x_0 is different from the one being scanned in x_1; or

   (b) x_b has been completely scanned, but some part of x_{1-b} is still unexplored; or

   (c) the scannings of x_0 and x_1 are completed simultaneously.

As soon as one of these situations arises for the first time, we are able to compare x_0 with x_1; in fact, in the respective cases we have:

(a) $x_b \lhd x_{1-b}$, if the smaller element which is being scanned lies in x_b;

(b) $x_b \lhd x_{1-b}$;

(c) $x_0 = x_1$.

We have thus shown that if $\lhd$ is a partial order, it is then a total order. Preliminary to the proof that it is indeed a partial ordering, let us introduce the notation

$$x = x_0 \heartsuit x_1 \heartsuit \cdots \heartsuit x_n$$

to indicate the following situation:

- the union of the x_is gives x, i.e. $x = x_0 \cup x_1 \cup \cdots \cup x_n$;

- for $0 \le i < j \le n$, every member of x_i precedes any member of x_j, i.e. if $y \in x_i$ and $z \in x_j$, then $y < z$.

It follows from the latter condition that the x_is are pairwise disjoint sets, i.e.

- $x_i \cap x_j = \emptyset$ for $0 \le i < j \le n$.

Exploiting this notation one can easily characterize the relation $\lhd$:

$x \lhd y$ holds if and only if one can either write

$$y = q \heartsuit x \quad \text{with} \quad q \neq \emptyset,$$

or one can decompose x and y as follows:

$$x = p \heartsuit \{v\} \heartsuit z, \; y = q \heartsuit \{w\} \heartsuit z, \quad \text{with} \; v < w.$$

It is immediately transparent from this characterization that one cannot have $x \lhd x$: it is impossible to have

$$x = q \heartsuit x \text{ with } q \neq \emptyset;$$

nor can one have

$$x = p \heartsuit \{v\} \heartsuit z = q \heartsuit \{w\} \heartsuit z \text{ with } v < w,$$

or else it would follow that

$$w \notin z, \; w \neq v, \; u < v < w \text{ for every } u \text{ in } p,$$

contradicting the assumption that $w \in x$.

Assuming now that $r \lhd s \lhd t$, we must prove that $r \lhd t$.

Case 1: $s = s_1 \heartsuit r$, $s_1 \neq \emptyset$;
$t = t_1 \heartsuit s$, $t_1 \neq \emptyset$.

Then
$$t = t_1 \heartsuit s_1 \heartsuit r, \text{ where } t_1 \heartsuit s_1 \neq \emptyset;$$
hence $r \lhd t$.

Case 2: $s = s_1 \heartsuit r$, $s_1 \neq \emptyset$;
$s = p \heartsuit \{v\} \heartsuit z$,
$t = q \heartsuit \{w\} \heartsuit z$, with $v < w$.

Subcase 2.1: $v \in r$, hence
$r = r_1 \heartsuit \{v\} \heartsuit z$,
$t = q \heartsuit \{w\} \heartsuit z$, with $v < w$;

hence $r \lhd t$.

Subcase 2.2: $v \in s_1$, hence
$z = z_1 \heartsuit r$,
$t = q \heartsuit \{w\} \heartsuit z_1 \heartsuit r$, where $q \heartsuit \{w\} \heartsuit z_1 \neq \emptyset$;

hence $r \lhd t$.

Case 3: $r = p \heartsuit \{v\} \heartsuit z$,
$s = q \heartsuit \{w\} \heartsuit z$, with $v < w$,
$t = t_1 \heartsuit s = t_1 \heartsuit q \heartsuit \{w\} \heartsuit z$;

hence $r \lhd t$.

Case 4: $r = p_1 \heartsuit \{v_1\} \heartsuit z_1$,
$s = q_1 \heartsuit \{w_1\} \heartsuit z_1$, with $v_1 < w_1$;
$s = p_2 \heartsuit \{v_2\} \heartsuit z_2$,
$t = q_2 \heartsuit \{w_2\} \heartsuit z_2$, with $v_2 < w_2$.

Subcase 4.1: $w_1 \in z_2$, hence
$z_2 = z_{21} \heartsuit \{w_1\} \heartsuit z_1$,
$t = q_2 \heartsuit \{w_2\} \heartsuit z_{21} \heartsuit \{w_1\} \heartsuit z_1$, with $v_1 < w_1$;

hence $r \lhd t$.

Subcase 4.2: $w_1 = v_2$, hence
$z_1 = z_2$, $q_1 = p_2$;
$t = q_2 \heartsuit \{w_2\} \heartsuit z_1$, with $v_1 < w_1 = v_2 < w_2$;

hence $r \lhd t$.

Subcase 4.3: $w_1 \in p_2$, hence
$z_1 = z_{11} \heartsuit \{v_2\} \heartsuit z_2$,
$r = p_1 \heartsuit \{v_1\} \heartsuit z_{11} \heartsuit \{v_2\} \heartsuit z_2$,
$t = q_2 \heartsuit \{w_2\} \heartsuit z_2$, with $v_2 < w_2$;

hence $r \lhd t$.

Assuming now that S is a set and that $<$ is a well-ordering, let us prove that $\lhd$ is a well-ordering too. Without loss of generality, in view of Theorem 2.1, we can assume that F is the set of *all* finite subsets of S. We consider a descending chain

$$\cdots \lhd f_2 \lhd f_1 \lhd f_0,$$

and proceed to prove that the length of this chain is finite. If $\emptyset = f_i$ for some i, then the chain has length $i + 1$, because $\emptyset$ is of course the minimum of F with respect to $\lhd$. Otherwise, we can write

$$f_0 = \{ s_0, \ldots, s_m \} \text{ with } s_m < s_{m-1} < \cdots < s_0,$$

and proceed by transfinite induction on

$$\gamma = ord(\{ s \in S : s < s_0 \}, <).$$

For each γ, we consider the following two cases:

Case A: s_0 is the maximum, with respect to $<$, in each of the sets f_i. This is the only possible case when $\gamma = 0$. Then, by deleting s_0 from each f_i, we obtain another anti-lexicographic chain,

$$\cdots \lhd f_2' \lhd f_1' \lhd f_0',$$

which has the same length as the first. If $\emptyset$ occurs somewhere in this chain, we are finished; otherwise we can exploit the induction hypothesis, because the maximum s_0' of f_0' precedes s_0 in $(S, <)$, and hence

$$ord(\{ s \in S : s < s_0' \}, <) \in ord(\{ s \in S : s < s_0 \}, <).$$

Case B: There is some f_i whose maximum s_0'' is different from s_0. Plainly, $s_0'' < s_0$. The induction hypothesis applies to the descending subchain that originates with f_i; hence the chain is finite.   ∎

Remark 2.1 The above lemma can be further strenghtened. To illustrate this, we point out a more general theorem, which is very useful in termination proofs (cf. [DM79]).

We begin by defining

$$\mathcal{M}(S) = \bigcup_{R \in Pow(S)} \{M : M \text{ is a function from } R \text{ into } \omega \setminus \{0\}\},$$

for any class S. The members of $\mathcal{M}(S)$ are called *multisets over* S. If $M \in \mathcal{M}(S)$ and $x \in S \setminus Dom(M)$, we convene to write $M x = 0$.

Suppose then that $(S, <)$ is a partially ordered class, and that F is a class consisting of finite multisets over S. Define $\lhd$ on F as follows: $X \lhd Y$ holds iff $X \neq Y$ and, furthermore,

> to every r in S such that $Y r \in X r$ there corresponds some other s in S, with $r < s$, for which $X s \in Y s$.

This $\lhd$ is a partial ordering of F. Moreover, $\lhd$ is total (respectively: well-founded) if $<$ is total (respectively: well-founded).

In order to obtain Lemma 2.1 from this statement, it will suffice to identify each $R \subseteq S$ with the singleton multiset $M_R : R \to \{1\}$.   □

Consider a finitely transitive, totally ordered class $(S, <)$. In the manner seen in Lemma 2.1, $<$ induces an ordering $\lhd$ among the finite subsets of S.

Lemma 2.2 *Let x, y be finite subsets of S. One has that:*

(1) if $x \subsetneq y$, then $x \lhd y$;

(2) if neither $x \subseteq y$ nor $y \subseteq x$, then:

$$x \lhd y \text{ holds if and only if } max(x \setminus y) < max(y \setminus x).$$

Proof. Concerning (1), assume that $x \subsetneq y$. If we scan the two sets x, y in parallel, proceeding from larger to smaller elements, one of the following will eventually occur:

(a) the element s which is being scanned in x is different from the one, u, being scanned in y;

(b) x has been completely scanned, but some part of y is still unexplored.

As soon as either (a) or (b) arises for the first time, we are able to compare x with y. Of course $x \lhd y$ if (b) is the case. In case (a), we have

$$x = v \heartsuit w, \quad y = z \heartsuit \{u\} \heartsuit w \text{ with } \emptyset \neq v \subseteq z;$$

where u is larger than any member of z, so that in particular $s = max\, v < u$. Again $x \lhd y$.

Concerning (2): only (a) can occur under the assumption that $x \not\subseteq y \not\subseteq x$; hence we have

$$x = t \heartsuit \{s\} \heartsuit w, \quad y = z \heartsuit \{u\} \heartsuit w.$$

If $s < u$, then $x \lhd y$ and also $u = max(y \setminus x)$, while $max(x \setminus y) \leq s$; hence $max(x \setminus y) < max(y \setminus x)$. Symmetrically, if $u < s$, then $y \lhd x$ and $max(y \setminus x) < max(x \setminus y)$. ∎

Let F be the set of those finite sets that belong to S

$$F = \{s \in S : |s| \in \omega\}.$$

By definition, $F \subseteq S$; moreover, every member of F is a subset of S, because we have assumed that S is finitely transitive. We can therefore consider the restrictions of both $<$ and $\lhd$ to F.

Definition 2.2 $(S, <)$ *is* anti-lexicographic *iff* $(F, <)$ *and* $(F, \lhd)$ *are the same, i.e. the equivalence*

$$x < y \text{ if and only if } x \lhd y$$

holds for all x, y in F. □

From Lemma 2.2 we easily deduce:

Corollary 2.1 (S, $<$) *is anti-lexicographic if and only if, for all* x, y *in* F:

(1) *if* $x \subsetneq y$, *then* $x < y$;

(2) *if* $x \not\subseteq y$ *and* $y \not\subseteq x$, *then:*
   $x < y$ *holds if and only if* $max(x \setminus y) < max(y \setminus x)$.

In particular, if $<$ *is anti-lexicographic and* $\emptyset \in S$, *then* $\emptyset$ *precedes every other finite set in* S.

Proof. Note that (1) and (2) together give a criterion for comparing any two distinct sets x, y in F; in fact, either $x \subsetneq y$ or $y \subsetneq x$ or $x \not\subseteq y \not\subseteq x$ when $x \neq y$. Hence there is at most one total ordering $\prec$ of F such that analogues of (1) and (2) hold, with '$x \prec y$' in place of '$x < y$'. The corollary follows, in view of the fact that $\lhd$ is such an ordering. ∎

Corollary 2.2 *Both* (ω, $\in$) *and* (Ord, $\in$) *are anti-lexicographic.*

Proof. We have already noted in Section 2.3 that

$$x \in y \text{ if and only if } x \subsetneq y,$$

when x, y are natural (or even ordinal) numbers. Condition (1) of Corollary 2.1 is therefore satisfied.

Moreover, for x, y in Ord, it is always the case that either

$$x = y, \text{ or } x \in y, \text{ or } y \in x;$$

hence it never happens that $x \not\subseteq y$ and $y \not\subseteq x$ hold together. Condition (2) of Corollary 2.1 is hence vacuously satisfied. ∎

Corollary 2.3 *Suppose that* (S, $<$) *is anti-lexicographic and that* $s_0, \ldots, s_n \in F$ *are non-empty and pairwise disjoint, with* $s_0 < s_1 < \cdots < s_n$. *For all* $K \subseteq \{0, \ldots, n\}$, *define*

$$S_K = \bigcup_{k \in K} s_k.$$

If I, $J \subseteq \{0, \ldots, n\}$ *are such that* S_I, $S_J \in F$, *then the following four conditions are equivalent:*

(i) $S_I < S_J$;

(ii) $\{s_i : i \in I\} \lhd \{s_j : j \in J\}$;

(iii) $\{max\ s_i : i \in I\} \lhd \{max\ s_j : j \in J\}$;

(iv) I *anti-lexicographically precedes* J *with respect to* (ω, $\in$).

Proof. Since the sets s_i are non-empty and pairwise disjoint, we get from Corollary 2.1:

$$s_i < s_j \quad \text{if and only if} \quad max\, s_i < max\, s_j\,, \text{ hence}$$
$$i \in j \quad \text{if and only if} \quad max\, s_i < max\, s_j\,,$$

for i, j in $\{0,\ldots,n\}$. This shows the equivalence between (ii), (iii), and (iv). Continuing to use Corollary 2.1, we obtain:

$$S_I < S_J \text{ if and only if either } S_I \subsetneq S_J \text{ or}$$
$$S_I \not\subseteq S_J \not\subseteq S_I \text{ and } max(\,S_I \setminus S_J\,) < max(\,S_J \setminus S_I\,)\,.$$

Note that

$$S_K = \emptyset \text{ if and only if } K = \emptyset\,,$$
$$max\, S_K = max_{k \in K}\, max\, s_k = max\, s_{max\, K} \text{ if } K \ne \emptyset\,,$$
$$S_K \setminus S_L = S_{K \setminus L}\,,$$

for all K, $L \subseteq \{0,\ldots,n\}$, and, therefore,

$$
\begin{aligned}
S_K \subseteq S_L \quad &\text{iff} \quad S_K \setminus S_L = \emptyset \\
&\text{iff} \quad S_{K \setminus L} = \emptyset \\
&\text{iff} \quad K \setminus L = \emptyset \\
&\text{iff} \quad K \subseteq L\,,
\end{aligned}
$$

$$S_K = S_L \quad \text{iff} \quad K = L\,,$$

$$
\begin{aligned}
max\, S_K < max\, S_L \quad &\text{iff} \quad max\, s_{max\, K} < max\, s_{max\, L} \\
&\text{iff} \quad max\, K \in max\, L\,.
\end{aligned}
$$

We conclude that: $S_I < S_J$ if and only if either $I \subsetneq J$, or $I \not\subseteq J \not\subseteq I$ and $max(\,I \setminus J\,) \in max(\,J \setminus I\,)$; i.e., if and only if I anti-lexicographically precedes J. This shows the equivalence between (i) and (iv). ∎

A lemma concerning well-orderings:

Lemma 2.3 (Concatenation of well-orderings) *Let $(I, <)$ and $(S_i, <_i)$, with i in I, be well-ordered sets. Suppose that the S_is are pairwise disjoint, i.e.*

$$S_i \cap S_j = \emptyset \text{ when } i \ne j\,.$$

A well-ordering $\lhd$ of the set $S = \bigcup_{i \in I} S_i$ can be defined by putting $x \lhd y$ iff either

> *there is some i in I for which x, $y \in S_i$ and $x <_i y$, or*
> *there are i, j in I for which $x \in S_i$, $y \in S_j$, and $i < j$.*

We call $\lhd$ the $<$-concatenation of the $<_i$s.

Proof. For no x in S can $x \lhd x$ hold, since this would imply that either $x <_i x$ for some i, or $x \in (S_i \cap S_j)$ with $i \neq j$. If $x \lhd y$ and $y \lhd z$, where $x \in S_i$, $y \in S_j$, and $z \in S_k$, then either

$$i < j < k, \text{ or}$$
$$i = j < k \text{ and } x <_i y, \text{ or}$$
$$i < j = k \text{ and } y <_j z, \text{ or}$$
$$i = j = k \text{ and } x <_i y <_i z.$$

In each of these four cases, one has $x \lhd z$ according to the definition. If $x, y \in S$, then there are i, j such that $x \in S_i$, $y \in S_j$. If $i = j$, then either $x \lhd y$ or $x = y$ or $y \lhd x$ according to whether $x <_i y$ or $x = y$ or $y <_i x$; if $i \neq j$, then either $x \lhd y$ or $y \lhd x$ according to whether $i < j$ or $j < i$.

Consider a chain

$$\cdots \lhd x_2 \lhd x_1 \lhd x_0.$$

Suppose that $x_h \in S_{i_h}$ for $h = 0, 1, 2, \ldots$. Since each $<_{i_h}$ is a well-ordering, we cannot have an infinite sequence

$$\cdots = i_{h+2} = i_{h+1} = i_h;$$

for this would in fact imply that

$$\cdots <_{i_h} x_{h+2} <_{i_h} x_{h+1} <_{i_h} x_h$$

is an infinite sequence. We can therefore divide the chain

$$\cdots \lhd x_2 \lhd x_1 \lhd x_0$$

into subchains

$$x_{h_0} \lhd \quad \cdots \qquad \lhd x_1 \lhd x_0,$$
$$x_{h_1} \lhd \quad \cdots \qquad \lhd x_{h_0+2} \lhd x_{h_0+1},$$
$$x_{h_2} \lhd \quad \cdots \qquad \lhd x_{h_1+2} \lhd x_{h_1+1},$$
$$\cdots$$
$$\cdots,$$

with

$$\cdots < i_{h_2} < i_{h_1} < i_{h_0},$$

each of which is finite and a subset of the set $S_{i_{(h_j)}}$. Since $<$ is a well-ordering, the chain

$$\cdots < i_{h_2} < i_{h_1} < i_{h_0}$$

must have finite length, and therefore

$$\cdots \lhd x_2 \lhd x_1 \lhd x_0$$

is finite too. ∎

2.5 The von Neumann universe and its ordering

Let us define

$$
\begin{aligned}
V_0 &= \emptyset\,; \\
V_{\alpha+1} &= Pow(V_\alpha)\,, \quad \text{for each ordinal } \alpha\,; \\
V_\alpha &= \textstyle\bigcup_{\beta\in\alpha} V_\beta\,, \quad \text{for each limit ordinal } \alpha\,.
\end{aligned}
$$

Consider the union, taken over all natural numbers,

$$
\mathcal{H} = \bigcup_{\alpha\in\omega} V_\alpha
$$

of these sets. The members of the resulting family are called *hereditarily finite sets* (justification for this name will emerge in Chapter 4). If the union is taken over all ordinals, the resulting class

$$
\mathcal{V} = \bigcup_{\alpha\in Ord} V_\alpha
$$

is called the *von Neumann universe* of sets. It is generally assumed that $\mathcal{V}$ exhausts the totality of sets: *every set belongs to* $\mathcal{V}$. Hence, if we succeed in well-ordering $\mathcal{V}$, then we will be allowed to compare any two sets, and to extract the least member from each class. The existence of the desired well-ordering of $\mathcal{V}$ will be proved exploiting the well-known Zermelo principle:

'every set is well-orderable',

which is equivalent to the axiom of choice (see, e.g. [Abi65]). In turn, the existence of a well-ordering of $\mathcal{V}$ readily implies the Zermelo principle; we must therefore abandon from the beginning any hope to define the ordering of $\mathcal{V}$ constructively. Due to the non-constructive character of our definition of the ordering, we will not be able to characterize it uniquely, although we will constrain it very heavily.

As an application of the principle of transfinite induction, let us begin by introducing the notion of rank, and by proving its basic properties. We have assumed that every set s belongs to $\mathcal{V}$; hence, there is some β such that $s \in V_\beta$. Moreover, since the class Ord is well-ordered by $\in$, there is a minimum such β. Due to its minimality, β cannot be a limit ordinal; hence it has the form $\beta = \alpha + 1$.

Definition 2.3 *The* rank *of any set s, denoted $rk\ s$, is the least ordinal α such that $s \subseteq V_\alpha$, i.e. $s \in V_{\alpha+1}$.* $\qquad\square$

Lemma 2.4 *(i) If $s \in V_\alpha$, then $s \subseteq V_\alpha$ (otherwise stated: V_α is transitive);*

(ii) if $\alpha \in \beta$, then $V_\alpha \subsetneq V_\beta$;

(iii) $rk(x \cup y) = max\{rk\,x, rk\,y\};$

(iv) $rk \bigcup_{i \in I} x_i = \bigcup_{i \in I} rk\,x_i;$

(v) $rk\,\emptyset = 0$ *and* $rk\{s_0, \ldots, s_n\} = (max_{i=0}^n rk\,s_i) + 1;$

(vi) $rk\,s = \bigcup_{r \in s}(rk\,r + 1);$

(vii) *if* $r \in s$, *then* $rk\,r \in rk\,s;$

(viii) $V_\alpha = \{s : rk\,s \in \alpha\};$

(ix) $rk\,V_\alpha = rk\,\alpha = \alpha.$

Proof. Throughout this proof, we take the freedom of writing $\alpha < \beta$ and $\alpha \le \beta$ instead of $\alpha \in \beta$ and $\alpha \subseteq \beta$ respectively, for α, β in Ord.

(i) We proceed by transfinite induction on α. If $\alpha = 0$, then (i) holds vacuously. If $\alpha = \gamma + 1$ and $s \in V_\alpha$, then $s \subseteq V_\gamma$, according to the definition of V_α; hence $r \in V_\gamma$ for all r in s. By the induction hypothesis, it follows that $r \subseteq V_\gamma$, i.e. $r \in V_\alpha$, for all r in s; we therefore conclude that $s \subseteq V_\alpha$. If α is a limit ordinal and $s \in V_\alpha$, then $s \in V_\gamma$ for some γ in α, according to the definition of V_α. By the induction hypothesis, $s \subseteq V_\gamma$ and hence $s \subseteq V_\alpha$, because $V_\gamma \subseteq V_\alpha$.

(ii) We proceed by transfinite induction on β. If $\beta = 0$, then (ii) holds vacuously. If $\beta = \gamma + 1$ and $\alpha \in \beta$, then either $\alpha = \gamma$ or $\alpha \in \gamma$. In the first case, $V_\alpha = V_\gamma \in Pow(V_\gamma) = V_{\gamma+1} = V_\beta$; since V_β is transitive by (i), this implies that $V_\alpha \subseteq V_\beta$; but $V_\alpha \notin V_\alpha$ and hence $V_\alpha \ne V_\beta$. If $\alpha \in \gamma$ and $\beta = \gamma + 1$, then $V_\alpha \subsetneqq V_\gamma$ by the induction hypothesis, and $V_\gamma \subsetneqq V_\beta$ as we have just proved; hence $V_\alpha \subsetneqq V_\beta$. If β is a limit ordinal and $\alpha \in \beta$, then $\alpha + 1 \in \beta$ and $V_\alpha \subsetneqq V_{\alpha+1}$ by the induction hypothesis; moreover $V_{\alpha+1} \subseteq V_\beta$ according to the definition of V_β, and hence $V_\alpha \subsetneqq V_\beta$.

(iii) This is a special case of (iv)—see proof below—with $I = \{0, 1\}$, $x_0 = x$, and $x_1 = y$.

(iv) By the definition of rank, and by (ii), $x_i \subseteq V_{rk\,x_i} \subseteq V_{\bigcup_{i \in I} rk\,x_i}$ holds for all i in I. It follows that $\bigcup_{i \in I} x_i \subseteq V_{\bigcup_{i \in I} rk\,x_i}$, hence $rk \bigcup_{i \in I} x_i \le \bigcup_{i \in I} rk\,x_i$. Conversely, since $\bigcup_{i \in I} x_i \subseteq V_{rk \bigcup_{i \in I} x_i}$, one has that $x_i \subseteq V_{rk \bigcup_{i \in I} x_i}$; hence $rk\,x_i \le rk \bigcup_{i \in I} x_i$, for all i in I, which implies that $\bigcup_{i \in I} rk\,x_i \le rk \bigcup_{i \in I} x_i$.

(v) It is trivial from the definition of rank that $rk\,\emptyset = 0$. We now prove that $rk\{x\} = rk\,x + 1$; in view of (iv), it will follow that

$$rk\{s_0, \ldots, s_n\} = (max_{i=0}^n rk\,s_i) + 1,$$

because $\{s_0, \ldots, s_n\} = \bigcup_{i=0}^n \{s_i\}$. Since $x \in V_{rk\,x+1}$, it follows that $\{x\} \subseteq V_{rk\,x+1}$, hence $rk\{x\} \le rk\,x + 1$. Moreover, since $\{x\} \subseteq V_{rk\{x\}}$, i.e. $x \in V_{rk\{x\}}$, it follows that $rk\,x < rk\{x\}$; hence $rk\{x\} = rk\,x + 1$.

(vi), (vii) By (iv), since $s = \bigcup_{r \in s} \{r\}$, we have $rk\,s = \bigcup_{r \in s} rk\{r\}$. In view of (v), this implies that $rk\,s = \bigcup_{r \in s} (rk\,r + 1)$. Thus $rk\,r + 1 \leq rk\,s$, i.e. $rk\,r < rk\,s$, for all r in s.

(viii) If $rk\,s \in \alpha$, then $s \in V_{rk\,s+1} \subseteq V_\alpha$. Therefore $\{\,s : rk\,s \in \alpha\,\} \subseteq V_\alpha$. Conversely, if $s \in V_\alpha$, then $rk\,s \in \alpha$, by the minimality of $rk\,s$.

(ix) Exploiting (vi), we get

$$rk\,\alpha = \bigcup_{\beta < \alpha} (rk\,\beta + 1) = \bigcup_{\beta < \alpha} (\beta + 1) = \alpha\,,$$

if we inductively assume that $\beta = rk\,\beta$ holds for every β preceding α. Suppose now that $rk\,V_\alpha = \gamma$. Then $\alpha \subseteq V_\alpha \subseteq V_\gamma$ and hence $\alpha \subseteq V_\gamma$, $rk\,\alpha \leq \gamma$, $\alpha \leq \gamma$. Conversely, $\gamma \leq \alpha$ because $V_\alpha \subseteq V_\alpha$. The desired conclusion $\alpha = \gamma$ follows immediately. ∎

Example 2.1 Properties (v) and (vi) in the preceding lemma suggest the following intuitive interpretation of the rank function. Roughly speaking, $rk\,v$—which is transfinite in most cases—is an exact count of how nested the braces would be in a fully explicit representation of v (if the latter could be carried out effectively). For instance

$rk\,\emptyset = rk\,\{\} = 0;$

$rk\,\{\{\emptyset\}, \{\emptyset, \{\emptyset\}\}\} = 3;$

$rk\,X = X$, for every ordinal X and, in particular:
for every cardinal X;
for every natural number X
(identified here with the set $\{0, 1, \ldots, X - 1\}$);

$rk\,\{3, 4, 5, \ldots\} = \omega,$

$rk\,\{\{1, 2, 3, \ldots\}, 1, 2, 3, \ldots\} = \omega \cup \{\omega\},$

where $\omega = \{0, 1, 2, \ldots\}$ by definition. □

Lemma 2.5 *$rk\,X$ is a successor ordinal if and only if there exists an x in X with highest rank in X, i.e. such that*

$$rk\,y \subseteq rk\,x \quad \text{for every } y \text{ in } X\,.$$

Proof. If there is an x in X with highest rank in X, then

$$y \subseteq V_{rk\,y} \subseteq V_{rk\,x}\,,$$

which implies $y \in V_{rk\,x+1}$, for every y in X. It follows that $X \subseteq V_{rk\,x+1}$, hence $rk\,X \le rk\,x + 1$; on the other hand, $rk\,x \in rk\,X$, i.e., $rk\,x + 1 \le rk\,X$, because $x \in X$. We conclude that $rk\,X = rk\,x + 1$, a successor ordinal.

Conversely, if $rk\,X = \alpha + 1$, then recalling that $rk\,X = \bigcup_{y \in X}(rk\,y + 1)$, we infer that $X \neq \emptyset$ and $rk\,y + 1 \le \alpha + 1$, i.e. $rk\,y \le \alpha$, for every y in X. If $rk\,x = \alpha$ for some x in X, then x has highest rank in X and we are finished. As a matter of fact this must be the case, else we would have

$$rk\,y < \alpha, \text{ hence } rk\,y + 1 \le \alpha, \text{ for every } y \text{ in } X.$$

It would then follow that

$$rk\,X = \bigcup_{y \in X}(rk\,y + 1) \le \alpha,$$

contradicting our initial assumption.   ∎

Lemma 2.6 *Let α be an ordinal such that there exists an infinite set X with $rk\,X = \alpha$. Then, for every k in ω, there is an infinite set X_k such that*

$$rk\,X_k = \alpha + k.$$

Moreover, there is a set X_ω such that

$$\alpha + k < rk\,X_\omega \text{ for all } k \text{ in } \omega.$$

Proof. Incidentally, we note that since $\alpha = rk\,\alpha$ for every ordinal α (see Lemma 2.4(ix)), the hypothesis

'there exists an infinite set X with $rk\,X = \alpha$'

could be stated simply as

'α is an infinite ordinal'.

Also recall that by the notation $\alpha + k$ we indicate the ordinal

$$\alpha + 1 + \cdots + 1,$$

where the successor operation $\beta \overset{+1}{\longmapsto} \beta \cup \{\beta\}$ has been applied k consecutive times.

We define:

$$X_0 = X\,;$$
$$X_{h+1} = X_h \cup \{X_h\}, \text{ for all } h \text{ in } \omega\,;$$
$$X_\omega = \bigcup_{k \in \omega} X_k\,.$$

An easy induction then shows that

$$X \subseteq X_k \subseteq X_\omega,$$

(so that X_k and X_ω are infinite) and that

$$rk\, X_k = rk\, X + k = \alpha + k, \text{ for all } k \text{ in } \omega.$$

Since $X_k \subseteq X_\omega$, and hence

$$\alpha + k = rk\, X_k \leq rk\, X_\omega, \quad \text{for all } k \text{ in } \omega,$$

it follows that

$$\alpha + k < rk\, X_\omega, \quad \text{for all } k \text{ in } \omega.$$

∎

Theorem 2.3 *There exists an anti-lexicographic well-ordering $<$ of V that agrees with rank comparison, in the following sense: for every two sets x, y such that $rk\, x \in rk\, y$, one has $x < y$. Moreover, if x is finite, y is infinite, and $rk\, x = rk\, y$, then $x < y$.*

Proof. We will obtain $<$ in stages, by inductively defining the restriction $<_\alpha$ of $<$ to V_α, for every ordinal α. Recalling that

$$V = \bigcup_{\alpha \in Ord} V_\alpha,$$

and that

$$V_\alpha \subseteq V_\beta \text{ when } \alpha \in \beta,$$

we are forced to put

$$< = \bigcup_{\alpha \in Ord} <_\alpha.$$

In our construction we need to consider, together with each $<_\alpha$, the binary relation $<^\alpha$ on the finite sets of rank α that is anti-lexicographically induced by $<_\alpha$ (see Lemma 2.1).

We will prove, using simultaneous transfinite induction, that the following five statements hold for every α:

(1) $<_\alpha$ is a well-ordering of V_α;

(2) if $\gamma \in \alpha$, then $<_\gamma$ is the restriction of $<_\alpha$ to V_γ;

(3) if $rk\, r \in rk\, s \in \alpha$, then $r <_\alpha s$;

(4) if $rk\, r = rk\, s \in \alpha$, r is finite, and s is infinite, then $r <_\alpha s$;

(5) $<_\alpha$ is anti-lexicographic on the finite sets s with $rk\, s$ in α.

In order to make sure that (5) is meaningful, we note that the finite sets s with $rk\,s$ in α constitute a finitely transitive set, because if $s = \{\,s_1,\ldots,\,s_n\,\}$ and $rk\,s \in \alpha$, then $rk\,s_i \in rk\,s \in \alpha$ and hence $rk\,s_i \in \alpha$. By Lemma 2.1, (1) implies the following condition:

(6) $<^\alpha$ is a well-ordering of the finite sets of rank α.

Our construction begins with $\alpha = 0$, for which we put $<_0 = \,<^0 = \;\emptyset$, because $V_0 = \emptyset$. For $\alpha = 0$: (2), (3), and (4) are vacuously true; also (1), (5), and (6) are trivial, because $\emptyset$ is a well-ordering of any set that has fewer than two elements. Suppose next that $\alpha = \beta + 1$. We can decompose V_α as the disjoint union

$$V_\alpha = V_\beta \cup W_\alpha \cup Z_\alpha,$$

where

$$W_\alpha = \{\,x \in V_\alpha \setminus V_\beta : x \text{ is finite }\},$$
$$Z_\alpha = \{\,x \in V_\alpha \setminus V_\beta : x \text{ is infinite }\}.$$

According to the inductive hypotheses (4) and (6):

whenever $rk\,r = rk\,s \in \beta$ and r is finite but s is infinite, one has $r <_\beta s$;

and, moreover,

$$<^\beta \text{ is a well-ordering of } W_\alpha.$$

A well-ordering $<^\beta_\infty$ of Z_α also exists, by the Zermelo principle; hence we can define $<_\alpha$ to be the concatenation of $<_\beta$ with $<^\beta$ and with $<^\beta_\infty$, which automatically ensures that (1) and (4) are satisfied. Incidentally note that a great deal of indeterminacy in the overall definition of $<$ originates from the absolute freedom that we have here, in choosing the well-ordering of Z_α.

Concerning (2) with $\alpha = \beta + 1$, note that $r, s \in V_\beta$ when $r, s \in V_\gamma$ and $\gamma \in \alpha$. Therefore $r <_\alpha s$ if and only if $r <_\beta s$. Moreover, either because $\gamma = \beta$ or by the inductive hypothesis (2), if $\gamma \in \beta$, we have that $r <_\beta s$ if and only if $r <_\gamma s$. Hence $r <_\alpha s$ if and only if $r <_\gamma s$, and (2) is proved.

Concerning (3) with $\alpha = \beta + 1$, note that $rk\,r \in \beta$, i.e. $r \in V_\beta$, when $rk\,r \in rk\,s \in \alpha$. If also $rk\,s \in \beta$, we can apply the inductive hypothesis (3) to deduce that $r <_\beta s$, whence $r <_\alpha s$ follows by definition. If $s \in V_\alpha \setminus V_\beta$ instead, then again $r <_\alpha s$ by definition, because $<_\alpha$ is the concatenation of a well-ordering of V_β with a well-ordering of $V_\alpha \setminus V_\beta$. Thus (3) is proved.

Concerning (5) with $\alpha = \beta + 1$, note first that since $\emptyset$ is the minimum in $(V_\beta, <_\beta)$, $\emptyset$ is the minimum in $(V_\alpha, <_\alpha)$ too, according to the definition of $<_\alpha$. Suppose next that

$$rk\,\{\,r_0,\ldots,\,r_n\,\},\, rk\,\{\,s_0,\ldots,\,s_m\,\} \in \alpha$$

with

$$r_n <_\alpha r_{n-1} <_\alpha \cdots <_\alpha r_0 ,$$

$$s_m <_\alpha s_{m-1} <_\alpha \cdots <_\alpha s_0 ,$$

and

$$\{ r_0 , \ldots , r_n \} <_\alpha \{ s_0 , \ldots , s_m \} .$$

We must ascertain that $\{ r_0 , \ldots , r_n \}$ anti-lexicographically precedes $\{ s_0 , \ldots , s_m \}$, i.e., either

(i) $n \in m$ and $r_i = s_i$ for $i = 0 , \ldots , n$; or

(ii) there is some j such that $n , m \notin j$ and

$$r_0 = s_0 , \ldots , r_{j-1} = s_{j-1}, r_j <_\alpha s_j .$$

Since

$$rk \{ s_0 , \ldots , s_m \} \notin rk \{ r_0 , \ldots , r_n \}$$

by (1) and (3), we are left with the following three cases to consider.
Case 1: $rk \{ r_0 , \ldots , r_n \} \in rk \{ s_0 , \ldots , s_m \}$.
Since

$$r_i , r_{i+1} , s_j , s_{j+1} \in V_\beta ,$$

we have

$$r_{i+1} <_\beta r_i , \quad s_{j+1} <_\beta s_j$$

according to the definition of $<_\alpha$; it follows from the inductive hypotheses (1) and (3) that

$$rk\, r_i \notin rk\, r_{i+1} \text{ and } rk\, s_j \notin rk\, s_{j+1} ,$$

for $i = 0 , \ldots , n - 1$ and $j = 0 , \ldots , m - 1$. As a consequence,

$$rk \{ r_0 , \ldots , r_n \} = rk\, r_0 + 1 ,$$

$$rk \{ s_0 , \ldots , s_m \} = rk\, s_0 + 1 ,$$

so that

$$rk\, r_0 \in rk\, s_0 \in \beta \in \alpha .$$

It follows, by the inductive hypothesis (3), that $r_0 <_\beta s_0$, so that (ii) holds with $j = 0$.
Case 2: $rk \{ r_0 , \ldots , r_n \} = rk \{ s_0 , \ldots , s_m \} = \beta$.
Then we have

$$\{ r_0 , \ldots , r_n \} <^\beta \{ s_0 , \ldots , s_m \} ,$$

by the definition of $<_\alpha$; hence (i) or (ii) holds by the definition of $<^\beta$ and by the implication

$$\text{if } r <^\beta s , \text{ then } r <_\alpha s .$$

Case 3: $rk\{r_0,\ldots,r_n\} = rk\{s_0,\ldots,s_m\} \in \beta$ or $rk\{r_0,\ldots,r_n\} \in rk\{s_0,\ldots,s_m\}$. Then we have

$$\{r_0,\ldots,r_n\} <_\beta \{s_0,\ldots,s_m\};$$

hence (i) or (ii) holds by the inductive hypothesis (5) and by the implication

$$\text{if } r_j <_\beta s_j, \text{ then } r_j <_\alpha s_j.$$

The proof of (5) is now complete.

It remains to consider the case when α is a limit ordinal. In this case (6) holds vacuously, because there are no finite sets of rank α. We are going to prove that (1) through (5) hold if one defines

$$<_\alpha = \bigcup_{\beta \in \alpha} <_\beta.$$

Dealing with the class Ord of all ordinals as if it were a limit ordinal, one can prove analogues of (1) through (5) with $<$ in place of $<_\alpha$; the theorem will then follow.

Concerning (1), when α is a limit ordinal, we can exploit the inductive hypothesis that $<_\beta$ is a well-ordering of V_β, for all β in α. Let us prove first that if $r <_\alpha s$, then $r, s \in V_\alpha$ and $r \neq s$. In fact, $r <_\beta s$ for some β in α; hence $r, s \in V_\beta \subseteq V_\alpha$ and $r \neq s$. Let us prove next that $r <_\alpha s <_\alpha t$ implies that $r <_\alpha t$. Assuming that $r <_\beta s <_\gamma t$ with β, γ in α, we have $r <_\delta s <_\delta t$, where $\delta = \beta \cup \gamma$, by the inductive hypothesis (2). Hence $r <_\delta t$, and therefore $r <_\alpha t$ by definition. In the third place, let us prove that if $r, s \in V_\alpha$, then either $r <_\alpha s$ or $r = s$ or $s <_\alpha r$. If $r \neq s$, take $\beta = (rk\,r \cup rk\,s) + 1$. Then $\beta \in \alpha$ because β is a non-limit ordinal whose predecessor belongs to α, hence we have either $r <_\beta s$ or $s <_\beta r$. Correspondingly, $r <_\alpha s$ or $s <_\alpha r$, by definition. To complete the proof of (1), we must still prove that every descending chain

$$\cdots <_{\beta_2} s_2 <_{\beta_1} s_1 <_{\beta_0} s_0$$

with $\beta_0, \beta_1, \beta_2, \ldots \in \alpha$ has finite length. In fact, by the inductive hypothesis (3), $rk\,s_i \notin rk\,s_{i+1}$ for $i = 0, 1, 2, \ldots$. Let $\mu = rk\,s_0 + 1$. By the inductive hypothesis (2), one has $s_{i+1} <_\mu s_i$ for $i = 0, 1, 2, \ldots$. Hence the chain cannot be infinite, because $<_\mu$ is a well-ordering.

Concerning (2), when α is a limit ordinal: if $\gamma \in \alpha$ and $r <_\gamma s$ then $r <_\alpha s$ by definition. Conversely, if $\gamma \in \alpha$, $r, s \in V_\gamma$ and $r <_\alpha s$, then we have $r <_\beta s$ for some β in α, hence $r, s \in V_\beta$. From the inductive hypothesis (2) it follows that

$$r <_\gamma s \text{ if and only if } r <_\beta s;$$

hence $r <_\gamma s$.

Concerning (3), when α is a limit ordinal, suppose that

$$rk\,r \in rk\,s \in \alpha,$$

so that, for $\gamma = rk\,s + 1$:

$$rk\,r \in rk\,s \in \gamma \in \alpha,$$

whence $r <_\gamma s$ by the inductive hypothesis (3). We conclude that $r <_\alpha s$.

Concerning (4), when α is a limit ordinal, suppose that $rk\,r = rk\,s \in \alpha$, r is finite, and s is infinite. Then, for $\gamma = rk\,s + 1$, $rk\,r = rk\,s \in \gamma \in \alpha$; hence $r <_\gamma s$ by the inductive hypothesis (4). We conclude that $r <_\alpha s$.

Concerning (5), when α is a limit ordinal, suppose that

$$\{r_1,\ldots,r_n\}, \{s_1,\ldots,s_m\} \in V_\alpha,$$
$$\gamma = (rk\,r_1 \cup \cdots \cup rk\,r_n \cup rk\,s_1 \cup \cdots \cup rk\,s_m) + 1 + 1,$$

so that $\gamma \in \alpha$ and, by the inductive hypothesis (5),

$\{r_1,\ldots,r_n\} <_\gamma \{s_1,\ldots,s_m\}$ holds if and only if
$\{r_1,\ldots,r_n\}$ anti-lexicographically precedes $\{s_1,\ldots,s_m\}$
with respect to $<_\gamma$.

We have already proved (2); therefore we get

$\{r_1,\ldots,r_n\} <_\gamma \{s_1,\ldots,s_m\}$ holds if and only if
$\{r_1,\ldots,r_n\} <_\alpha \{s_1,\ldots,s_m\}$;

on the other hand

$\{r_1,\ldots,r_n\}$ anti-lexicographically precedes $\{s_1,\ldots,s_m\}$
with respect to $<_\gamma$ if and only if it so does with respect to $<_\alpha$.

One readily concludes the proof of (5):

$\{r_1,\ldots,r_n\} <_\alpha \{s_1,\ldots,s_m\}$ holds if and only if
$\{r_1,\ldots,r_n\}$ anti-lexicographically precedes $\{s_1,\ldots,s_m\}$
with respect to $<_\alpha$. ∎

Corollary 2.4 *Every infinite set X has infinitely many infinite subsets Y with $rk\,Y = rk\,X$. For no infinite ordinal α there exists a maximum finite set of rank α.*

Proof. For all i in ω, define

$$x_i = min\,(X \setminus \{x_j : j < i\}),$$
$$Y_i = X \setminus \{x_i\},$$

where the 'minimum' operation is referred to an ordering $<$ like the one that we have constructed in the theorem. That is, x_i is the $(i+1)$-st member of X, and Y_i is X deprived of its $(i+1)$-st member. It is obvious that the x_is (respectively, the Y_is) are pairwise distinct. Since

$$rk\, X \;=\; \bigcup_{x \in X} (rk\, x \,+\, 1),$$

$$rk\, Y_i \;=\; \bigcup_{y \in X \setminus \{x_i\}} (rk\, y \,+\, 1),$$

(see Lemma 2.4(vi)), and moreover $rk\, x_i \,\leq\, rk\, x_{i+1}$, so that $rk\, x_i \,+\, 1 \,\leq\, rk\, x_{i+1} \,+\, 1$ and hence $rk\, x_i \,+\, 1 \,\subseteq\, rk\, x_{i+1} \,+\, 1$ (with $x_{i+1} \in Y_i$), it follows that $rk\, Y_i \,=\, rk\, X$, for all i in ω, which proves the first claim.

For the second claim, let us assume that finite sets of rank α exist, so that α is a successor ordinal $\alpha \,=\, \beta + 1$ with $\omega \,\leq\, |\beta| \,=\, \beta \,=\, rk\, \beta$. By the preceding claim, there are distinct subsets b_0, b_1, $b_2, \ldots$ of β with $rk\, b_i \,=\, \beta$ for all i. For every finite set x of rank α, $\overline{x} \,=\, x \cup \{\, b_i \,:\, i \in \{0, \ldots, |x|\}\,\}$ has rank α, is finite and includes x as a proper subset; hence $x \,<\, \overline{x}$ as desired.∎

2.6   Well-founded graphs and their realizations. The principle of transfinite recursion

In this section we give the definition and a few examples of *well-founded graphs*. Our definition is not very strict, because we allow the collection V of nodes of a well-founded graph to be a proper class. If f is a function defined on a set V, then a particular well-founded graph whose set of nodes is V can be defined by taking as edges those pairs $(v, w) \in V \times V$ such that $fv \in fw$ (see Example 2.2 below). Conversely (see Lemma 2.11 below), we will show that if the class V of nodes of a well-founded graph is a set, then there exists a function $f : V \rightarrow Z$, to be called a *realization* of the graph, such that (v, w) is an edge if and only if $fv \in fw$. In proving this, we will exploit an extremely powerful inductive principle, called the *principle of transfinite recursion*. We will in fact define the concept of *grafting* of a well-founded graph G in such a way that every function f that is related to G in the manner just explained, is induced, via transfinite recursion, by a suitable grafting of G (see Example 2.7).

It is postulated in [Acz88] that a correspondence can be established between arbitrary graphs (possibly containing edges of the form (x, x)) and non-well-founded sets, in a way that generalizes the way we will establish, in this section, a correspondence between well-founded graphs and ordinary sets.

Definition 2.4 *A (directed) graph is a class V—perhaps a proper one— paired with a function $v \mapsto \dot{v}$ whose values are subclasses of V; such that:*

(1) $v \notin \dot{v}$, for any v in V; and

(2) each class $\{ v \in V : w \in \dot{v} \}$, with w in V, is a set.

The members of V are called nodes *(or 'vertices') of the graph; the non-empty classes of the form $\dot{v}$ are called its* places; *and the ordered pairs (v, w) with w in $\dot{v}$ are called its* edges. *If there is no* infinite descending path, *i.e. no infinite sequence*

$$\dots, (v_3, v_2), (v_2, v_1), (v_1, v_0)$$

of edges, the graph is then said to be well-founded. *If there is no* cycle, *i.e., no finite sequence*

$$(v_0, v_1), (v_1, v_2), \dots, (v_{m-1}, v_m), (v_m, v_0)$$

of edges, with $m \geq 1$ and $v_i \neq v_j$ for $0 \leq i < j \leq m$, the graph is then said to be acyclic. *We abbreviate the two phrases 'well-founded graph' and 'directed acyclic graph' into* WFG *and* DAG, *respectively.*

If a graph G has a node r such that for every other node v there is exactly one *path leading from v to r, i.e. a sequence*

$$(v_0, v_1), (v_1, v_2), \dots, (v_{n-1}, v_n), (v_n, v_{n+1})$$

of edges where v_0 is v and v_{n+1} is r, then G is called a tree, *and r is said to be the* root *of G.*

If $G = (V, \dot{\ })$ is a WFG and $W \subseteq V$, then a well-founded *subgraph Γ of G is obtained by taking W and*

$$\{(x, y) : x, y \in W \text{ and } y \in \dot{x}\}$$

as the sets of nodes and edges of Γ, respectively. Γ is called the restriction *of G to W.*

If $G = (V, \dot{\ })$ is a WFG and $f : V \to Z$ is a function such that the equivalence

$$w \in \dot{v} \text{ if and only if } fv \in fw$$

holds for every pair v, w of nodes (i.e. the edges are the pairs (v, w) with fv in fw), then we say that G originates *from f or, equivalently, that f is a* realization *of G.* □

An example of DAG can be obtained from any finite partially ordered set $(V, \prec)$ by putting, for all v in V:

$$\dot{v} = \{w \in V : v \prec w \text{ and there is no } z \text{ in } V \text{ with } v \prec z \prec w\}.$$

Note that if a graph $G = (V, \dot{\ })$ with a finite number of nodes is given, one can easily check whether G is well-founded, by use of the fact:

Lemma 2.7 *When the set V of nodes is finite,*

$$G \text{ is a WFG if and only if } G \text{ is a DAG}.$$

Proof. If there is a cycle (v_0, v_1), (v_1, v_2), ..., (v_{m-1}, v_m), (v_m, v_0) then $\dots, (v_{m-1}, v_m), (v_m, v_0), (v_0, v_1), (v_1, v_2), \dots, (v_{m-1}, v_m)$ is an infinite descending path; that is, if G fails to be a DAG, then G also fails to be a WFG. Conversely, if $\dots, (v_3, v_2), (v_2, v_1), (v_1, v_0)$ is an infinite descending path, then some node w must occur more than once in the infinite sequence $(v_0, v_1, \dots, v_i, \dots)$, because there are only finitely many nodes available. Let h be the smallest subscript for which $v_h = v_k$ for some $k < h$. Then k is uniquely determined by h, and $(v_h, v_{h-1}), \dots, (v_{k+1}, v_k)$ is a cycle, according to our definition. ∎

Corollary 2.5 *It is possible to generate all WFGs that have a pre-established finite set V of nodes, in a finite amount of time.*

Proof. Suppose we want to output each graph $G = (V, {}^\cdot)$ by supplying the set E of its edges, i.e. those pairs v, w in V for which $w \in \dot{v}$. If V consists of n nodes, then there are $2^{(n^2 - n)}$ sets of distinct pairs, and from among these graphs we want to discard only those that have cycles. Since a cycle cannot involve more than n edges, we can check whether a graph has cycles by inspecting all sequences consisting of $2, 3, \dots, n$ edges and by checking, for each of these sequences, whether or not it is a cycle. ∎

Example 2.2 What conditions are to be satisfied by a function $f : V \to Z$ (where V, Z are classes) in order that a WFG originates from f ? Let $\dot{v} = \{w \in V : fv \in fw\}$ for every v in V (see Figure 2.1); in the next paragraph we show that $(V, {}^\cdot)$ is a graph, and even a WFG, provided that the class $\{v \in V : fv \in fw\}$ is a set for every w in V. This is the case, for instance, when V is a set, or when $V = Z$ and f is the identity function $fv = v$. On the other hand, if $V = \mathcal{V}$ is the class of all sets, $Z = \{\emptyset, \{\emptyset\}\}$, and we define $f\emptyset = \{\emptyset\}$, $fv = \emptyset$, for every non-empty set v, then

$$\{v \in V : fv \in f\emptyset\} = V \setminus \{\emptyset\}$$

is a proper class.

We must prove that:

(i) $v \notin \dot{v}$, for any v in V;

(ii) there exists no infinite sequence $(v_0, v_1, v_2, \dots)$ with $v_i \in \dot{v}_{i+1}$ for all i in ω.

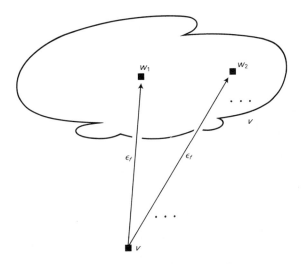

Fig. 2.1. The arrows indicate that $f_v \in f_{w_i}$

Indeed, $v \in \dot{v}$ would imply that $fv \in fv$, while the existence of a sequence like the one in (ii) would imply that $\cdots \in fv_2 \in fv_1 \in fv_0$; and in both cases the well-foundedness of membership would be violated. We conclude that $(V, \dot{})$ is a WFG. □

Example 2.3 Our definition of graph is so tolerant that a WFG $(V, \dot{})$ can be defined by choosing V to be the class Ord of all ordinals, and each v to be the class of all ordinals w greater than v. This is in fact a particular case of the preceding example, in view of von Neumann's equivalence (valid for all ordinals v, w)

$$v < w \text{ if and only if } v \in w.$$

□

One may wonder why, in defining the notion of a graph (see Definition 2.4), we have demanded that $\{v \in V : w \in \dot{v}\}$ be a set, for every node w. The reason is that, with this condition, an inductive principle even more powerful than the one already presented in Section 2.3 becomes available. We borrow such a principle, with some changes in terminology, from [Mos69].

Principle 2.2 (Transfinite recursion) *Let:*

(1) $G = (V, \dot{})$ be a WFG;

(2) V be the class of all sets; and

(3) $g : V \times V \to V$ be an arbitrary function.

Then there exists a unique function h satisfying, for every w in V, the identity

$$hw = g(\{hv : v \in V \text{ and } w \in \dot{v}\}, w).$$

We say that g induces *h on G.*   □

The following induction principle also holds for WFGs:

Lemma 2.8 *Let V be the set of nodes of a WFG, and P(w) be a sentence such that, for all w in V:*

   if P(v) holds for every edge (v, w), then P(w) is also true.

Under these assumptions, P(v) holds for every v in V.

Proof. Suppose that $P(v_0)$ could be false for some v_0 in V. Inductively, since $P(v_i)$ is false, $P(v_{i+1})$ is false for some edge (v_{i+1}, v_i). Thus we discover an infinite descending path $\ldots$, (v_2, v_1), (v_1, v_0) in G, which, instead, was supposed to be well-founded.   ■

Example 2.4 The function H induced on G by

$$g(s, v) = \bigcup_{r \in s} (r \cup \{r\})$$

is called *height*. By definition,

$$Hw = \bigcup_{w \in \dot{v}} (Hv \cup \{Hv\});$$

hence if w is a node such that Hv is an ordinal for every edge (v, w), then Hw is an ordinal too (indeed, it is the least upper bound of $\{Hv + 1 : (v, w) \text{ is an edge}\}$—see Section 2.3). By Lemma 2.8, it follows that Hv is an ordinal for every node v.   □

Example 2.5 In Example 2.2, let us take

$$V = \mathcal{V} = \text{the class of all sets}; \text{ and}$$
$$\dot{v} = \{w \in V : v \in w\}.$$

The height of the nodes, in this case, is just the *rank* function (see Lemma 2.4(vi)).   □

Example 2.6 If $u \mapsto \ddot{u}$ is a function is defined on a set U, and

$$g(s, v) = s \cup \bigcup_{u \in U, v \in \ddot{u}} u,$$

then the induced function will often be written (leaving $\ddot{\ }$ implicit) $v \mapsto v'$. According to the definition, we have

$$w' = \{v' : w \in \dot{v}\} \cup \bigcup_{w \in \ddot{u}} u$$

for every node w.

A well-known special case of this construction, called the *Mostowski contraction* of the WFG, is obtained when U is empty, in which case the recurrence formula simplifies into $w' = \{ v' : w \in \dot{v} \}$. □

Definition 2.5 *In the construction of Example 2.6, the members of $Un(U)$ are called* scions. *If the members of U are non-empty pairwise disjoint sets, and for every u in U one has*

$$\emptyset \neq \ddot{u} \subseteq V, \text{ and}$$
$$v' \notin u \text{ for any } v \text{ in } V,$$

then $(U, \ddot{\ })$ is called a grafting *of the WFG and the sets of the form $\ddot{u}$ are called the* places *of the grafting. (Note that the word 'place' now has two different meanings—recall Definition 2.4). A grafting $(U, \ddot{\ })$ of a WFG $(V, \dot{\ })$ is said to be* extensional *iff $\dot{v} = \dot{w}$ holds for every pair v, w of nodes belonging to the same places (of the WFG or of the grafting), i.e., $\dot{v} = \dot{w}$ holds whenever both*

$$v \in \dot{z} \text{ if and only if } w \in \dot{z}, \text{ for all } z \text{ in } V, \text{ and}$$
$$v \in \ddot{u} \text{ if and only if } w \in \ddot{u}, \text{ for all } u \text{ in } U.$$

□

Example 2.7 If the function $f : V \to Z$ is defined on a set V, and $G = (V, \dot{\ })$ is the WFG originating from f as in Example 2.2, then a particular grafting of G can be defined as follows. We take

$$D = \left(\bigcup_{v \in V} fv \right) \setminus f[V];$$

that is, D is the family of all those sets which belong to some image $f v$ with v in V, but do not, themselves, have the form $f w$ with w in V. For a, b in D we write

$$a \sim_f b \text{ iff } \{ v \in V : a \in fv \} = \{ v \in V : b \in fv \};$$

then we consider a partition U of D, compatible with $\sim_f$. That is to say, we require that

$$u \neq \emptyset \text{ for all } u \text{ in } U;$$
$$u_0 \cap u_1 = \emptyset \text{ for all } u_0, u_1 \text{ in } U;$$
$$Un(U) = D;$$
$$a \sim_f b \text{ whenever } a, b \text{ belong to the same } u \text{ in } U.$$

(The two extreme admissible values for U are:

$$U = \{\{d\} : d \in D\},$$
$$U = \{\{a : a \sim_f d\} : d \in D\}.)$$

Finally, for every u in U, we put

$$\ddot{u} = \{v \in V : d \in fv\},$$

where d is any member of u. In order to prove that $(U, \ddot{\ })$ is a grafting of G, it will suffice to show that $v' = fv$ for every v in V, because, as an immediate consequence of the definitions of D, U, the fact $fv \notin u$ holds for all u in U, v in V. The identity $v' = fv$ is proved using the induction principle stated in Lemma 2.8. As a matter of fact,

$$\begin{aligned}
w' &= \{v' : w \in \dot{v}\} \quad \cup \textstyle\bigcup_{w \in \ddot{u}} u \\
&= \{v' : fv \in fw\} \cup \{d \in D : d \in fw\} \\
&= \{fv : fv \in fw\} \cup \{d \notin f[V] : d \in fw\} = fw,
\end{aligned}$$

if we inductively assume that $v' = fv$ for every edge (v, w).

Slightly stretching the terminology, we could concisely describe the situation that G originates from the function $v \mapsto v'$ and that this function is in turn induced by a grafting of G, by saying:

'*G is realized by one of its graftings*'.

It is easy to show that the above-defined grafting $(U, \ddot{\ })$ is extensional. In fact, if v, w are nodes belonging to the same places, then

$$\begin{aligned}
fv = v' &= \{z' : v \in \dot{z}\} \cup \textstyle\bigcup_{v \in \ddot{u}} u \\
&= \{z' : w \in \dot{z}\} \cup \textstyle\bigcup_{w \in \ddot{u}} u = w' = fw,
\end{aligned}$$

so that

$$\dot{v} = \{z \in V : fv \in fz\} = \{z \in V : fw \in fz\} = \dot{w}.$$

∎

Lemma 2.9 *If the function $f : V \to Z$ is defined on a set V, so that f originates a WFG $G = (V, \dot{\ })$, then there exists a unique one-to-one grafting of G that induces the function f. This grafting is extensional.*

Proof. The existence of an extensional grafting that induces f has been proved in Example 2.7; moreover the grafting constructed there turns out easily to be one-to-one if one chooses

$$U = \{\{a : a \sim_f d\} : d \in D\}.$$

Only uniqueness hence remains to be proved.

Suppose that $(U, \ddot{\ })$ is a one-to-one grafting whose induced function $v \mapsto v'$ coincides with f. For every node v, we have

$$fv = v' = \{z' : v \in \dot{z}\} \cup \bigcup_{v \in \ddot{u}} u,$$

and hence

$$fv = \{fz : v \in \dot{z}\} \cup \bigcup_{v \in \ddot{u}} u,$$

where unions are disjoint by the very definition of graftings. It follows (taking into account that $\emptyset \neq \ddot{u} \subseteq V$ for all u in U) that

$$\left(\bigcup_{v \in V} fv\right) \setminus f[V] = \bigcup_{v \in V} \bigcup_{v \in \ddot{u}} u = Un(U),$$

where the left-hand side is the set D of scions, already seen in Example 2.7.

Assuming now that $u \in U$, $d \in u$, we must prove that $\ddot{u} = \{v \in V : d \in fv\}$ and $u = \{a : a \sim_f d\}$. In fact, if $v \in \ddot{u}$, then we have $d \in v' = fv$. Consequently, $\ddot{u} \subseteq \{v \in V : d \in fv\}$. Conversely, if $d \in fv$, then it can only be so by virtue of the fact that $v \in \ddot{u}$; hence we have $\ddot{u} = \{v \in V : d \in fv\}$ and the fact that this equality holds no matter how d is selected proves that U is compatible with $\sim_f$, so that $\{a : a \sim_f d\} = u \cup u_1 \cup \cdots \cup u_n$, where unions are disjoint. However n must be zero in the latter equality, else we would have $\ddot{u} = \ddot{u}_1$, contradicting the assumption that $\ddot{\ }$ is one-to-one. ∎

We now want to prove that Example 2.2 is the most general conceivable example of a WFG whose class of nodes is a set. That is, assuming that V is a set and that $G = (V, \dot{\ })$ is a WFG, we intend to show how to construct a function $f : V \to Z$ from which G originates. In our construction, f will be the function $v \mapsto v'$ induced (as in Example 2.6) by an extensional grafting of G; the following result is therefore important for our current purposes.

Lemma 2.10 *In an extensional grafting, for every pair v, w of nodes one has that:*

(1) *$v' = w'$ if and only if v and w belong to the same places, so that (by extensionality) $\dot{v} = \dot{w}$;*

(2) *$v' \in w'$ if and only if $w \in \dot{v}$. Otherwise stated, the WFG is realized by the grafting.*

Proof. We prove (1) by induction on the height $K = max\{Hv, Hw\}$ (see Example 2.4). Assuming that v and w do not belong to the same places, our aim is to show that the inequality $v' \neq w'$ holds. Let us therefore begin by choosing a place p such that

$$v \in \dot{p} \text{ if and only if } w \notin p.$$

If p has the form $\ddot{u}$ (as is the only possibility when $K = 0$), then we immediately deduce

$$d \in v' \text{ if and only if } d \notin w',$$

for any d in u, and hence $v' \neq w'$.

Alternatively, p has the form $\dot{z}$, i.e. there is a node z such that either $v \in \dot{z}$ or $w \in \dot{z}$, but not both. For definiteness, assume that

$$v \in \dot{z}, \, w \notin \dot{z}$$

(the case when $w \in \dot{z}$, $v \notin \dot{z}$ is treated in an entirely symmetrical fashion). We will prove that, while trivially $z' \in v'$, one instead has $z' \notin w'$, so that $v' \neq w'$.

If w' contains no members of the form z_1', we are already finished. Otherwise, assuming that $w \in \dot{z}_1$, one has

$$max\{Hz, \, Hz_1\} < K$$

because

$$Hz + 1 \subseteq Hv \leq K \text{ and } Hz_1 + 1 \subseteq Hw \leq K.$$

If z' could equal z_1', then we would have $\dot{z} = \dot{z}_1$ by our induction hypothesis, hence $w \in \dot{z}$, contradicting one of our temporary assumptions.

Let us now prove (2). Trivially, if $w \in \dot{v}$ then $v' \in w'$. Conversely, if $v' \in w'$, then it must be $v' = z'$ for some node z such that $w \in \dot{z}$. By (1), this implies $\dot{v} = \dot{z}$; hence $w \in \dot{v}$.   ∎

Preliminary to the construction of an extensional grafting for any given WFG, we define the extensional sets of nodes:

Definition 2.6 *Assume that V is a set and that $G = (V, \, \dot{} \,)$ is a WFG. An extensional set of nodes of G is a subset S of V such that $\dot{v} = \dot{w}$ for every pair v, w of nodes in S with*

$$\{x : v \in \dot{x}\} = \{y : w \in \dot{y}\}.$$

□

According to this definition, the empty set of nodes, as well as any singleton set of nodes, is extensional. More generally, S is extensional whenever it satisfies the following property:

(i) for every pair v, w of distinct nodes in S which have the same height, there is a node z in V such that exactly one of the pairs (z, v), (z, w) is an edge of G, i.e.

$$\text{either } v \in \dot{z} \text{ or } w \in \dot{z}, \text{ but not both.}$$

In fact, whenever $\{x : v \in \dot{x}\} = \{y : w \in \dot{y}\}$, it follows that v and w have the same height, so that, by (i), v and w coincide, and hence $\dot{v} = \dot{w}$ as desired. Note that (i) implies also that

(ii) at most one node of height zero belongs to S;

in fact, whenever v and w have height 0, one has that:

$$\{x : v \in \dot{x}\} = \{y : w \in \dot{y}\} = \emptyset.$$

Lemma 2.11 *If S is an extensional set of nodes for $G = (V, \dot{})$, then there exists a grafting $(U, \ddot{})$ of G (see Definition 2.5) such that:*

(0) all members of U are singleton and have the same rank > 1.

(1) $u \mapsto \ddot{u}$ is a one-to-one correspondence between U and $\{\{v\} : v \in V \setminus S\}$. It is an extensional grafting and puts

　(1a) exactly one scion in each set v' with v in $V \setminus S$;

　(1b) no scion in any set w' with w in S;

　(1c) no scion in more than one set of the form z'.

(2) $w' = \emptyset$ if and only if w has height zero and belongs to S, for all node w.

(3) For every pair v, w of distinct nodes such that $v' = w'$, one has

　(3a) $v, w \in S$, and

　(3b) $\{z : v \in \dot{z}\} = \{z : w \in \dot{z}\}$, so that $\dot{v} = \dot{w}$ by the definition of an extensional set.

Note that thanks to the extensionality of the grafting, Lemma 2.10 applies, yielding the corollary:

(4) G originates from the function $v \mapsto v'$.

Also note that when S satisfies condition (i) preceding this lemma, one can simplify (3) into

(3') $v' = w'$ implies that $v = w$.

Proof. Since V is a set, the two classes

$$\alpha = \bigcup_{v \in V} Hv,$$
$$\beta = |V \setminus S|$$

are both ordinal numbers. Put

$$\gamma = \alpha \cup \beta,$$
$$D = \{\{\delta, \gamma\} : \delta \in \beta\},$$

so that D turns out to be a set with

$$rk\, d \,=\, \gamma + 1 \text{ for all } d \text{ in } D$$

and

$$|D| \,=\, |U| \,=\, \beta,$$

where $U \,=\, \{\{d\} \,:\, d \,\in\, D\}$. We can therefore establish a one-to-one correspondence $u \,\mapsto\, \ddot{u}$ between U and the singleton subsets of $V \setminus S$. For notational convenience, we will also write $\ddot{d} = v$ when $u = \{d\}$ and $\ddot{u} = \{v\}$, for some u in U.

In order to see that $(U, \ddot{\ })$ is a grafting (whence (0) and (1c) will follow immediately), we now prove that, for every w in V

$$\text{either } rk\, w' \,=\, Hw \,\in\, \gamma + 1 \text{ or } \gamma + 1 \,\in\, rk\, w',$$

so that in neither case w' can belong to any u in U. Here by Hw we indicate the height of w, as usual (see Example 2.4). Recalling that

$$w' \,=\, \{v' \,:\, w \,\in\, \dot{v}\} \,\cup\, \{d \,\in\, D \,:\, w \,=\, \ddot{d}\},$$

we apply the induction hypothesis to all v for which $w \,\in\, \dot{v}$. If, for each such v, we have $rk\, v' \,=\, Hv$, then

$$\text{either } rk\, w' \,=\, Hw \text{ or } \gamma + 1 \,\in\, rk\, w',$$

according to whether $\{d \,:\, w \,=\, \ddot{d}\}$ is empty or not. If, on the contrary, $\gamma + 1 \,\in\, rk\, v'$ for some v with w in $\dot{v}$, then $\gamma + 2 \,\in\, rk\, w'$. We can now write, for all w in S:

$$w' \,=\, \{v' \,:\, w \,\in\, \dot{v}\};$$

and, for all w in $V \setminus S$:

$$w' \,=\, \{v' \,:\, w \,\in\, \dot{v}\} \,\cup\, \{d_w\},$$

where d_w is the scion for which $\ddot{d}_w \,=\, w$. Thus (1a) and (1b) follow immediately; moreover (2) is obvious since, by the definition of the height function (see Example 2.4), a node w has height zero if and only if $w \,\in\, \dot{v}$ holds for no v.

In order to prove the extensionality of the grafting, we note that when two distinct nodes v, w belong to the same places, they must both belong to a singleton place $\ddot{u}$). Moreover they must satisfy the identity

$$\{z \,:\, v \,\in\, \dot{z}\} \,=\, \{z \,:\, w \,\in\, \dot{z}\},$$

because they only belong to places of the form $\dot{z}$. Since S is an extensional set, this identity implies that $\dot{v} \,=\, \dot{w}$, as we had to prove. Now (3) becomes an obvious consequence of Lemma 2.10(1).   ∎

Corollary 2.6 *For every WFG G whose nodes form a set, there exists a well-ordering $\lhd$ of the nodes such that $v \lhd w$ whenever (v, w) is an edge of G.*

Proof. Let S be $\emptyset$ in the preceding lemma, so that by (1a) and (1c) we have $v' \neq w'$ when $v \neq w$. An induction similar to the one used in the lemma shows that $rk\,v' = \gamma + 2 + Hv$ for every node v. If we now put $v \lhd w$ iff $v' < w'$ holds in the well-ordering of all sets, we clearly obtain a well-ordering of the nodes. When (v, w) is an edge, then $Hv < Hw$, hence $v' < w'$, implying that $v \lhd w$. ∎

Remark 2.2 Note that although the construction of the grafting $(U, \ddot{\ })$ in the proof of Lemma 2.11 may seem very abstract, it can be carried out effectively when the set V of nodes is finite. In this case one is able to calculate the height of each node (which is a natural number), hence the value of $\alpha = max_{v \in V} Hv$. One is also able to determine which sets S of nodes are extensional and, for each such set, to calculate the values of β, γ, and U; the correspondence $u \mapsto \ddot{u}$ can be established, generally, in various ways. □

Corollary 2.7 *If S is an extensional set of nodes for $G = (V, \dot{\ })$ whose set of nodes V has finitely many members, then G has an extensional grafting $(U, \ddot{\ })$ as in Lemma 2.11, such that*

$$Hv \leq rk\,v' \leq max\,\{\alpha, \beta - 1\} + 2 + Hv$$

for every node v, where

$$\alpha = max_{v \in V} Hv,$$
$$\beta = |V \setminus S|.$$

Proof. The construction of the grafting $(U, \ddot{\ })$ can be carried out in almost the same manner as in Lemma 2.11, with only a minor change in the definition of D when $\alpha < \beta$. If $\alpha < \beta$ we define in fact

$$D = \{\{\beta - 1\}\} \cup \{\{\delta, \beta - 1\} : \delta \in \beta - 1\}.$$

Let us put

$$\gamma = max\,\{\alpha, \beta - 1\},$$

so that this γ and the one appearing in the proof of the lemma are different when $\alpha < \beta$, while it remains true in all cases that

$$rk\,d = \gamma + 1 \text{ for all } d \text{ in } D$$

and

$$|D| = |U| = \beta.$$

An induction similar to the one used in the lemma shows that, for every w in V, either $rk\,w' \;=\; H\,w \;<\; \gamma + 1$ or $\gamma + 1 \;<\; rk\,w' \;\leq\; \gamma + 2 + H\,w,$

thus providing the desired conclusion, since trivially $H\,w \leq \alpha \leq \gamma$. ∎

Lemma 2.12 *If a DAG G has only a finite number n of nodes, then G has*

(i) *an extensional grafting in which*

$$rk\,v' \;=\; n + 1 + H\,v \;\leq\; 2 \cdot n \;\; \text{for all } v\,;$$

(ii) *an extensional grafting in which*

$$rk\,v' \;<\; n + 1 + H\,v \;\leq\; 2 \cdot n \;\; \text{for all } v\,;$$

(iii) *an extensional grafting such that, assuming that the nodes are $v_1, \ldots, v_n$ with $H\,v_i \leq H\,v_j$ for $0 < i < j \leq n$, one has*

(1) $rk\,v'_i \;=\; 2 \cdot i - 2$ *for $i = 1, \ldots, n$;*

(2) *exactly one scion d_i belongs to each v'_i with $1 < i$;*

(3) $rk\,d_i \;=\; 2 \cdot i - 3$ *for $i = 2, \ldots, n$, hence d_i is the member of highest rank in v'_i.*

Proof. (i) It suffices to take $S = \emptyset$ in the preceding corollary, and to observe that $H\,v \leq n - 1$ plainly holds for every node v, so that

$$\gamma \;=\; max\,\{\,\alpha\,,\,\beta - 1\,\} \;=\; \beta - 1 \;=\; n - 1.$$

The inductive hypothesis of the corollary is then strengthened into

$$rk\,w' \;=\; \gamma + 2 + H\,w\,,$$

to obtain the desired conclusion.

(ii) In this case we take a singleton S consisting of a node of height 0. If there is a node of height 0 also in $V \setminus S$, then we have $H\,v \leq n - 2 = \beta - 1$ for every node v, and hence the inequality

$$rk\,v' \;\leq\; max\,\{\,\alpha\,,\,\beta - 1\,\} + 2 + H\,v$$

in the preceding corollary becomes $rk\,v' \leq n + H\,v$, which gives the desired result.

If G has only one node z of height 0, then $rk\,z' = 0 = H\,z$, which allows us to strengthen the inductive hypothesis of the corollary into

$$rk\,w' \;<\; \gamma + 2 + H\,w\,,$$

where $\gamma \leq n - 1$ and $H\,w \leq n - 1$, so that our thesis again follows.

(iii) We take $v'_1 = \emptyset$. By induction on $i = 2, \ldots, n$ we then define the sets d_i, v'_i, putting $d_i = \{\,\bigcup_{j<i} v'_j\,\}$ and $v'_i = \{\,v'_j : v_i \in \dot{v}_j\,\} \cup \{\,d_i\,\}$. Note that

this definition is non-circular, because we have $Hv_j < Hv_i$, and hence $j < i$, when $v_i \in \dot{v}_j$. Trivially, one has $d_2 = \{\emptyset\}$ and $rk\, v_1' = 0$. Moreover, from the assumptions $rk\, d_j = 2 \cdot j - 3$, $rk\, v_j' = 2 \cdot j - 2$ with $j = 2, \ldots, i-1$ one inductively derives that:

$$rk\, d_i = rk\, v_{i-1} + 1 = 2 \cdot i - 3,$$

d_i has highest rank in v_i', and hence

$$rk\, v_i' = 2 \cdot i - 2.$$

We therefore get the conclusions (1), (2), and (3), whence it follows that $d_i \dot{\mapsto} \{\, v_i\,\}$ is indeed a grafting inducing the function $v \mapsto v'$ on the nodes. Since no two distinct nodes belong to the same places, this grafting is extensional. ∎

Remark 2.3 From Lemma 2.11(4), recalling that $S = \emptyset$ is an extensional set, we obtain that every WFG whose nodes constitute a set originates from at least one of its graftings. Does it originate from *all* of its graftings? The answer is 'no' in general, as shown by the example in Figure 2.2, where an arrow leading from node A to node B indicates that (A, B) is an edge. We have

$$V = \{\, v, x, z, w\,\},$$
$$\dot{v} = \{\, z\,\},$$
$$\dot{x} = \{\, z, w\,\},$$
$$\dot{z} = \dot{w} = \emptyset.$$

A grafting, in this example, is found by taking

$$u = \{\emptyset\}, \quad U = \{\, u\,\},$$
$$\ddot{u} = \{\, v, x\,\},$$

so that

$$v' = x' = \{\emptyset\},$$
$$z' = w' = \{\{\emptyset\}\},$$

and hence

$$v' \in w' \text{ but } w \notin \dot{v}.$$

We conclude that if G is a WFG whose nodes constitute a set, then:

(1) There exist functions from which G originates (see Lemma 2.11(4)).

(2) Every function f from which G originates is induced by a suitable grafting of G (see Example 2.7). Moreover there exists a unique one-to-one grafting that induces f (see Lemma 2.9).

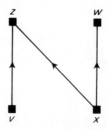

Fig. 2.2. A WFG

(3) There may exist a function induced by a grafting of G, from which G does not originate (see preceding paragraph); however this is never the case for an extensional grafting (see Lemma 2.10(2)).   □

3. THE VALIDITY PROBLEM

3.1 Introduction

In this chapter we present an interpreted formal language upon which an axiomatic theory of sets (see [Jec78, Man77]) could be based, along lines hinted at in [Omo88]. Unlike the meta-theory which has been informally described in Chapter 2, this formal theory would deal with sets only, avoiding reference to proper classes; in the syntax of our language, two restrictions reflect this decision:

(1) we have only one sort of free variables, each variable being supposed to range over all sets;

(2) we will not allow ourselves to employ abstraction terms of the kind $\{ x : p(x) \}$ unless the formula $p(x)$ fulfils severe syntactical constraints.

Our language differs from those most commonly used to express set theories, in that we have two choice operators available: η and H (lower case and upper case eta). The first, when applied to a term t that denotes a non-empty set s, 'extracts' a member ηt of s. The second can be used to bind a free variable x appearing in a formula p; the resulting term $H x\, p$ either denotes a singleton $\{s\}$ such that p holds when x is interpreted as s, or denotes $\emptyset$ (in case no s satisfying p exists). In the semantics just outlined for both η and H, the choice of s is *minimal* with respect to the well-ordering $<$ of the von Neumann universe (see Section 2.5). The expressive power of H is so great, that we can dispense with existential or universal quantifiers as primitive constructs of our language (indeed, even the propositional connectives are redundant, as we will see in Section 3.5). Closely related to our η and H are Bernays' σ-symbol and Hilbert's ε-symbol (cf. [Lei69]). A weaker connection exists between them and the Peano-Russell definite description (see [Car70]).

Let us remark that the quantifying constructs of the object language (namely: H, $\forall$, $\exists$ and abstraction terms) are very rarely employed or treated in the book, save in this chapter and in Chapter 7, where satisfiability decision tests are presented for formulae that involve restricted quantifiers. On the other side, the reader must be warned that several unquantified constructs (e.g. those representing domain, range and rank), not considered in this chapter, will make their first appearance in the chapters where they are respectively studied. Our justification for this state of affairs is that we intend to keep this chapter limited in size, while giving an illustration of the appropriate level of insight at which the overall matter of the book should be discussed if one wished to be very formal. Our expectation and hope is that the ongoing illustration, no matter how incomplete, is rich enough to let the reader who has previously been exposed to symbolic logic feel at ease with the rest of the book.

After introducing the syntax of our language, we supply designation rules that assign meaning to the well-formed expressions (i.e. terms and formulae). In particular, each formula will denote a truth value, or 'bit', while each term will denote a set. Our designation rules are patterned after [Mos69]. We make a few notational conventions for abbreviating terms and formulae, and note that various primitive constructs of our language are redundant, in the sense that even if they were not present to begin with, we could nevertheless introduce them as shorthand notations. Some notational conventions make sense—and can be useful—in connection with the meta-theory too; when this is the case, they will be tacitly 'exported', i.e., made available at both levels.

Later in this chapter, we discuss the dual notions of satisfiability and validity, and several variants of them: some of these variants (e.g. finite satisfiability) would also make sense in the absence of choice operators; some others come into play due to the presence of η and H. It must be noted, in fact, that a certain indeterminacy surrounds the validity notion, because the construction carried out in Section 2.5 did not uniquely characterize the well-ordering of the von Neumann universe: this lack of uniqueness also propagates to η and H, and seems to be unavoidable.

We will be ready, at this point, to carry into the scene the main problem of this book: namely, the problem of determining whether a formula meeting specific syntactical constraints is valid, or satisfiable. An automatic procedure for this determination is called a *decision test* if it is able to always produce a yes/no answer (provided the input is inside the appropriate collection of formulae). Semi-decision tests are weaker—in a sense that will be explained— but their study is useful nonetheless, because it often paves the way towards the discovery of a decision test, or towards the proof that no such procedure exists for a given collection of formulae.

By the end of the chapter, we describe a fundamental technique for normalizing formulae, which will be often exploited throughout the book in order to simplify various instances of the satisfiability decision problem.

A simple quantifier elimination procedure concludes the chapter.

3.2   Language

Definition 3.1 *The* primitive symbols *of our language are:*

(a) a denumerable infinity ξ_0, ξ_1 ξ_2, ... of variables;

(b) the operators

$$\emptyset \qquad\qquad \textit{(nullary)};$$
$$\cap, \setminus, \cup, \times \qquad \textit{(binary)};$$
$$Un, \ Pow, \ \eta \qquad \textit{(unary)};$$

(c) the relators

$$=, \subseteq, \in \textit{ (binary)};$$
$$Finite \quad \textit{(unary)};$$

(d) the (propositional) connectives

$$\&, \vee, \rightarrow, \leftrightarrow \textit{ (binary)};$$
$$\neg \qquad\qquad \textit{(unary)};$$

(e) the descriptor

$$H;$$

(f) and the auxiliary symbols:

$$\textit{parentheses }) \textit{ and } (\ ;$$
$$\textit{braces } \quad \} \textit{ and } \{ \ ;$$
$$\textit{comma} \quad , \qquad ;$$
$$\textit{colon} \quad : \qquad .$$

$\square$

Definition 3.2 *The* well-formed expressions *(or* wfes*) of our formalized set theory are subdivided into* terms *and* formulae *(or* wffs*). They are construed by means of the following inductive formation rules:*

(1) the constant $\emptyset$ is a term;

(2) each variable x is a term;

(3) if s, t are terms, $\otimes$ is a binary operator and $\ominus$ is a unary operator, then

$$(s \otimes t), \quad \ominus t,$$

are (compound) *terms;*

(4) if $t_0, \ldots, t_n$ are terms, then

$$\{ t_0, \ldots, t_n \}$$

is a (compound) *term, for all n in ω;*

(5) if s, t are terms, $\otimes$ is a binary relator and $\ominus$ is a unary relator, then

$$(s \otimes t), \quad \ominus t,$$

are formulae, called atomic *formulae;*

(6) if p, q are formulae, then

$$\neg p, (p \& q), (p \vee q), (p \to q), (p \leftrightarrow q)$$

are (compound) *formulae;*

(7) if x, y are distinct variables and t is a term, then

$$\{t : x \in y\}$$

is a (quantified) *term;*

(8) if x is a variable and p is a formula, then

$$Hx\,p$$

is a (quantified) *term.*

A finite sequence of primitive symbols is a wfe if and only if it can be shown to be a wfe on the basis of these eight rules. $\Box$

Remarks 3.1 (1) It is important to notice that it is possible to determine algorithmically whether or not any given finite sequence of primitive symbols is a wfe. Furthermore, each wfe admits only one construction ('unique readability theorem'; see, e.g., [End72]). The steps of this construction can be detected algorithmically too.

(2) We will usually employ the lower case letters (with or without subscripts or superscripts):

$$u, v, w, x, y, z$$

to denote variables,

$$p, q, r, a, b, c$$

to denote formulae, and

$$s, t, d, e, f, g$$

to denote terms. All the remaining lower case letters denote non-negative integers. The subscript b is sometimes used to denote a 'bit', i.e. 0 or 1. $\Box$

Wfes can be variously classified, on the basis of their syntactical structure. For instance, a formula of the form $\neg p$ (respectively, $(p \& q)$) is called a *negation* (respectively, a *conjunction*); a term of the form $Hx\,p$ (respectively, $\{t : x \in y\}$) is called a *description* (respectively, an *abstraction term*). We omit defining disjunctions, implications, etc., because this terminology is quite widespread in the literature. We recall only a few definitions here.

Definition 3.3 *A negated or un-negated atomic formula, i.e., a formula of one of the forms*

$$(s = t), \quad (s \subseteq t), \quad (s \in t), \quad Finite\ s,$$

$$\neg (s = t), \ \neg (s \subseteq t), \ \neg (s \in t), \ \neg\, Finite\ s,$$

is called a literal. *Any wfe whose construction requires use of rule (7) or (8) is said to be* quantified; *each occurrence of x within a wfe is said to be* bound *or* free *according to whether or not it belongs to a well-formed sub-expression of the form* $\{t : x \in y\}$ *or* Hxp. *An unquantified formula is a* disjunctive normal form *iff it is either a literal, or a conjunction of literals, or a disjunction* $(q_0 \vee q_1)$ *where both* q_0 *and* q_1 *are disjunctive normal forms.* □

3.3 Semantics

In the following we assume that $<$ is a well-ordering of the class $\mathcal{V}$ of all sets, which is anti-lexicographic on finite sets and such that

$$rk\, X \in rk\, Y \text{ implies that } X < Y,$$

for every two sets X and Y. Moreover,

if $rk\, X = rk\, Y$, X is finite, and Y is infinite, then $X < Y$.

We denote by $\Phi(E)$, where E is a wfe, the set of all those variables that have at least one free occurrence in E. An *interpretation of E* is a function φ from $\Phi(E)$ into $\mathcal{V}$. We are now going to provide *designation rules* which assign a value E^φ to E, for every interpretation φ of E. E^φ will be a set when E is a term, and one of the truth values *FALSE*, *TRUE* when E is a formula. For notational convenience, we identify here:

$$FALSE \text{ with } \emptyset,$$
$$TRUE \text{ with } \{\emptyset\}.$$

The set E^φ is said to be *denoted by E under* φ; or simply *denoted by E*, when the reference to φ is unnecessary.

Definition 3.4 *The* designation rules *are as follows:*

(1′) *If E is* $\emptyset$, *then* $E^\varphi = \emptyset$.

(2′) *If E is a variable x, then* $E^\varphi = \varphi x$.

(3′) *If E is* $(s \otimes t)$, *where* $\otimes$ *stands for one of the signs* $\cap, \setminus, \cup, \times$; *and* ψ, χ *are the restrictions of* φ *to* $\Phi(s)$ *and to* $\Phi(t)$ *respectively, then* $E^\varphi = s^\psi \otimes t^\chi$.

$(3'')$ If E is $\ominus t$, where $\ominus$ stands for one of the signs Pow, Un, then $E^\varphi = \ominus t^\varphi$.

$(3''')$ If E is ηt, then E^φ is the minimum, with respect to $<$, of the set t^φ. When t^φ is $\emptyset$, we put $E^\varphi = \emptyset$.

$(4')$ If E is $\{t_0, \ldots, t_n\}$ and φ_i is the restriction of φ to $\Phi(t_i)$, for $i = 0, \ldots, n$, then $E^\varphi = \{t_0^{\varphi_0}, \ldots, t_n^{\varphi_n}\}$.

$(5')$ If E is $(s \otimes t)$, where $\otimes$ stands for one of the signs $=, \subseteq, \in$; and ψ, χ are the restrictions of φ to $\Phi(s)$ and to $\Phi(t)$ respectively, then $E^\varphi = $ TRUE if and only if $s^\psi \otimes t^\chi$.

$(5'')$ If E is Finite s, then $E^\varphi = $ TRUE if and only if s^φ has finite cardinality.

$(6')$ If E is $\neg p$, then $E^\varphi = \{\emptyset\} \setminus p^\varphi$.

$(6'')$ If E is $(p \,\&\, q)$, and ψ, χ are the restrictions of φ to $\Phi(p)$ and to $\Phi(q)$ respectively, then $E^\varphi = p^\psi \cap q^\chi$.

$(6''')$ If E is $(p \vee q)$, and ψ, χ are the restrictions of φ to $\Phi(p)$ and to $\Phi(q)$ respectively, then $E^\varphi = p^\psi \cup q^\chi$.

$(6'''')$ If E is $(p \rightarrow q)$, and ψ, χ are the restrictions of φ to $\Phi(p)$ and to $\Phi(q)$ respectively, then $E^\varphi = q^\chi \cup (\{\emptyset\} \setminus p^\psi)$.

$(6''''')$ If E is $(p \leftrightarrow q)$, and ψ, χ are the restrictions of φ to $\Phi(p)$ and to $\Phi(q)$ respectively, then $E^\varphi = (q^\chi \cup (\{\emptyset\} \setminus p^\psi)) \cap (p^\psi \cup (\{\emptyset\} \setminus q^\chi))$.

$(7')$ If E is $\{t : x \in y\}$, then for every set S let φ_S be the function such that:

    (i) the domain of φ_S is $\Phi(t)$;

    (ii) $\varphi_S z = \varphi z$, for every variable z in $\Phi(t) \setminus \{x\}$;

    (iii) $\varphi_S x = S$, if $x \in \Phi(t)$.

Then $E^\varphi = \{t^{\varphi_S} : S \in \varphi y\}$.

$(8')$ If E is $H x p$, then for every set S let φ_S be the function such that:

    (i) the domain of φ_S is $\Phi(p)$;

    (ii) $\varphi_S z = \varphi z$, for every variable z in $\Phi(p) \setminus \{x\}$;

    (iii) $\varphi_S x = S$, if $x \in \Phi(p)$.

Consider the class $K = \{S : p^{\varphi_S} = $ TRUE $\}$. Then

$$E^\varphi = \emptyset \text{ if } K = \emptyset;$$
$$E^\varphi = \{M\}, \text{ where } M \text{ is the minimum with respect to } <$$
of K, if $K \neq \emptyset$.

$\square$

Remarks 3.2   (a) The seeming circularity in some of the above designation rules is due to the fact that the same signs are used at times as relators (or operators), at times instead as relations (or operations). For instance, when we write

$$(s \cup t)^\varphi \; = \; s^\psi \cup t^\chi,$$

the sign $\cup$ that appears on the left side is an operator (i.e. a symbol of our formal language for set theory), while the one that appears on the right side indicates the actual union operation, applied to the two sets s^ψ and t^χ. Similar considerations apply to $\emptyset$, which is at times a symbol, at times a set.

(b) If E is a wfe and V is a set of variables such that

$$\Phi(E) \subseteq V,$$

then we will freely call any function

$$\varphi : V \to \mathcal{V}$$

an interpretation of E. If we wanted to be mercilessly precise, however, when the above inclusion is proper we should have to consider the restriction ψ of φ to $\Phi(E)$. This ψ is indeed an interpretation of E; being somewhat sloppy, we will also write E^φ to indicate E^ψ.

(c) We anticipate, as a consequence of propositions that will be proved in the next chapter, that the truth value of any unquantified variable-free formula p can be calculated. Indeed, each term appearing in p denotes a hereditarily finite set and can—after some calculations—be replaced by a 'normalized' term. One will then evaluate the atomic formulae appearing in p. Completing the evaluation of p is then a matter of routine, because the semantics of the propositional connectives is entirely constructive.   □

Without proof, we recall a classical proposition:

Theorem 3.1 *For every unquantified formula p, there exists a disjunctive normal form q which involves the same atomic formulae as p (so that $\Phi(p) = \Phi(q)$), and for which*

$$p^\varphi \; = \; q^\varphi$$

in every interpretation φ of p.   □

3.4   Shorthand notations

Rigorously speaking, only those combinations of symbols that can be shown to be wfes on the basis of the formation rules (1) through (8) in Definition

3.2 should be regarded as wfes. However, we will often employ illegal expressions (called *abbreviations*) as alternative notations for well-formed expressions. Abbreviations are not necessarily shorter than the wfes they stand for: the motivation for their adoption is that they convey meaning more clearly, or belong to some widely accepted notational tradition. We will freely use the words 'term' and 'formula' for the 'abbreviated' notations for terms and for formulae.

CONVENTIONS FOR DROPPING PARENTHESES Unless it occurs as part of a longer formula, a formula of the form $(p \otimes q)$, where $\otimes$ is a propositional connective, is often abbreviated $p \otimes q$. Under the same circumstances (respectively, if it appears in a context of the form $\{ (s \otimes t) \cdots \}$ or of the form $\{ t_0, \ldots, t_i, (s \otimes t) \cdots \}$), the formula (respectively, term) $(s \otimes t)$, where $\otimes$ is a relator (respectively, an operator), becomes $s \otimes t$. We have another rule for dropping parentheses: namely, we assume that

$$\eta, \ Un, \ Pow, \ \cap, \ \backslash, \ \cup, \ Finite, \ =, \ \subseteq, \ \in, \ \neg, \ \&, \ \vee, \ \rightarrow, \ \leftrightarrow$$

have decreasing strengths (or decreasing 'precedences'). Thus, for instance,

$$x = y \vee \neg x \in y \,\& \, z = y \cap x \cup z$$

stands for

$$((x = y) \vee (\neg (x \in y) \,\& \, (z = ((y \cap x) \cup z)))).$$

When no precedence rule applies, we assume associativity to the left; thus, for instance,

$$p \rightarrow q \rightarrow r \rightarrow a$$

stands for

$$(((p \rightarrow q) \rightarrow r) \rightarrow a).$$

□

JOINING SYMBOLS

$s \notin t$	stands for	$\neg s \in t,$
$s \neq t$	stands for	$\neg s = t,$
$s \nsubseteq t$	stands for	$\neg s \subseteq t,$
$s \subsetneq t$	stands for	$s \subseteq t \,\& \, s \neq t,$
$s \underset{=}{\in} t$	stands for	$s \in t \vee s = t,$
$s \underset{\neq}{\in} t$	stands for	$s \in t \,\& \, s \neq t.$

□

EXISTENTIAL AND UNIVERSAL QUANTIFIERS.

$\exists x \, p$	stands for	$\emptyset \neq Hx \, p,$
$\exists x \in y \, p$	stands for	$\exists x \, (x \in y \,\& \, p),$
$\forall x \, p$	stands for	$\neg \exists x \, \neg p,$
$\forall x \in y \, p$	stands for	$\neg \exists x \in y \, \neg p.$

The quantifiers $\exists x \in y$, $\forall x \in y$ are called *restricted quantifiers*. They are defined only when x and y are distinct variables.   □

EXPONENTIATING SYMBOLS For $n = 0, 1, 2, \ldots$:

$$\neg^n p \text{ stands for } \neg \cdots \neg p,$$

with n consecutive occurrences of the symbol $\neg$;

$$\ominus^n t \text{ (where } \ominus \text{ is a unary operator) stands for } \ominus \cdots \ominus t,$$

with n consecutive occurrences of the symbol $\ominus$;

$$s \in^{n+1} t \text{ stands for } \exists x_0 \exists x_1 \cdots \exists x_{n+1} (x_0 = s \, \& \, x_{n+1} = t$$
$$\& \, x_0 \in x_1 \, \& \, x_1 \in x_2 \, \& \, \cdots \, \& \, x_n \in x_{n+1}),$$

where the variables $x_0, x_1, \ldots, x_{n+1}$ do not occur in s or in t.   □

REVERSING RELATORS

$$s \supseteq t \quad \text{stands for} \quad t \subseteq s,$$
$$s \ni t \quad \text{stands for} \quad t \in s,$$
$$s \supsetneq t \quad \text{stands for} \quad t \subsetneq s,$$

and so on.   □

CHAINING RELATORS We often shrink a conjunction

$$t_0 \otimes_0 t_1 \, \& \, t_1 \otimes_1 t_2 \, \& \, \cdots \, \& \, t_n \otimes_n t_{n+1}$$

of two or more literals, where each $\otimes_i$ is either a primitive relator (e.g. $\in$ or $\subseteq$) or a defined relator (e.g. $\subsetneq$ or $\in^m$), into a single 'chain'

$$t_0 \otimes_0 t_1 \otimes_1 t_2 \otimes_2 \cdots \otimes_n t_{n+1}.$$

Thus, for instance,

$$t \in s \subseteq d \quad \text{stands for} \quad t \in s \, \& \, s \subseteq d;$$
$$\emptyset \not\ni \{s, t\} = d \quad \text{stands for} \quad \emptyset \not\ni \{s, t\} \, \& \, \{s, t\} = d.$$

In order to avoid confusion, we do not treat propositional connectives in the same manner. Thus, for example,

$$p \leftrightarrow q \leftrightarrow r$$

should *not* be regarded as an abbreviation for

$$((p \leftrightarrow q) \, \& \, (q \leftrightarrow r)),$$

but rather for

$$((p \leftrightarrow q) \leftrightarrow r).$$

□

ORDERING RELATORS

$$s < t \quad \text{stands for} \quad t \neq \eta\{s, t\}$$
$$s \leq t \quad \text{stands for} \quad s < t \vee s = t.$$

□

These definitions of the relators $<$, $\leq$ do not deserve particular comments, because they simply reflect our interpretation of η, according to which $(\eta\{s, t\})^\varphi$ denotes the smaller of s^φ and t^φ in any interpretation φ. We could not define $s < t$ as

$$s = \eta\{s, t\},$$

because it might well be the case that $s^\varphi = t^\varphi$ in some interpretation φ, in which case

$$s^\varphi = (\eta\{s, t\})^\varphi$$

would also be satisfied.

BINDING OPERATORS We have already met a few binding operators, namely the primitive descriptor H and the quantifiers $\exists$ and $\forall$. Another binding construct that we have seen is the abstraction term

$$\{t : x \in y\}.$$

We now introduce, as shorthand notations, new descriptors and a new kind of abstraction term:

$$\varepsilon x p \quad \text{stands for} \quad \eta H x p;$$
$$I x p \quad \text{stands for} \quad H y \forall x (p \leftrightarrow x = y),$$

where y is a variable not appearing in p;

$$\iota x p \quad\quad\quad\quad \text{stands for} \quad \eta I x p;$$
$$\{t : x \in y : p\} \quad \text{stands for} \quad Un\{Iz (p \& z = t) : x \in y\},$$

where z is a variable not appearing in p or in t.   □

Let us give suggestions about how to read the binding constructs:

$\forall x p$ can be read: 'p holds for all x';
$\exists x p$ can be read: 'p holds for some x';
$\varepsilon x p$ can be read: 'some x such that p';
$H x p$ can be read: 'some $\{x\}$ such that p holds for x';
$\iota x p$ can be read: 'the x such that p';
$I x p$ can be read: 'the $\{x\}$ such that p holds for x';

$\{t : x \in y\}$ can be read: 'the set of all those values that t takes on, as x varies in y';

$\{t : x \in y : p\}$ can be read: 'the set of all those values that t takes on, corresponding to values of x which belong to y and satisfy p'.

Note that:

1. When no set x satisfying p exists, then $\exists x\, p$ is false, and $\varepsilon\, x\, p$, $H x\, p$, $\iota\, x\, p$, $I x\, p$ all denote $\emptyset$.

2. When exactly one set S exists such that p holds for $x = S$, then: $\exists x\, p$ is true, both $\varepsilon\, x\, p$ and $\iota\, x\, p$ take the value S, and both $H x\, p$ and $I x\, p$ take the value $\{S\}$.

3. When more than one set x satisfying p exists, then: $\exists x\, p$ is true, both $\iota\, x\, p$ and $I x\, p$ denote $\emptyset$, $\varepsilon\, x\, p$ denotes the least value S such that p holds for $x = S$, and $H x\, p$ denotes $\{S\}$.

In some sense, the descriptions

$$H x\, p, \quad I x\, p$$

are less ambiguous than the corresponding descriptions

$$\varepsilon\, x\, p, \quad \iota\, x\, p,$$

because $H x\, p = \emptyset$ (respectively, $I x\, p = \emptyset$) indicates the non-existence (respectively, non-existence or lack of uniqueness) of an x satisfying p, whereas $\varepsilon\, x\, p = \emptyset$ (respectively, $\iota\, x\, p = \emptyset$) can also be an indication that p holds for $x = \emptyset$ (respectively, that p holds for $x = \emptyset$ and for no other value of x).

Formalized presentations of set theory (see e.g. [Jec78]) usually include an axiom scheme called the 'substitution axiom' or the 'replacement axiom'. This scheme generally has some form equivalent to

$$\exists z\, \forall u\, (u \in z \leftrightarrow \exists v \in x\, \forall y\, (p \leftrightarrow y = u)),$$

where p is a formula not involving z or u. In typical applications, p involves both v and y. In our language we can explicitly denote the set z whose existence is asserted by the substitution axiom, by means of the following term:

$$\{\iota\, y\, p : v \in x : \emptyset \neq I y\, p\}.$$

Analogous remarks apply to the so-called 'foundation' (or 'regularity') axiom and to the axiom of choice, which, stated in words, respectively say that

every non-empty set x has a member y with $\emptyset = x \cap y$,

and that

whenever the members of a set w are pairwise disjoint non-empty sets, there exists a set z which has a singleton intersection with each member of w.

Concerning the first of these, it follows from the semantics of η that

$$x \neq \emptyset \;\rightarrow\; \eta\, x \in x \,\&\, \emptyset = x \cap \eta\, x$$

is true in any interpretation, so that $\eta\, x$ is the desired y. Concerning the second, if we assume the members of w to be pairwise disjoint non-empty sets, then the desired z is simply

$$z = \{\, \eta\, v : v \in w \,\}.$$

3.5　Redundancies

Our language is highly redundant. For instance, if propositional connectives were absent from our repertoire of primitive symbols (see Definition 3.1), then of course we would be forced to abandon the formation rule (6) in Definition 3.2. We could nonetheless re-introduce the connectives into the language, by abbreviations.

Note, indeed, that if x does not appear in p then, in any interpretation φ of p (which is also an interpretation of $Hx\,p$), the term $Hx\,p$ takes either the value $\emptyset$ or the value $\{\emptyset\}$ according to whether p^{φ} is false or true. We could therefore define:

$$\textit{Truth Value}(p) \text{ stands for } Hx\,p\,,$$

with x not occurring in p;

$$
\begin{aligned}
\neg\, p \quad &\text{stands for} \quad \emptyset = \textit{Truth Value}(p); \\
\textit{False} \quad &\text{stands for} \quad \emptyset \in \emptyset; \\
\textit{True} \quad &\text{stands for} \quad \neg\, \textit{False}.
\end{aligned}
$$

An equally good definition—not involving $\in$—for *False* would be:

$$\textit{False} \text{ stands for } \emptyset = Hx\,\emptyset = x\,.$$

Observe next that $\&$ and $\vee$ operate on truth values in the same manner $\cap$ and $\cup$ operate on the sets $\emptyset$, $\{\emptyset\}$. We could therefore define:

$$
\begin{aligned}
(p \,\&\, q) \text{ stands for } &\textit{Truth Value}(p) \cap \textit{Truth Value}(q) \neq \emptyset, \\
(p \vee q) \text{ stands for } &\textit{Truth Value}(p) \cup \textit{Truth Value}(q) \neq \emptyset, \\
(\textit{If}\,p \textit{ Then}\,q \textit{ Else}\,r) \text{ stands for } &(p \,\&\, q) \vee (\neg\, p \,\&\, r), \\
(p \,\rightarrow\, q) \text{ stands for } &\textit{If}\,p \textit{ Then}\,q \textit{ Else True}, \\
(p \,\leftrightarrow\, q) \text{ stands for } &\textit{If}\,p \textit{ Then}\,q \textit{ Else}\,\neg\, q.
\end{aligned}
$$

Also the operator $\cap$ and the relators $\subseteq$, $\in$ are redundant, because one could define

$$x \cap y \quad \text{stands for} \quad x \setminus (x \setminus y),$$
$$x \subseteq y \quad \text{stands for} \quad \emptyset = x \setminus y,$$
$$x \in y \quad \text{stands for} \quad \emptyset = \{x\} \setminus y.$$

The construct $\{\bullet, \ldots, \bullet\}$ is redundant too. Terms of the form

$$\{t_0, \ldots, t_n\}$$

can in fact be regarded as shorthand notations, as follows:

$$\{t\} \text{ stands for } Hx \, x = t,$$

where x is a variable not appearing in t;

$$\{s_0, \ldots, s_m, t\} \text{ stands for } \{s_0, \ldots, s_m\} \cup \{t\},$$

for $m = 0, 1, 2, \ldots$. In a subtler sense, η is redundant too. In this case we are unable to regard terms of the form ηt as abbreviations for other terms not involving η. Nonetheless, one can recursively define a function $p \mapsto p'$ sending formulae into formulae, so that, for every formula p:

- the operator η does not occur in p';

- $\Phi(p) = \Phi(p')$, i.e., the free variables in p and in p' are the same;

- $p^\varphi = p'^\varphi$ in every interpretation φ of p.

The basic remark here is that the formula $x = \eta t$ can be regarded as an abbreviation for

$$\emptyset = x = t \lor x \in Hy \, y \in t,$$

provided that x, y do not appear in t. Assuming that $\eta t_1, \ldots, \eta t_n$ are the distinct terms of the form ηt that appear in p, we choose distinct variables $x_1, \ldots, x_n, y_1, \ldots, y_n$ not occurring in p, and define p' to be the formula

$$\exists x_1 \cdots \exists x_n \left(p^* \, \& \, \&_{i=1}^n \left((\emptyset = x_i = t_i)' \lor x_i \in Hy_i (y_i \in t_i)' \right) \right),$$

where p^* is obtained from p by replacing each outermost occurrence of a term of the form ηt_j by the corresponding variable x_j.

To conclude, let us show how the relator $Finite$ can be eliminated:

$$s \, Swells \, t \quad \text{stands for} \quad \eta s \underset{\neq}{\subseteq} t \in s,$$
$$Finite \, t \quad \text{stands for} \quad \neg \exists v \, \exists w \, (v \, Swells \, w \subseteq t),$$

where v, w are distinct variables not appearing in t. $\square$

Concerning the relator $Swells$, observe in fact that if s $Swells$ t holds in some interpretation φ, then t^φ must be infinite. For, assuming to the contrary that t^φ is finite, let us show that

$$t^\varphi \;<\; X \text{ whenever } t^\varphi \;\subsetneqq\; X\,.$$

Indeed, if X is finite then $t^\varphi \subsetneqq X$ implies $t^\varphi < X$ by anti-lexicographicity. If X is infinite and $t^\varphi \subsetneqq X$, then either $rk\,t^\varphi < rk\,X$ or $rk\,t^\varphi = rk\,X$; in both cases $t^\varphi < X$, because t^φ is finite. It is therefore impossible to have

$$\eta\,s^\varphi \;\subsetneqq\; t^\varphi \;\in\; s^\varphi\,,$$

or else we would have

$$\eta\,s^\varphi \;\in\; s^\varphi \text{ and } t^\varphi \;\in\; s^\varphi\,, \text{ but } t^\varphi \;<\; \eta\,s^\varphi\,,$$

contradicting the minimality of $\eta\,s^\varphi$ in s^φ.

It follows from the preceding paragraph that if

$$\neg\,Finite\,t$$

holds in some interpretation φ, then t^φ must be infinite. In fact, from the existence of sets V, W such that

$$V\,Swells\,W \;\subseteq\; t^\varphi\,,$$

one deduces that both W and t^φ are infinite. Conversely, assuming that t^φ is infinite, let us prove that $\neg\,Finite\,t$ holds in φ. We can in fact choose a sequence

$$U_0\,,\,U_1\,,\,U_2\,,\ldots$$

of distinct members of t^φ. Putting

$$W_i \;=\; \{\,U_i\,,\,U_{i+1}\,,\,U_{i+2}\,,\ldots\}$$

for every i in ω, and

$$W_j \;=\; min\,\{\,W_i \,:\, i \in \omega\,\}\,,$$

$$V \;=\; \{\,W_j\,,\,W_{j+1}\,\}\,,\quad W \;=\; W_{j+1}\,,$$

we have

$$t^\varphi \;\supseteq\; W_j \;=\; min\,V \;\subsetneqq\; W \in V\,,$$

so that

$$V\,Swells\,W$$

and
$$W \subseteq t^\varphi .$$

Incidentally, the above remarks show one way in which one can express, in our language, the existence of infinite sets; this is by means of the formula

$$\exists\, v\, \exists\, w\; v\; Swells\; w\, .$$

3.6   Restricted notions of satisfiability and validity

Consider a formula p.

Definition 3.5 *A* finite interpretation *of p is an interpretation of p (i.e., a function φ from $\Phi(p)$ into V) whose values are finite sets; a* hereditarily finite interpretation *of p is one whose values are hereditarily finite sets. If p is unquantified, and t^φ is finite for every term t occurring in p (not only when t is a variable of p), then φ is said to be an* easy *interpretation of p. The formula p is said to be:*

valid	*iff p holds true in each of its interpretations;*
satisfiable	*iff p holds true in some of its interpretations.*

*The definitions of (*hereditarily*) finite validity and satisfiability and those of easy validity and satisfiability are similar, but they refer to (hereditarily) finite interpretations and to easy interpretations respectively. For instance:*

> *p is* easily satisfiable *iff there is some easy interpretation φ of p such that p^φ = TRUE.*

'Unsatisfiable' *is synonymous with 'not satisfiable'. An interpretation φ such that p^φ = TRUE, is called a* model *of p.*   □

Trivially:

- if p is easily satisfiable, or hereditarily finitely satisfiable, then p is finitely satisfiable;

- if p is finitely satisfiable, then p is satisfiable;

- p is valid if and only if $\neg p$ is unsatisfiable;

- p is easily (respectively, (hereditarily) finitely) satisfiable if and only if $\neg p$ fails to be easily (respectively, (hereditarily) finitely) valid;

- if p is valid, then p is finitely valid;

- if p is finitely valid, then p is both easily valid and hereditarily finitely valid.

Note that the distinction between finite and easy satisfiability vanishes when p is unquantified and the operators Un, η do not appear in p. Indeed, $(\eta t)^\varphi$ and $(Un\, t)^\varphi$ may fail to be finite when t^φ is finite (for instance when $t^\varphi = \{\omega\}$), whereas the remaining operators $\cap$, $\setminus$, $\cup$, $\times$ and Pow always yield finite sets when applied to finite sets.

A useful characterization of easy satisfiability is the following:

Lemma 3.1 *Let p be an unquantified formula, and, for each term t occurring in p, let x_t be a variable uniquely associated with t. Suppose that none of the variables x_t occurs in p. Then p is easily satisfiable if and only if the formula*

$$(\star) \qquad\qquad p \,\&\, \&_{t\,\text{occurs in}\,p}\, x_t = t$$

is finitely satisfiable.

Proof. If φ is an easy interpretation of p that makes p true, then we obtain ψ by enlarging the domain of φ as follows:

$\psi\, x \,=\, \varphi\, x$, for every variable x in $\Phi(p)$;
$\psi\, x_t \,=\, t^\varphi$, for every term t occurring in p.

Thus

$p^\psi \,=\, p^\varphi \,=\, TRUE$, and
$x_t^\psi \,=\, t^\varphi \,=\, t^\psi$ is finite for every term t occurring in p.

Consequently, $(\star)$ is finitely satisfied by ψ. Conversely, by restricting any finite model of $(\star)$ to $\Phi(p)$, one gets an easy model of p. $\qquad\square$

It will turn out, as a consequence of results proved in Chapter 8, that any satisfiable formula not involving the constructs *Finite*, η, Un, or $\times$ is hereditarily finitely satisfiable too. On the contrary, when any one of the constructs η, Un, $\times$ comes into play it is not hard to find examples of formulae that are satisfiable without even being finitely satisfiable. One such example, discussed in the preceding section, is the formula $v\ Swells\ w$; another one, to be encountered in Chapter 10, is $u = Un(u)\ \&\ v \in u$; a third example is $x \times y \subseteq x \neq \emptyset\,\&\, y \neq \emptyset$, whose analysis is left to the reader. When new constructs will be added to the language (e.g. an operator designating domain of an arbitrary set, in Chapter 9), new fashions of expressing the existence of infinite sets by means of unquantified formulae will make their appearance.

3.7   Absolute notions of satisfiability and validity. Decision tests

Let $<$ be a well-ordering of the class $\mathcal{V}$ of all sets, which is anti-lexicographic and agrees with rank comparison. As explained in Section 3.3, $<$ induces semantics for η and H. To be more specific, consider the term x, consisting of an isolated variable. Let φ_S be the function, defined on the singleton domain $\{x\}$, that sends x into S; hence φ_S is an interpretation of ηx, for every set S. The choice function K defined on all of $\mathcal{V}$ by putting

$$KS = \eta x^{\varphi_S},$$

selects the minimum member of S with respect to $<$, for every set S. It can be regarded as the 'meaning' of η, and uniquely determines $<$, because we have

$$S < T \text{ if and only if } K\{S, T\} \neq T.$$

Definition 3.6 *The choice function*

$$KS = \eta x^{\varphi_S}$$

induced by an anti-lexicographic, rank-preserving, well-ordering $<$ of $\mathcal{V}$, is called a version *of η.*   □

The notions of satisfiability and validity that we have discussed in the preceding section, are related to the meaning of the choice operators, hence indirectly related to the version of η. It is quite likely, one may argue, that some formula involving η or H is satisfiable for a particular version of η (i.e., for a particular choice of $<$), but unsatisfiable for some other version. We now want to introduce new satisfiability and validity notions that are *absolute* in the sense that they are not affected by the partial indeterminacy in the meaning of η.

In what follows we denote by C a privileged version of η that, for some reason, we want to regard as its 'legitimate' version; for instance, if η was also included among the operators of our meta-theory, then C would be the version such that

$$CS = \eta S$$

for every set S. By K we instead denote some arbitrary version of η. We now explicitly speak of K-models, where we simply spoke of models before.

Definition 3.7 *A formula p is said to be:*

- weakly satisfiable *iff p admits a K-model for some version K of η;*

- satisfiable *iff p admits a C-model;*

- strongly satisfiable *iff p admits a K-model for every version K of η;*

- uniformly satisfiable *iff p admits a C-model which is also a K-model for every version K of η.*

The three notions of strong validity, validity, and weak validity are dual to those of weak satisfiability, satisfiability, and strong satisfiability, respectively. Thus, in particular, a formula is weakly valid if and only if its negation fails to be strongly satisfiable. The adverbs 'finitely', 'hereditarily finitely', and 'easily' can be used to modify these notions in the same manner we have explained in Section 3.6. □

Consider some fragment F of set theory: i.e. F is a recursive collection of set-theoretic formulae, which we assume to be closed under negation.

Definition 3.8 *A procedure T is said to be:*

- a complete *(respectively,* sound*) satisfiability test iff every formula p in F such that T produces the output 'yes' (respectively, 'no') on the input p, is (respectively, fails to be) satisfiable;*

- a total *satisfiability test iff, corresponding to every input p in F, T issues a yes/no answer;*

- a semi-decision *procedure for F iff T is both sound and complete; and a* decision *procedure iff T is also total. A semi-decision satisfiability procedure that answers 'no' to every unsatisfiable p in F is sometimes called a* quasi-decision *procedure; one that answers 'yes', to every satisfiable p in F is said to be* strongly complete. □

Obviously, every satisfiability test can be converted into a validity test.

Weak, strong, and uniform soundness or completeness are defined by analogy with the above, in terms of weak, strong, and uniform satisfiability. If we succeed in finding a total satisfiability test which is both weakly sound and strongly complete, then we can conclude that F is decidable since, trivially, for every formula p in F, the first three notions of satisfiability (as well as their dual notions of validity) will coincide.

One method for proving the strong completeness of a test T is by showing that some modified version of T will produce, for every formula p for which T gives the answer 'yes', an interpretation of p making p true for any version of η. If this is the case, then T is even uniformly complete. Typical 'uniform' models of p, when p is unquantified, will be the hereditarily finite models, as we will soon explain.

In the absence of any restriction on the use of η, finite satisfiability of a formula p is not easier to detect than simple satisfiability. For, if p_η is obtained from p by replacing each free variable x by the corresponding term ηx, then:

p is satisfiable if and only if p_η is finitely satisfiable.

In fact, if φ is an interpretation of p such that $p^\varphi = TRUE$, then $p_\eta^\psi = TRUE$ in the finite interpretation ψ such that $\psi x = \{\varphi x\}$ for every free variable x of p. Conversely, if ψ is an interpretation of p_η such that $p_\eta^\psi = TRUE$, then $p^\varphi = TRUE$ in the interpretation φ such that $\varphi x = C\psi x$ for every free variable x of p. (Here C continues to stand for the privileged version of η.) Conclusion:

Lemma 3.2 *If a fragment F of set theory contains p_η whenever it contains p, then any algorithm that is able to determine, for every formula q in F, whether or not q is finitely satisfiable, can be converted into a decision algorithm for F.* □

We will prove in the next chapter that the notion of hereditarily finite satisfiability, although, *a priori*, a relative notion, is absolute when applied to unquantified formulae. The following is in fact a semi-decision procedure to test such a formula p for hereditarily finite satisfiability:

> Systematically replace the variables that occur in p by normalized terms denoting hereditarily finite sets (recall Remark 3.2(c)), in all possible ways. For each replacement, check whether p has become true: if this is the case, then p is hereditarily finitely satisfiable and the procedure terminates. Otherwise, another replacement is to be tried.

Of course, this procedure can go on for ever, because there are infinitely many hereditarily finite sets, hence infinitely many replacements to be tried.

The soundness of this test is entirely trivial, as it does not even attempt to prove p unsatisfiable. Completeness is also obvious, because the replacement $x \mapsto t_x$ that led to the output 'p is satisfiable' corresponds indeed to a model of p: to the one which interprets each variable x as the hereditarily finite set denoted by t_x. A sort of strong completeness can also be proved, namely that:

> p is hereditarily finitely satisfiable only if the test is able to detect this fact.

Indeed, we will prove through Lemma 4.5 and Corollary 4.8 that every hereditarily finite set is denoted by a normalized hereditarily finite term; hence, in doing the replacements, we are not overlooking any potential hereditarily finite model.

We continue to suppose that p is unquantified. If p is hereditarily finitely satisfiable, then p is easily satisfiable, because every term in p denotes a hereditarily finite (hence a finite) set when the variables are interpreted as hereditarily finite sets, as will result from the next chapter. Conversely, one often succeeds in proving, for formulae p in a *certain* unquantified fragment F of set theory, that:

> p is easily satisfiable only if p is hereditarily finitely satisfiable.

Such a discovery will make the above semi-decision test for hereditarily finite satisfiability also applicable as a test for *easy* satisfiability.

Does there exist a *decision* test for hereditarily finite satisfiability of unquantified formulae? In Chapters 5, 6, and 8 we will describe several decision algorithms which can determine whether *or not p* is hereditarily finitely satisfiable, when p belongs to various broad collections of unquantified formulae. However, we are as yet unable to answer the question in general, without imposing any restrictions on the form of p.

3.8   Normalization of formulae

In order to facilitate the study of the satisfiability decision problem for a given portion F of set theory, we will usually make simplifying assumptions regarding the form of the input; moreover we will sometimes refer to a variant of the satisfiability notion, called *injective satisfiability*, which is introduced below. Generality will not be affected in either case; in particular the simplifying assumptions regarding the input formula p will be legitimated by defining a *normalization process* for F, i.e. a translation $p \mapsto p'$ where p' also belongs to F but has a very regular syntactic structure, so that:

- p' is satisfiable if and only if p is satisfiable;

- a model for p is easily extracted from any model of p'.

All examples of normalization given in this book will be adaptations of the basic normalization algorithm described in this section. This prototype algorithm is applicable to *any* unquantified formula p.

Definition 3.9 *Let p be any formula of set theory. An assignment φ of sets to the variables of p is said to be an* injective interpretation *of p if $\varphi x \neq \varphi y$ for any pair x, y of distinct variables in p. If p admits an injective model, then p is said to be* injectively satisfiable.   □

When F contains equality and is closed with respect to the propositional connectives, the following holds:

Lemma 3.3 *The satisfiability problem is reducible to the injective satisfiability problem, for every given collection F of set-theoretic formulae. Otherwise stated, a sound and complete satisfiability decision test can be obtained from any sound and complete decision test for injective satisfiability.*

Proof.   Given disjoint non-empty sets $X_1, \ldots, X_n$ of variables with $X_1 \cup \cdots \cup X_n = \Phi(p)$, we select a representative variable x_i from each X_i in the partition $\pi = \{ X_1, \ldots, X_n \}$ and obtain the formula p_*^π by replacing each variable y in p by the representative x_i with y in X_i. Then we put

$$p^\pi = (\, p_*^\pi \ \& \ \&_{i=1}^{n} \ \&_{x,y \in X_i} \ x = y \,).$$

Since there are finitely many partitions π of $\Phi(p)$, also the number of formulae of the form p^π is finite. We claim that p is satisfiable if and only if p^π is injectively satisfiable for some π. Indeed, any model of p^π clearly satisfies p too. Conversely, any model φ of p uniquely determines a partition $\pi_\varphi = \{ X_1, \ldots, X_n \}$ such that

$$\varphi x = \varphi y \text{ if and only if } x, y \text{ belong to the same } X_i,$$

for all x, y in $\Phi(p)$. Thus φ is an injective model for p^{π_φ}.   ■

In order to describe a normalization process with sufficient generality, we will assume that the language of the class F of formulae consists of:

  (i) a denumerable infinity of variables;

 (ii) the constant $\emptyset$;

(iii) the boolean set operators $\cap$, $\backslash$, $\cup$;

(iv) additional operators $f_1, \ldots, f_k$ (which, for simplicity, we assume all of arity 1);

 (v) the relators $\in$, $\subseteq$, $=$;

(vi) the propositional connectives $\neg$, $\&$, $\vee$, $\rightarrow$, $\leftrightarrow$.

Given a formula p of F, the process ν to be described below will produce a formula $\nu(p)$ such that:

- p and $\nu(p)$ are *equi-satisfiable* (i.e. p is satisfiable if and only if $\nu(p)$ is satisfiable);

- $\nu(p)$ is a disjunction $q_1 \vee \cdots \vee q_n$ in which each q_i is a conjunction of positive atoms of type

$$
\begin{array}{ll}
(=) & x = y \cup z, \quad x = y \backslash z, \\
(f) & x = f_i(y), \\
(\in) & x \in y,
\end{array}
$$

where x, y, z are variables.

Algorithm 3.1 (Normalization process ν)

- Initialize Ψ to the empty set of formulae.

- Replace each occurrence in p of literals of type $t_1 \subseteq t_2$ by the atomic formula $\emptyset = t_1 \backslash t_2$.

- Introduce a new variable y_0, not otherwise occurring in p, replace each occurrence in p of $\emptyset$ by the new variable y_0 and put the literal $y_0 = y_0 \backslash y_0$ in Ψ.

WHILE p contains terms of type $t_1 \cap t_2$ **DO**

- let $t_1 \cap t_2$ be a term in p containing exactly one occurrence of the boolean set operator $\cap$;
- choose a new variable z not otherwise occurring in p or Ψ and add the formula $z = t_1 \cap t_2$ to Ψ;
- replace each occurrence in p of the term $t_1 \cap t_2$ by the term $t_1 \setminus (t_1 \setminus t_2)$.

END WHILE

[At this point the constant $\emptyset$ and the set operator $\cap$ have been eliminated.]

WHILE p contains compound terms **DO**

- let τ be a compound term in p not containing any compound sub-term;
- choose a brand new variable z_τ (i.e. one not occurring in p or Ψ) and add the formula $z_\tau = \tau$ to Ψ;
- replace each occurrence in p of the term τ by the variable z_τ.

END WHILE

[At this point compound terms have been eliminated.]

- Bring p to disjunctive normal form $p_1 \vee \cdots \vee p_h$, where each p_i is a conjunction of literals (of type $x = y$, or $x \neq y$, or $x \in y$, or $x \notin y$);

FOR ALL $i = 1, \ldots, h$ **DO**

    WHILE p_i contains literals of type $x \notin y$ **DO**

-     * let $x \notin y$ be one of the literals of p;
-     * choose brand new variables $u_{x,y}, w_{x,y}$ and replace each occurrence in p of the literal $x \notin y$ by the conjunction $x \in u_{x,y} \ \& \ u_{x,y} = w_{x,y} \setminus y$;

    END WHILE

[At this point, all literals of type $x \notin y$ have been eliminated.]

    WHILE p_i contains literals of type $x \neq y$ **DO**

-     * let $x \neq y$ be one of the literals of p_i;
-     * choose new variables $v_{x,y}, v_{y,x}, z_{x,y}$ and replace each occurrence in p_i of the literal $x \neq y$ by the disjunction

$$(z_{x,y} \in v_{x,y} \ \& \ v_{x,y} = x \setminus y) \vee (z_{x,y} \in v_{y,x} \ \& \ v_{y,x} = y \setminus x).$$

    END WHILE

[At this point, all inequalities of the form $x \neq y$ have been eliminated.]

WHILE p_i contains literals of type $x = y$ **DO**

    * let $x = y$ be one of the literals of p_i;

    * replace each occurrence in p_i of $x = y$ by the atom $x = y \cup y$.

END WHILE

[At this point, all equalities $x = y$ have been eliminated.]

– Let $q_{i,1} \vee \ldots \vee q_{i,m_i}$ be a disjunctive normal form of p_i & $\&_{q \in \Psi} q$;

    [Each $q_{i,j}$ is a conjunction of positive literals of type $x = y \cup z$, $x = y \setminus z$, $x = y$, $x = f_i(y)$, $x \in y$.]

END FOR ALL

RETURN $\bigvee_{i=1}^{h} \bigvee_{j=1}^{m_i} q_{i,j}$.

                                                             □

Remark 3.3 Atoms of the form $x \in y$ are easily eliminated from $\nu(p)$ when one of the f_is is $\{ \bullet \}$. As a matter of fact, $x \in y$ can be replaced by

$$x' = \{ x \} \,\&\, y_0 = x' \setminus y,$$

where x' is a new variable and y_0 occurs in a literal $y_0 = y_0 \setminus y_0$—so that y_0 denotes $\emptyset$.   □

3.9   Quantifier elimination

A formula

$$Q_1 x_1 \cdots Q_n x_n \, p \,,$$

where p is unquantified and each Q_i is either $\exists$ or $\forall$, is said to be in *prenex normal form*. It is well-known that any formula q can be brought to prenex normal form q_*, so that $\Phi(q) = \Phi(q_*)$ and $q^\varphi = q_*^\varphi$ in every interpretation φ (for a proof, see e.g. [CL73]—minor additional difficulties arise with our formalism, due to the presence of abstraction terms). Many classical studies concerning the validity problem in first-order predicate calculus (cf. [Ack62, DG79, Lew79]) have therefore mainly focused on prenex formulae.

The instances of the validity problem treated in this book refer, typically, to families of unquantified formulae. Nevertheless, the applicability of the decision tests can often be extended to restricted classes of prenex formulae, thanks to the following lemma taken from [CC88].

Lemma 3.4 *If p is in prenex normal form, and every bound occurrence x of a variable in p belongs to an atom of the form $x = t$ or of the form $x \in t$,* *with no occurrences of x in t, then p is logically equivalent to an unquantified formula p_* in the same language of p.*

This is to say, $\Phi(p) = \Phi(p_*)$ and $p^\varphi = p_*^\varphi$ in every interpretation φ, for a suitable formula p_*, algorithmically obtainable from p. □

Since $\exists$ and $\forall$ have been introduced in terms of H, the proof of this lemma can be based upon the following definition:

Definition 3.10 *A formula p is in* prenex normal form *iff p is unquantified or p has either one of the forms $H x\,r = \emptyset$, $H x\,r \neq \emptyset$, where r is in prenex normal form.*

In the first case, p is said to be its own matrix; *in the other two cases, the matrix of r is said to be the matrix of p too. Moreover, in the first case p is said to be* left, *while in the other two cases p is* left *iff r is left and every occurrence of x in p belongs to an atom of the form $x = t$ or $x \in t$, where t does not involve x.*

The definition of $\in$-left prenex formulae is entirely analogous, except that atoms of the form $x = t$, with x bound, are disallowed. □

Proof of the lemma. Without loss of generality, since $x = t$ can be rewritten as $x \in \{t\}$, we may assume that p contains no atoms of the form $x = t$, with x bound in p. Moreover, it will suffice to consider the case when p is of the form $H x\,r \neq \emptyset$, i.e. $\exists x\,r$.

Recursively, r is brought to unquantified form r_* As will be clear at the end, if x is bound in r, then x vanishes in the translation of r to r_*; differently, the atoms involving x in r_* and in p are the same, and each of them is of the form $x \in t$.

Next, r_* is brought to disjunctive normal form (on the basis of Theorem 3.1), giving

$$\bigvee_{i=1}^{n} (\&_{j=1}^{h_i}\, d_{ij}\ \&\ \&_{j=h_i+1}^{k_i}\, x \in t_{ij}\ \&\ \&_{j=k_i+1}^{m_i}\, x \notin t_{ij}),$$

logically equivalent to r, without occurrences of x in any d_{ij}. In turn, p will be equivalent to

$$\bigvee_{i=1}^{n} \left(\&_{j=1}^{h_i}\, d_{ij}\ \&\ \exists x\, (\&_{j=h_i+1}^{k_i}\, x \in t_{ij}\ \&\ \&_{j=k_i+1}^{m_i}\, x \notin t_{ij}) \right),$$

obviously reducible to the unquantified formula

$$\bigvee_{i=1}^{n} \left(\&_{j=1}^{h_i}\, d_{ij}\ \&\ \emptyset \neq \left(\bigcap_{j=h_i+1}^{k_i} t_{ij} \right) \setminus \left(\bigcup_{j=k_i+1}^{m_i} t_{ij} \right) \right)$$

(or simply to $\bigvee_{i=1}^{n} (\&_{j=1}^{h_i}\, d_{ij})$, if $k_i = h_i$). ■

The above lemma is based on the implicit assumption that the language to which p belongs includes—in addition to membership, equality, and boolean operators both propositional and set-theoretic—the *singleton* construct. The latter in fact has been exploited in order to eliminate atoms of the form $x = t$, with x bound. Without this assumption, only the following weaker version of the lemma holds.

Corollary 3.1 *There is a procedure that transforms any given $\in$-left prenex formula p into a logically equivalent unquantified formula p_* in the same language to which p belongs. No terms of the form $\{t_0, \ldots, t_N\}$ occur in p_*, unless such terms also appear in p.* □

4. A PARTIAL SOLUTION TO THE FINITE SATISFIABILITY PROBLEM

4.1 Introduction

In this chapter we present two semi-decision procedures which detect the *finite* satisfiability of p, for all p in a certain extensive collection S of formulae. Being interested in finite satisfiability only, we temporarily disable use of the *Finite* relator. Apart from this restriction, S includes all unquantified formulae that do not involve *Un* or η; moreover, η is allowed to appear in certain favourable contexts. The members of S will be called *safe* formulae. It will turn out that when a safe formula p is finitely satisfiable, then a suitable substitution of *hereditarily finite sets* (see Section 2.5) for the variables of p makes p true. This fact is the key of a semi-decision method applicable to safe formulae, although not the main key to prove semi-decidability. We will see, in fact, that a set X is hereditarily finite if and only if:

- every X_i is finite in any chain of the form $X_n \in X_{n-1} \in \cdots \in X_0 = X$ with n in ω; and, moreover,

- there is a maximum integer n for which such a chain exists.

It follows from this characterization that X is hereditarily finite if and only if it is the value of a variable-free term belonging to the smallest collection of terms that contains $\emptyset$ and is closed with respect to the construct $\{\bullet,...,\bullet\}$. Symbolically performing set-theoretic operations on terms of this kind is possible (indeed, there are programming languages, such as SETL [SDDS86], that *do* perform them); hence, it will be possible to evaluate any safe formula p in any substitution of hereditarily finite sets for the variables of p. It is also possible to systematically attempt all substitutions of hereditarily finite sets for the variables of p, but this requires a perpetual process, because the family $\mathcal{H}$ of hereditarily finite sets is infinite. This is why we obtain a semi-decision algorithm, instead of a decision algorithm.

Essential to our analysis of finite satisfiability is the introduction of a class of hierarchic structures, constructed in a way similar to $\mathcal{H}$ and to the von Neumann universe. These new structures are defined as follows:

Definition 4.1 *Let B_0 be a finite, possibly empty, family of infinite sets. Put:*

$$B_1 = B_0 \cup \{\emptyset\};$$

$$B_{i+1} = B_i \cup Pow(B_i),$$

for $i = 1, 2, 3, \ldots$; and

$$\mathcal{B} = (\bigcup_{i \in \omega} B_i) \setminus B_0,$$

where ω is the set of all natural numbers. The set $\mathcal{B}$ is called the superstructure *over B_0.* □

It will turn out that $\mathcal{B} = \mathcal{H}$ if and only if $B_0 = \emptyset$. The fact that the members of $\mathcal{H}$ are denoted by terms involving only $\emptyset$ and the construct $\{\bullet,...,\bullet\}$ generalizes to an arbitrary superstructure $\mathcal{B}$, as follows. We take n distinct variables $x_1, \ldots, x_n$, where $n = |B_0|$, and call n-*terms* the terms composed by means of the construct $\{\bullet,...,\bullet\}$ starting from $\emptyset$ and from the variables x_i. Indicating by χ the interpretation that sends the i-th of these variables, x_i, into the i-th member of B_0 (with respect to the ordering $<$ of $\mathcal{V}$—see Section 2.5), we will prove that $\{ t^\chi : t$ is an n-term $\} = \mathcal{B} \cup B_0$.

To the same member Y of $\mathcal{B}$, there will generally correspond several n-terms t with $t^\chi = Y$. In defining *normalized n-terms*, however, we will manage to insure that exactly one normalized n-term satisfies this equality for each Y. While the definition of n-term will uniquely depend on n, the definition of normalized n-term will also depend on the restriction of $<$ to $\mathcal{B}$. Properly speaking, we should use the locution 'normalized with respect to $\mathcal{B}$' instead of the single word 'normalized'. It will turn out, fortunately, that a certain tuple $(k_2, \ldots, k_n)$—which is empty for $n = 0$ or $n = 1$— with each k_i belonging to $\omega \cup \{\omega\}$, entirely determines the collection of all normalized n-terms. Even, knowing this tuple, we can calculate, for each n-term t, the normalized n-term t_* for which $t_*^\chi = t^\chi$. The tuple $(n, k_2, \ldots, k_n)$ will be called the *characteristic tuple* of $\mathcal{B}$.

We will prove that every tuple of the form $(n, k_2, \ldots, k_n)$, with k_i in $\omega \cup \{\omega\}$ for $i = 2, \ldots, n$, is the characteristic tuple of a suitable superstructure $\mathcal{B}$. Knowing this tuple, it will become possible to evaluate p_*^χ, where p_* is any formula resulting from a safe formula p by replacing each variable by an n-term, not a variable, which is normalized with respect to $\mathcal{B}$.

With this outline in mind, we now outline our basic semi-decision procedure:

'Generate all pairs m, $(n, k_2, \ldots, k_n)$, consisting of a positive integer and a characteristic tuple. For each such pair, let us denote by B_0 and B_m the bottom and the m-th stage of a superstructure $\mathcal{B}$ whose characteristic tuple is $(n, k_2, \ldots, k_n)$. Systematically replace the variables that occur in the given safe formula p by normalized n-terms t with t^χ in $B_m \setminus B_0$, in all possible ways. If, for some replacement, the resulting formula p_* is such that $p_*^\chi = TRUE$, then p is finitely satisfiable and the procedure terminates. Otherwise, the process continues'.

This procedure bears some resemblance to the Herbrand procedure for predicate calculus (cf. [CL73]). As a matter of fact, it systematically tries all possible substitutions of variables by terms, drawing the terms from an infinite family of 'ground' terms. Each substitution leads to a 'ground' formula which can be directly evaluated. A major difference should be noted, however: the Herbrand procedure is a prototype of the process of proving theorems by contradiction and halts—with an unsatisfiability diagnosis for the negation of the theorem—when the evaluation of the ground formula gives the result 'UNSATISFIABLE'. Our procedure is a prototype of the search for a finite model or counter-example: it halts when the evaluation gives the result 'TRUE'. Following the terminology that we have introduced in Section 3.7, we can say that our semi-decision procedure is strongly complete (as far as finite satisfiability is concerned), while the Herbrand procedure is a quasi-decision test.

The more refined semi-decision test hinted at at the beginning avoids choosing the characteristic tuple $(n, k_2, \ldots, k_n)$; this is in fact assumed to be invariably (0), i.e. the characteristic tuple of $\mathcal{H}$. Only m remains to be chosen. In order to legitimate this simplification, we will show how to construct, for any given $(n, k_2, \ldots, k_n)$ and m, a one-to-one function $v \mapsto v'$ from B_m into $\mathcal{H}$ that satisfies the conditions:

$$x \in y \text{ if and only if } x' \in y',$$
$$x < y \text{ if and only if } x' < y',$$
$$\{x_1, \ldots, x_j\}' = \{x'_1, \ldots, x'_j\},$$

for all $x, y, \{x_1, \ldots, x_j\}$ in B_m, with j an arbitrary natural number. The construction of such a $'$ was carried out in [CFO88] for the first time, while the less refined finite satisfiability test described above is taken from [Omo84].

4.2 Closure properties of superstructures. Safe formulae

Throughout this chapter, the notations B_0, B_i, $\mathcal{B}$ are used consistently with Definition 4.1. We also indicate $\mathcal{B} \cup B_0$ by $\overline{\mathcal{B}}$. The following lemma shows that every b in a superstructure $\mathcal{B} = \bigcup_{i \in \omega} (B_i \setminus B_0)$ is finite. Its corollary proves that every hereditarily finite set is finite.

Lemma 4.1 *Each B_i and every b in $B_i \setminus B_0$ is finite; moreover $i + |B_0| \leq |B_i|$, for all i in ω.*

Proof. $B_0 \setminus B_0$ is empty and $\emptyset$, which is the sole member of $B_1 \setminus B_0$, is finite. Moreover B_0 and B_1 are finite, and hence the induction hypothesis holds for $i = 0, 1$. For $i = j + 2$, we inductively assume that B_{j+1} is finite, whence it follows that $Pow(B_{j+1})$ and $B_i = B_{j+1} \cup Pow(B_{j+1})$ are finite, as well as each member of $Pow(B_{j+1})$ (hence of $B_i \setminus B_0$). In order to see that $i + |B_0| \leq |B_i|$, it suffices to observe that when $1 < i$ then $B_i \supseteq B_{i-1}$ and $B_{i-1} \in B_i \setminus B_{i-1}$. ∎

Corollary 4.1 $\mathcal{B} = \mathcal{H}$ *if and only if* $B_0 = \emptyset$. $\mathcal{B} = \mathcal{H}$ *if and only if* $\mathcal{B}$ *is closed with respect to* Un, *i.e.*, $Un(b) \in \mathcal{B}$ *for every* b *in* $\mathcal{B}$.

Proof. From Lemma 2.4(ii) it follows that $V_\alpha \subsetneq V_{\alpha+1} = Pow(V_\alpha)$ for all ordinal α. Hence $\mathcal{H} = V_\omega = \bigcup_{i \in \omega} V_i$ is the superstructure over $\emptyset$ (see Definition 4.1). In a superstructure $\mathcal{B}$ which has $B_0 \neq \emptyset$, taking $x \in B_0$ we will have $x \in \{x\} \in B_2 \setminus B_0 \subseteq \mathcal{B}$. Since x is infinite, $x \notin \mathcal{B}$. Therefore $\mathcal{B}$ is not transitive and hence $\mathcal{B}$ differs from $\mathcal{H}$, which is transitive by Lemma 2.4(i). The first statement is proved at this point.

Concerning the second statement, we first observe that when $B_0 \neq \emptyset$, then taking x in B_0 we will have $\{x\} \in \mathcal{B}$, but $Un(\{x\}) = x \notin \mathcal{B}$. Hence $\mathcal{B}$ is not closed with respect to Un when $\mathcal{B} \neq \mathcal{H}$. Consider now an arbitrary h in $\mathcal{H}$. If $h = \emptyset$, then $Un(h) = h \in \mathcal{H}$. Otherwise, $h = \{g_0, \ldots, g_Q\}$ where $Q \in \omega$ and each g_i belongs (by transitivity) to $\mathcal{H}$, so that g_i and $rk\,g_i$ are both finite (recall Lemma 2.4(viii)). Hence

$$rk\,Un(h) = rk(g_0 \cup \cdots \cup g_Q) = max\{rk\,g_0, \ldots, rk\,g_Q\} < \omega,$$

where the second equality holds by Lemma 2.4 (iii). This shows that $Un(h) \in \mathcal{H}$. ∎

The rest of this section is devoted to the study of the closure properties enjoyed by superstructures.

Lemma 4.2 *For all* i, n *in* ω:

(1) $x \in B_{i+2} \setminus B_0$ *if and only if* $x \subseteq B_{i+1}$;

(2) $x_1, x_2, \ldots, x_n \in B_{i+1}$ *if and only if* $\{x_1, x_2, \ldots, x_n\} \in B_{i+2} \setminus B_0$;

(3) *If* $x \in B_{i+1} \setminus B_0$ *and* $u \subseteq x$, *then* $u \in B_{i+1} \setminus B_0$; *in particular* $x \cap z$, $x \setminus z$, *belong to* $B_{i+1} \setminus B_0$ *for any set* z;

(4) $x, y \in B_{i+1} \setminus B_0$ *if and only if* $x \cup y \in B_{i+1} \setminus B_0$;

(5) $x \in B_{i+1} \setminus B_0$ *if and only if* $Pow(x) \in B_{i+2} \setminus B_1$;

(6) $x, y \in B_{i+2} \setminus B_1$ *if and only if* $x \times y \in B_{i+4} \setminus B_1$.

Proof. (1) We proceed by induction on i. Let first $i = 0$. If $x \in B_2 \setminus B_0$, then either $x \in B_1 \setminus B_0 = \{\emptyset\}$ or $x \in Pow(B_1) \setminus B_1$. It is clear that in either case $x \subseteq B_1$. Conversely, if $x \subseteq B_1$ then $x \in Pow(B_1) \subseteq B_2$, whence it follows that $x \in B_2 \setminus B_0$, because B_1 is finite. Now suppose inductively that (1) is true for $i = j$, and let $i = j + 1$. If $x \in B_{j+3} \setminus B_0$ then either $x \in B_{j+2} \setminus B_0$ or $x \in Pow(B_{j+2})$. In either case, by induction, $x \subseteq B_{j+2} = B_{i+1}$. Conversely, if $x \subseteq B_{j+2}$, since x is finite by Lemma 4.1, it follows immediately that $x \in Pow(B_{j+2}) \setminus B_0 \subseteq B_{j+3} \setminus B_0$, concluding the proof of (1).

(2) Recall that B_{i+1} is finite by Lemma 4.1; hence (2) is just a rephrasing of (1).

(3) Assuming that $u \subseteq x \in B_{i+1} \setminus B_0$, one has either $x = \emptyset$ or $0 < i$ and $x \in B_{(i-1)+2} \setminus B_0$. Hence, by (1), either $u = \emptyset$ or $u \subseteq x \subseteq B_{(i-1)+1}$; and hence $u = \emptyset$ or $u \in B_{(i-1)+2} \setminus B_0$, again by (1). In either case $u \in B_{i+1} \setminus B_0$.

(4) The statement is trivial if $x = \emptyset$ or $y = \emptyset$. Otherwise, we have $0 < i$, so that x, $y \in B_{i+1} \setminus B_0$ holds if and only if x, $y \in B_{(i-1)+2} \setminus B_0$, i.e. (by (1)) if and only if x, $y \subseteq B_{(i-1)+1} \setminus B_0$. Equivalently, $x \cup y \subseteq B_{(i-1)+1} \setminus B_0$, i.e. $x \cup y \in B_{(i-1)+2} \setminus B_0 = B_{i+1} \setminus B_0$.

(5) Is entirely straightforward, in view of (1).

(6) Is obtained in a straightforward manner by repeated applications of (1), if, in agreement with Kuratowski's convention, we identify any ordered pair (a, b) with the set $\{\{a\}, \{a,b\}\}$.   ∎

Corollary 4.2 *Let x, $y \in \mathcal{B}$, and z, $z_0, \dots, z_Q \in \overline{\mathcal{B}}$, $Q \in \omega$. Then $u \in \mathcal{B}$ whenever $u \subseteq x$, so that in particular $\emptyset$, $x \cap z$, $x \setminus z \in \mathcal{B}$. Moreover $\{z_0, \dots, z_Q\}$, $x \cup y$, $Pow(x)$, $x \times y \in \mathcal{B}$. Finally, $x \subseteq \overline{\mathcal{B}}$, so that $\eta x \in \overline{\mathcal{B}}$.*

Proof. The first statement is an immediate consequence of (3) of the preceding lemma. Likewise, the second statement follows from (2), (4), (5), and (6). To prove the third statement, consider $x \in B_j \setminus B_0$. If $j = 1$, then $x = \emptyset$, so that trivially $x \subseteq \overline{\mathcal{B}}$. By (1), the same inclusion holds when $j = i + 2$. ∎

Before we can state the next corollary of Lemma 4.2, we must define safe expressions:

Definition 4.2 *A safe term is a term which is construed according to the following rules:*

(1) $\emptyset$ is a safe term;

(2) each variable y is a safe term;

(3) if s, t are safe terms, then $(s \cap t)$, $(\eta s \cap t)$, $(t \cap \eta s)$, $(s \setminus t)$, $(s \setminus \eta t)$, $(s \cup t)$, $Pow\, s$, $(s \times t)$ are safe terms;

(4) if s_0, $s_1, \dots$, s_Q are safe terms and t_i is either s_i or ηs_i for $i = 0, 1, \dots, Q$, then

$$\{t_0, \dots, t_Q\}, \quad \eta\{s_0, \dots, s_Q\}$$

are safe terms. Here Q is an arbitrary natural number.

A safe formula *is an unquantified formula p in which every term t is either safe or of the form ηs, where s is safe.*   □

Note that the formula $s < t$, which abbreviates $\neg\eta\{s,t\} = t$, is safe when s, t are safe terms.

Corollary 4.3 *Let p be a safe formula, φ be a finite interpretation of p, and $X = \{ y^\varphi : y \in \Phi(p) \}$ be the set consisting of all images, via φ, of the variables of p. If B is any superstructure such that $B \supseteq X$, then $s^\varphi \in \overline{B}$ for every term s that occurs in p, and in particular $s^\varphi \in B$ when s is safe.*

Proof. Recall from Corollary 4.2 that iterated application of the operations

$$\cap, \setminus, \cup, \{\, \bullet, \ldots, \bullet \,\}, \, Pow, \, \times$$

to the members of B cannot lead outside B, while a single application of η can lead from B to B_0. A single application of Un, instead, could actually lead 'below the bottom'.

Trivially, $\emptyset^\varphi \in B$ and $y^\varphi \in B$ for every variable y in p; and we have noted that $(s^\varphi \cap t^\varphi)$, $(t^\varphi \cap \eta s^\varphi)$, $(s^\varphi \setminus t^\varphi)$, $(s^\varphi \setminus \eta t^\varphi)$, $(s^\varphi \cup t^\varphi)$, $Pow(s^\varphi)$, $(s^\varphi \times t^\varphi) \in B$, if $s^\varphi \in B$ and $t^\varphi \in B$. Assuming that $s_0^\varphi, \ldots, s_Q^\varphi \in B$ and that t_i is either s_i or ηs_i, one has that $t_i^\varphi \in B \cup B_0$ for $i = 0, 1, \ldots, Q$; hence $\{ t_0, \ldots, t_Q \}^\varphi \in B$, by Corollary 4.2. Moreover $(\eta \{ s_0, \ldots, s_Q \})^\varphi$, being equal to some s_i^φ, also belongs to B. ∎

4.3 König's lemma

Preliminary to the definitions of the *bottom* of an arbitrary set x, and of the superstructure *generated* by x, we prove a classical proposition, known in the literature as König's lemma (see e.g. [Bet69, Knu75, Lov78]). From this lemma we derive the important fact that the finitely transitive closure of any set x is finite.

Lemma 4.3 (König) *Suppose that the set $\{v : w \in \dot{v}\}$ formed by the immediate predecessors v of w is finite for every node w of a WFG $G = (V, \dot{\,})$. Moreover assume that G has a node r such that for every other node v at least one path leads from v to r. Then V is finite.*

Proof. By induction on the height Hw of an arbitrary node w, we proceed to show that G_w has finitely many nodes, where G_w is the restriction of G to the set

$$\{w\} \cup \{v \in V : \text{there are paths from } v \text{ to } w \text{ in } G\}$$

(see Definition 2.4). In particular this will hold for G_r, which coincides with G; hence $|V| < \omega$.

For $Hw = 0$, the set of nodes of G_w is $\{w\}$, which clearly is finite. When $0 < Hw$, then $|G_w| \le 1 + \sum_{w \in \dot{v}} |G_v|$; one easily sees, in fact, that the nodes of G_w are—apart from w—those belonging to some G_v with w in $\dot{v}$. Since there are only a finite number of vs contributing to the summation, and since $|G_v|$ is finite for any of these (by the induction hypothesis), one concludes that $|G_w| < \omega$. ∎

After recalling that a class Z is *finitely transitive* iff $y \in Z$ whenever $y \in f \in Z$ with $|f| < \omega$, we state:

Corollary 4.4 *Every set x belongs to a finitely transitive family F which is a subset of any other finitely transitive set Z with x in Z. This family has $|F| < \omega$. Moreover, if we take $B_0 = \{ y \in F : \omega \le |y| \}$, then $F \setminus B_0 \subseteq \mathcal{B}$, where $\mathcal{B}$ is the superstructure over B_0.*

Proof. As a preliminary remark we observe that for any class $\mathcal{F}$ whose members are finitely transitive sets, the intersection $F = \bigcap \mathcal{F}$ is finitely transitive. In particular, if we take $\mathcal{F} = \{ Z : x \in Z$ and Z is finitely transitive $\}$, then F will be finitely transitive. With this choice, F is a set because x belongs to $V_{rk\,x+1}$ which is a transitive set, in view of Lemma 2.4(viii),(i), so that $F \subseteq V_{rk\,x+1}$. The first part of the thesis is hence trivially satisfied.

Trivially, if x is infinite, then $F = \{x\}$. When x is finite, an alternative characterization of F is as follows. We put $x_0 = x$ and consider the class F_1 consisting of all ys such that there is a 'membership chain' $y = x_n \in \cdots \in x_1 \in x_0$ with n in ω and with $|x_i| < \omega$ for $i = 1, \dots, n-1$. Taking in particular $n = 0$, we note that $x \in F_1$. Notice also that $F_1 \subseteq F$ as a consequence of the fact that F is finitely transitive: indeed, we have $x_{i+1} \in F$ for $i = 0, \dots, n-1$ in any chain as above, because $x_{i+1} \in x_i$, where x_i is finite and belongs to F. One easily recognizes that F_1 is finitely transitive, whence, by the minimality of F, it follows that $F_1 = F$.

Consider now the WFG having nodes F and edges $\{ (v, w) \in F \times F : v \in w$ and $|w| < \omega \}$. This graph trivially meets the conditions of Lemma 4.3 (note that for every node $y \ne x$ there is a path from y to x, i.e. a chain $y = x_n \in \cdots \in x_1 \in x$). Hence F is finite. An easy induction on the height of each node w will show (taking Corollary 4.2 into account) that either $w \in B_0$ or $w \in \mathcal{B}$. The last statement of the thesis hence follows. ∎

Definition 4.3 *The* bottom *of a finite set x is the set $B_0 = \{ y \in F : \omega \le |y| \}$, where F is as in Corollary 4.4. The superstructure over B_0 is said to be* generated by x. □

The reader will easily recognize, at this point, that: (a) the bottom of x is empty if and only if x is hereditarily finite; (b) the bottom of a finite set $x = x_0$ consists of those infinite sets y for which there exists a membership chain $y \in x_n \in \cdots \in x_1 \in x_0$, with n in ω, in which every x_i is finite. Note also that since $x \in F \setminus B_0 \subseteq \mathcal{B}$, one has $x \subseteq \overline{\mathcal{B}}$ by Corollary 4.2.

Remark 4.1 The bottom of an infinite set x would be $\{ x \}$ if the above definition were extended in the most straightforward manner. This extension, however, seems useless. A more sensible extension is to recursively define the bottom of an *infinite* set x as the union of $\{ y \in x : \omega \le |y| \}$ with the union of the bottoms of the members of x. According to this definition, for example, the bottom of the superstructure $\mathcal{B}$ over B_0 is B_0, which also contains as a subset the bottom of any subset of $\mathcal{B}$. The bottom of ω is $\emptyset$, while the bottom of $\omega + 1$ is $\{\omega\}$.   □

It is possible, at this point, to strenghten Corollary 4.3:

Corollary 4.5 *Let p be a safe formula and let φ be a finite interpretation of p. Then there exists a superstructure $\mathcal{B}$ such that $s^\varphi \in \mathcal{B}$ for every safe term s that occurs in p and $t^\varphi \in \overline{\mathcal{B}}$ for every non-safe term t that occurs in p.*

Proof. It suffices to let $\mathcal{B}$ be the superstructure generated by the set $X = \{y^\varphi : y \in \Phi(p)\}$. Thus, $X \subseteq \overline{\mathcal{B}}$. It follows that $X \subseteq \mathcal{B}$, because every member of X is finite, whereas the members of $\overline{\mathcal{B}} \setminus \mathcal{B}$ are infinite. Hence we can apply Corollary 4.3.   ∎

4.4 n-terms

In defining, for all n in ω, the collection T_n of the n-terms, we will refer to a pre-established infinite sequence x_1, x_2, $x_3, \ldots$ of distinct set-theoretic variables.

Definition 4.4 *For every n in ω, let T_n be the smallest collection of formal expressions such that*

- *each of the variables $x_1, \ldots, x_n$ belongs to T_n;*
- *the one-symbol expression $\emptyset$ belongs to T_n;*
- *the term $\{t_0, \ldots, t_Q\}$ belongs to T_n whenever $t_0, \ldots, t_Q$ belong to T_n, for all Q in ω.*

The members of T_n are called n-terms.   □

Let B_0 be the bottom of the superstructure $\mathcal{B}$, where

$$B_0 = \{X_1, \ldots, X_n\},$$

with

$$|X_j| \notin \omega \text{ for } j = 1, \ldots, n$$

and with

$$X_1 < \cdots < X_n$$

in the ordering $<$ of $\mathcal{V}$ (see Section 2.5).

Definition 4.5 *The $\mathcal{B}$-interpretation $\chi = \chi_\mathcal{B}$ of the n-terms, is the correspondence $x_j \mapsto X_j$ defined for $j = 1, \ldots, n$.*   □

Lemma 4.4 *For every n-term t, one has* $t^X \in \mathcal{B} \cup B_0$.

Proof. As a matter of fact,

$$x_j^X = X_j \in B_0, \text{ for } j = 1, \ldots, n;$$

$$\emptyset^X = \emptyset \in B_1 \setminus B_0 \subseteq \mathcal{B};$$

and, moreover, by Corollary 4.2,

$$\{ t_0, \ldots, t_Q \}^X = \{ t_0^X, \ldots, t_Q^X \} \in \mathcal{B}$$

if we inductively assume that $t_0^X, \ldots, t_Q^X \in \mathcal{B} \cup B_0$.　　　　■

　　Conversely:

Lemma 4.5 *For every Y in* $\mathcal{B} \cup B_0$, *there is an n-term t such that* $t^X = Y$.

Proof. We proceed by induction on the least number i for which $Y \in B_i$. For $i = 0$, we have $Y = X_j = x_j^X$ for a suitable j in $\{ 1, \ldots, n \}$. For $i = 1$, we have $Y = \emptyset = \emptyset^X$. For $i > 1$, Y has the form $Y = \{ Z_0, \ldots, Z_Q \}$, where each Z_h belongs to B_{i-1}, so that $Z_h = t_h^X$ for a suitable n-term t_h, by inductive hypothesis. Thus $Y = \{ t_0^X, \ldots, t_Q^X \} = \{ t_0, \ldots, t_Q \}^X$, where $\{ t_0, \ldots, t_Q \}$ is an n-term.　　　　■

　　Lemmas 4.4 and 4.5 can be summarized by the identity

$$\mathcal{B} \cup B_0 = \{ t^X : t \text{ is an n-term} \},$$

where the correspondence $t \mapsto t^X$ between n-terms and sets clearly is *not* one-to-one. For example,

$$\{ \emptyset, \{ \emptyset \} \}^X = \{ \{ \emptyset \}, \emptyset \}^X = \{ \emptyset, \{ \emptyset \} \}.$$

In an attempt to restrict the domain of this correspondence in such a manner that the correspondence becomes one-to-one, its set of values remaining unaltered, we give the following definition.

Definition 4.6 *An n-term t is* normalized *(with respect to* $\mathcal{B}$*) iff either:*

- *t is one of the variables* x_j $(0 < j \le n)$; *or*
- *t is* $\emptyset$; *or*
- *t has the form* $\{ t_0, \ldots, t_Q \}$, *where each t is normalized and, moreover,* $t_0^X < \cdots < t_Q^X$.　　　　□

　　It will follow from Corollary 4.8 that this definition fulfils our intentions, i.e.,

- $\mathcal{B} \cup B_0 = \{ t^X : t \text{ is a normalized } n\text{-term} \};$
- $t_0^X \neq t_1^X$ unless the normalized terms t_0, t_1 coincide.

4.5   The characteristic tuple of a superstructure

As in the preceding section, we assume that $\mathcal{B}$ is the superstructure over the family

$$B_0 = \{ X_1, \ldots, X_n \}.$$

We continue to assume that the infinite sets X_j are increasing with their subscripts, i.e.

$$X_1 < X_2 < \cdots < X_n,$$

so that

$$rk\, X_1 \leq \cdots \leq rk\, X_n.$$

Definition 4.7 *The characteristic tuple of $\mathcal{B}$ is the ordered list*

$$(n, k_2, \ldots, k_n)$$

with $max\{ 1, n \}$ components, such that

$$n = |B_0|$$

and, for $i = 2, \ldots, n$:

  $k_i \in \omega \cup \{ \omega \}$;
  if $k_i \in \omega$, then $rk\, X_i = rk\, X_{i-1} + k_i$;
  if $k_i = \omega$, then $rk\, X_{i-1} + k < rk\, X_i$ for all k in ω.

  □

Lemma 4.6 *Every ordered list of the form*

$$(n, k_2, \ldots, k_n),$$

with $n \in \omega$ and $k_2, \ldots, k_n \in \omega \cup \{ \omega \}$, is the characteristic tuple of a suitable superstructure $\mathcal{B}$.

Proof. We will manage to define the bottom $B_0 = \{ X_1, \ldots, X_n \}$ of $\mathcal{B}$ so that each X_j is the $(m_j + 1)$-st infinite set of its rank, for a suitable number m_j. Note that no X_j will be the last infinite set of its rank in this manner; in fact, by Corollary 2.4, if there exist infinite sets of a certain rank α, then there exist infinitely many of them.

We begin by putting

$$X_1 = \text{first infinite set},$$

so that $rk\, X_1 = \omega$ and $m_1 = 0$.

For $j = 2, \ldots, n$, if $k_j = 0$, we will put

$$X_j = (m_{j-1} + 2)\text{-nd infinite set having rank } rk\, X_{j-1}.$$

Thus we will have $X_{j-1} < X_j$, $m_j = m_{j-1} + 1$, and

$$rk\, X_j \;=\; rk\, X_{j-1} \;=\; rk\, X_{j-1} + k_j$$

as desired.

For $j = 2, \ldots, n$, if $k_j \in \omega \setminus \{0\}$, we will put

$$X_j \;=\; \text{first infinite set having rank } rk\, X_{j-1} + k_j\,.$$

It is indeed possible to find infinite sets of rank $rk\, X_{j-1} + k_j$, by Lemma 2.4 (ix), since X_{j-1} is infinite. In this case we have $m_j = 0$.

For $j = 2, \ldots, n$, if $k_j = \omega$, we will select an ordinal β such that:

$$rk\, X_{j-1} + k \;<\; \beta$$

for all k in ω, and, moreover, there exist infinite sets with rank β (again by Lemma 2.4). The first infinite set X with $rk\, X = \beta$ will be X_j. Also in this case we have $m_j = 0$. ∎

It is convenient for our subsequent purposes to extend the definition of the k_js as follows. We put

$$X_0 \;=\; \emptyset\,,$$

$$k_1 \;=\; k_{n+1} \;=\; \omega\,.$$

Note that, provided that $n > 0$, one has

$$rk\, X_0 + k \;<\; rk\, X_1$$

for all k in ω, consistent with our choice of k_1; in fact, X_1 is infinite (hence X_1 has infinite rank), whereas $rk\, X_0 = 0$.

Assuming that we know the characteristic tuple of $\mathcal{B}$, our first goal is to find a criterion for comparing the ranks

$$rk\, t_0^\chi\,, \quad rk\, t_1^\chi$$

of any two n-terms t_0, t_1. As usual, χ indicates the $\mathcal{B}$-interpretation (see Definition 4.5). Informally, our idea is the following.

Let $r_0 < r_1 \leq \cdots \leq r_n$ be the non-decreasing sequence

$$rk\, X_0 \;<\; rk\, X_1 \;\leq\; \cdots \;\leq\; rk\, X_n\,,$$

and let

$$R_0 \;<\; \cdots \;<\; R_h$$

be its increasing subsequence such that

$$\{\, r_0\,, r_1, \ldots, r_n \,\} \;=\; \{\, R_0, \ldots, R_h \,\}\,.$$

For $i = 1, \ldots, h$, let $K_i \in \omega \cup \{\omega\}$ be a measure of the 'gap' existing between R_{i-1} and R_i, in the same sense in which k_i can be viewed as a measure of the gap existing between r_{i-1} and r_i. Also put $K_{h+1} = \omega$.

Now consider the pairs

$$(j, k) \text{ with } 0 \leq j \leq h \text{ and } 0 \leq k < K_{j+1};$$

we will show that there is an easy way, given any two of these pairs, say

$$(j_*, k_*), \ (j^*, k^*),$$

to determine whether or not

$$R_{j_*} + k_* \leq R_{j^*} + k^*.$$

In particular, equality holds only if both $k_* = k^*$ and $j_* = j^*$. We will also produce an algorithm which, for every n-term t, calculates one of these pairs (j, k), for which

$$rk\, t^X = R_j + k.$$

Thus our goal will be achieved.

We now begin to fill in the details of the plan that has been outlined in the last three paragraphs. In the first place, we must be able to determine the tuple $(K_1, \ldots, K_{h+1})$. This is easy:

Lemma 4.7 *The ordered list* $(K_1, \ldots, K_{h+1})$ *is obtainable from the* $(n+1)$*-tuple* $(k_1, \ldots, k_{n+1})$ *by simply dropping the null components, i.e. those k_is for which $k_i = 0$.*

Proof. Let us put

$$(K_1, \ldots, K_{h+1}) = (k_{i_1}, \ldots, k_{i_{h+1}}),$$

where $1 = i_1 < \cdots < i_{h+1} = n + 1$ and

$$\{k_{i_1}, \ldots, k_{i_{h+1}}\} = \{k_1, \ldots, k_{n+1}\} \setminus \{0\}.$$

Thus $K_{h+1} = \omega$; and we will have (retaining for the R_js the meaning that has been explained preceding this lemma):

$$(R_0, R_1, \ldots, R_h) = (0, rk\, X_{i_1}, \ldots, rk\, X_{i_h}),$$

with $R_{j-1} < R_j \notin \omega$ and

$$R_j = rk\, X_{i_j} = \cdots = rk\, X_{i_{j+1}-1},$$

for $j = 1, \ldots, h$. This chain of equalities also holds for $j = 0$, if we put $i_0 = 0$.

We claim that for $j = 1, \ldots, h$:

$$\text{if } K_j \in \omega, \text{ then } R_j = R_{j-1} + K_j;$$
$$\text{if } K_j = \omega, \text{ then } R_{j-1} + k < R_j,$$

for all k in ω.

For $j = 1$, we have:

$$i_j = 1, \ K_j = k_{i_j} = k_1 = \omega,$$

and

$$R_j = rk\,X_{i_j} = rk\,X_1 \notin \omega.$$

Furthermore, we have $R_{j-1} = R_0 = 0$, and hence

$$R_{j-1} + k = k < R_j$$

for all k in ω.

For $j = 2, \ldots, h$, we have

$$
\begin{aligned}
K_j &= k_{i_j}, \\
R_j &= rk\,X_{i_j}, \\
R_{j-1} &= rk\,X_{i_{j-1}},
\end{aligned}
$$

and we already know that:

$$
\begin{aligned}
&\text{if } \ k_{i_j} \in \omega, \quad \text{then} \quad rk\,X_{i_j} = rk\,X_{i_{j-1}} + k_{i_j}; \\
&\text{if } \ k_{i_j} = \omega, \quad \text{then} \quad rk\,X_{i_{j-1}} + k < rk\,X_{i_j}, \\
&\qquad\qquad\qquad\qquad\quad \text{for all } k \text{ in } \omega.
\end{aligned}
$$

These two implications are easily seen to be equivalent to our claim. ∎

The following is a technical definition, which will not be used beyond this section.

Definition 4.8 *A* rank-related pair *is a pair* (j, k) *with:*

$$0 \le j \le h, \quad 0 \le k < K_{j+1}.$$

□

Lemma 4.8 *Let* (j_*, k_*) *and* (j^*, k^*) *be rank-related pairs. Then the inequality*

$$R_{j_*} + k_* \le R_{j_*} + k^*$$

holds if and only if either

$$j_* < j^*, or$$
$$j_* = j^* \text{ and } k_* \le k^*;$$

moreover, it is satisfied as an equality if and only if

$$j_* = j^* \text{ and } k_* = k^*.$$

Proof. The first claim follows easily from the following two facts:

- $R_0 < R_1 < \cdots < R_j \leq R_j + k < R_{j+1} < \cdots < R_h \leq R_h + K$ if $0 \leq k < K_{j+1}$, $K \in \omega = K_{h+1}$, for $j = 0, \ldots, h - 1$.

- An inequality
$$R + k_* \leq R + k^*$$
with k_*, k^* in ω and R in Ord, holds if and only if $k_* \leq k^*$.

The second claim is an easy consequence of the first, because the equality
$$R_{j_*} + k_* = R_{j^*} + k^*$$

holds if and only if the two inequalities
$$R_{j_*} + k_* \leq R_{j^*} + k^*,$$
$$R_{j^*} + k^* \leq R_{j_*} + k_*$$

are simultaneously satisfied. ∎

Lemma 4.9 *If the characteristic tuple*
$$(n, k_2, \ldots, k_n)$$

of $\mathcal{B}$ is known, then, for each n-term t, it is possible to calculate a rank-related pair (j, k) such that
$$rkt^{\chi} = R_j + k,$$

where χ is the $\mathcal{B}$-interpretation. By Lemma 4.8, there is only one such pair for each t.

Proof. By Lemma 4.7, we can easily calculate both the tuple
$$(K_1, \ldots, K_{h+1}) = (k_{i_1}, \ldots, k_{i_{h+1}})$$

and the function
$$i \mapsto j_i$$

such that, for $i = 1, \ldots, n$, one has
$$1 \leq j \leq h,$$
$$i_{j_i} \leq i < i_{j_i+1}.$$

If t is x_i with $0 < i \leq n$ then t^{χ} is X_i, and therefore
$$rkt^{\chi} = R_{j_i} + 0.$$

If t is $\emptyset$, then t^X is $\emptyset$, and therefore

$$rk\, t^X = R_0 + 0.$$

If t is $\{t_0, \ldots, t_Q\}$, then we can recursively calculate rank-related pairs

$$(j_m, \ell_m)$$

such that

$$rk\, t_m^X = R_{j_m} + \ell_m$$

for $m = 0, \ldots, Q$. Exploiting Lemma 4.8, we can select, from among these pairs, one for which $R_{j_m} + \ell_m$ is largest; let it be (j_M, ℓ_M). Since

$$t^X = \{t_0^X, \ldots, t_Q^X\},$$

we will have

$$rk\, t^X = R_{j_M} + \ell_M + 1.$$

Recall from the definition of rank-related pair that

$$0 \le \ell_M < K_{j_M+1},$$

and hence

$$0 < \ell_M + 1 \le K_{j_M+1}.$$

If $\ell_M + 1 < K_{j_M+1}$, then the desired pair is

$$(j, \ell) = (j_M, \ell_M + 1).$$

This is the case, in particular, if $j_M = h$, because in this case $K_{j_M+1} = \omega$. If $\ell_M + 1 = K_{j_M+1}$ instead, then $R_{j_M} + \ell_M + 1 = R_{j_M+1}$, and hence the desired pair is $(j_M + 1, 0)$.   ∎

Recalling Lemma 4.5, we obtain:

Corollary 4.6 *The rank R of each member x of $\mathcal{B} \cup \mathcal{B}_0$ can be decomposed in a unique way into the form*

$$R = R_j + k,$$

where (j, k) is a rank-related pair.   ∎

We also have:

Corollary 4.7 *If the characteristic tuple $(n, k_2, \ldots, k_n)$ of $\mathcal{B}$ is known, then, for every pair t_0, t_1 of n-terms, it is possible to determine whether*

$$rk\, t_0^X < rk\, t_1^X, \ \ or \ rk\, t_0^X = rk\, t_1^X, \ \ or \ rk\, t_1^X < rk\, t_0^X.$$

Proof. In fact, by Lemma 4.9, we can calculate rank-related pairs (j_*, k_*), (j^*, k^*) such that

$$rk\, t_0^X = R_{j_*} + k_*, \ \ rk\, t_1^X = R_{j^*} + k^*.$$

Exploiting Lemma 4.8, we are able to compare $R_{j_*} + k_*$ with $R_{j^*} + k^*$.   ∎

4.6   Normalization of n-terms. Evaluation of safe formulae

From Lemma 4.9, exploiting the anti-lexicographicity of $<$, and the implications

$$rk\, x \; < \; rk\, y \text{ implies that } x \; < \; y;$$
$$rk\, x \; = \; rk\, y,\, x \text{ is finite, and } y \text{ is infinite implies that } x \; < \; y;$$

we now derive:

Lemma 4.10 *If the characteristic tuple*

$$(n,\, k_2,\, \ldots,\, k_n)$$

of the superstructure $\mathcal{B}$ is known, then, for any given finite collection S of n-terms, one can calculate the binary relations $\lhd$, $\sim$, on S defined by

$$s \lhd t \text{ iff } s^{\chi} < t^{\chi},$$

$$s \sim t \text{ iff } s^{\chi} = t^{\chi},$$

for all s, t in S, where χ is the $\mathcal{B}$-interpretation.

Proof. Let us recursively define the complexity $C(t)$ of an n-term t as follows:

$$C(\emptyset) \;=\; C(x_1) \;=\; \cdots \;=\; C(x_n) \;=\; 0,$$
$$C(\{t_0, \ldots, t_Q\}) \;=\; max_{h=1}^{Q}\, C(t_h) + 1.$$

This function C is, of course, a computable function with values in ω.

Let us also define:

$$I \;=\; max_{s \in S}\, C(s).$$

For $i = 0, \ldots, I$, let S_i be the family consisting of those terms t, with

$$C(t) \leq i,$$

such that t either belongs to S or occurs as a subterm in some term s belonging to S.

For explanatory purposes, it will be easier to define $\lhd$ and $\sim$ on all of the set S_I of which S is a subset. We inductively calculate the restrictions $\lhd_i$, $\sim_i$ of $\lhd$, $\sim$ to S_i, for $i = 0, \ldots, I$. At the end of this process, we will obtain $\lhd_I = \lhd$, $\sim_I = \sim$. For $i = 0$, the calculation of $\lhd_i$ is immediate, since we know that

$$S_i \subseteq \{\emptyset,\, x_1, \ldots, x_n\}$$

and that

$$\emptyset \lhd x_1 \lhd \cdots \lhd x_n.$$

For $i = j + 1$, assuming that s, $t \in S$, we are led to consider the following two cases:

(1) exactly one of s, t—say s for definiteness—belongs to S_0;

(2) neither s nor t belongs to S_0.

(The case when both s and t belong to S_0 has already been covered at the beginning of the induction).

In case (1), we can compare $rk\,s^\chi$ with $rk\,t^\chi$, as explained in Corollary 4.7. If $rk\,s^\chi < rk\,t^\chi$, then $s \lhd t$; if $rk\,t^\chi < rk\,s^\chi$, then $t \lhd s$. If s^χ and t^χ have the same rank, then $t \lhd s$: on the one hand, in fact, t^χ is finite; on the other hand s^χ is infinite, because t^χ is non-empty, hence s cannot be $\emptyset$, and hence s is a variable.

In case (2):

$$s \text{ has the form } \{\, s_0, \ldots, s_Q \,\};$$
$$t \text{ has the form } \{\, t_0, \ldots, t_K \,\};$$

where

$$C(s_h) < C(s) \le i, \text{ for } 0 \le h \le Q;$$
$$C(t_k) < C(t) \le i, \text{ for } 0 \le k \le K.$$

We are already able to make comparisons among the s_hs and the t_ks by means of $\lhd_{i-1}$ and of $\sim_{i-1}$, and hence we are able to determine whether

$$s \lhd_i t, \text{ or } s \sim_i t, \text{ or } t \lhd_i s,$$

exploiting anti-lexicographicity. ∎

Corollary 4.8 *For any given n-term t, there is exactly one normalized n-term s with $s \sim t$; this term s can be calculated, if one knows the characteristic tuple. Unless t is $\emptyset$ or a variable, s will be of the form $\{\, s_0, \ldots, s_K \,\}$.*

Proof. If t is a variable x_j or the constant $\emptyset$, then t is already normalized and we let s coincide with t. The identity $s^\chi = t^\chi$ could not hold for any other choice of s, because the images, via χ, of $\emptyset$, $x_1, \ldots, x_n$ are distinct and none of them is both finite and non-empty; every n-term different from these, instead, has the form $\{\, s_0, \ldots, s_Q \,\}$ and hence its value in the interpretation is a finite non-empty set.

If t has the form $\{\, t_0, \ldots, t_Q \,\}$, then, using the preceding lemma, we can determine whether $t_i \sim t_j$ for some j with $0 \le j < i$, for each $i = 1, \ldots, Q$. If this is the case, we remove the subterm t_i from t: this leaves the value of t unaffected in the interpretation χ. After completion of this elimination process, we will be left with a term $\{\, t_{i_0}, \ldots, t_{i_K} \,\} \sim t$. Using the same lemma again, we find the permutation $i_j \mapsto i'_j$ of the subscripts i_j ($j = 0, \ldots, K$) for which $t_{i'_0} \lhd \cdots \lhd t_{i'_K}$. Plainly, a normalized n-term s is equivalent to t (hence to $\{\, t_{i_0}, \ldots, t_{i_K} \,\}$ and hence to $\{\, t_{i'_0}, \ldots, t_{i'_K} \,\}$) if and only if s has the form $\{\, s_0 \,, \ldots,\, s_K \,\}$, with each s_j normalized and equivalent to the corresponding $t_{i'_j}$. Recursively, we calculate such terms s_js, being able to assume by induction that each of them is uniquely determined. In this manner, we simultaneously determine the desired s and prove its uniqueness.   ∎

The following corollary is only needed to prepare the ground for the subsequent Corollaries 4.10 and 4.11.

Corollary 4.9 *Let:*

(1) S be a safe term ;

(2) $y_1, \ldots, y_Q$ be distinct variables, including those that appear in S; and

(3) $t_1, \ldots, t_Q$ be n-terms, other than variables.

Also let S_* be the term that results from S after one replaces each variable y_i by the corresponding term t_i, all substitutions being performed simultaneously. If the characteristic tuple is known, then it is possible to calculate normalized n-terms $\overline{S}, \overline{\eta S}$ such that

$$\overline{S} \sim S_* \,, \quad \overline{\eta S} \sim \eta S_* \,,$$

i.e.

$$\overline{S}^\chi = S_*^\chi \,, \quad \overline{\eta S}^\chi = \eta S_*^\chi$$

in the B-interpretation χ. The term $\overline{S}$ will have either the form $\emptyset$ or the form $\{\, s_0 \,, \ldots,\, s_K \,\}$, while $\overline{\eta S}$ could also be of the form x_i, with $0 < i \leq n$.

Proof. Note that χ is indeed an interpretation of S_*, because all variables appearing in S_* are chosen from $x_1, \ldots, x_n$, and these constitute the domain of χ. We also have

$$S_*^\chi = S^\varphi \,,$$

where φ is the interpretation

$$y_i \overset{\varphi}{\mapsto} t_i^\chi \ (i = 1, \ldots, Q)$$

of S. A similar identity

$$s_*^\chi = s^\varphi$$

holds for every subterm s of S.

How can we be convinced from the beginning that a normalized term $\overline{S}$ equivalent to S_* will have either the form $\emptyset$ or the form $\{\, s_0 \,, \ldots,\, s_K \,\}$? We notice that

$$y_i^\varphi = (\, y_i \,)_*^\chi = t_i^\chi \in \mathcal{B} \,,$$

for $i = 1, \ldots, Q$, because no t_i is a variable. Hence, applying Corollary 4.3 to the safe formula $S = S$, we obtain

$$s^\varphi = s_*^\chi \in \mathcal{B}$$

for every safe subterm s of S. In particular, this identity will hold when s is S.

Let us explain why ηS had to be considered along with S, in the statement of the corollary. The normalized term $\overline{S}$ will be defined by recursively calculating $\overline{s}$ for every subterm s of S, and, from Definition 4.2, one sees that every subterm s, with the possible exception of the subterms ηt that appear in contexts of the form $\{\eta t \cdots\}$ or $\{\ldots, \eta t \cdots\}$, is safe. Even for the exceptional subterms ηt, however, the subterm t is safe, and this will make the recursive calculation of S possible.

We add the remark that once S_* has been normalized into the form $\emptyset$ or in the form $\{s_0, \ldots, s_K\}$, the normalized form of ηS_* will be found immediately: it will be either $\emptyset$ or s_0, in the respective cases.

If S is $\emptyset$, then S_* is also $\emptyset$ and we let $\overline{S}$ be $\emptyset$. If S is a variable y_i, then S_* is the n-term t_i, and hence $\overline{S}$ can be calculated as in the preceding corollary.

If S has the form $Pow\,t$, we first calculate $\overline{t}$. If $\overline{t}$ is $\emptyset$, then $\overline{S}$ will be the term $\{\emptyset, \{\emptyset\}\}$; if $\overline{t}$ is $\{t_0, \ldots, t_K\}$, then $\overline{S}$ will be the term

$$\{\emptyset, \{t_0\}, \{t_1\}, \{t_0, t_1\}, \ldots, \{t_0, \ldots, t_K\}\},$$

within which all terms $\{t_{i_0}, \ldots, t_{i_j}\}$ with $0 \le j \le K$, $0 \le i_0 < \cdots < i_j \le K$ appear exactly once. The term $\{t_{i_0}, \ldots, t_{i_j}\}$ will occur before $\{t_{i'_0}, \ldots, t_{i'_{j'}}\}$ within $\overline{S}$ if and only if the set $\{i_0, \ldots, i_j\}$ anti-lexicographically precedes $\{i'_0, \ldots, i'_{j'}\}$ with respect to $(\omega, \in)$.

Similar, but simpler, manipulations of the terms $\overline{t}_0$, $\overline{t}_1$ will enable us to calculate $\overline{S}$ when S is of the form $(t_0 \otimes t_1)$, where $\otimes$ is one of the binary operators $\cap$, $\setminus$, $\cup$. For example, if $\otimes$ is $\cup$, then, assuming that $\overline{t}_0$ is $\{t_0, \ldots, t_K\}$ and $\overline{t}_1$ is $\{t'_0, \ldots, t'_{K'}\}$, we will first

(a) eliminate duplicates from the list $t_0, \ldots, t_K, t'_0, \ldots, t'_{K'}$ of terms; then

(b) sort in increasing order the reduced list, exploiting Lemma 4.10; and, finally,

(c) combine the terms in the resulting list into the single normalized term $\overline{S}$.

We leave the details of the remaining cases to the reader. Note that, for each binary operator, the cases when $\overline{t}_0$ is $\emptyset$ or $\overline{t}_1$ is $\emptyset$ need a separate (but trivial) treatment.

Assuming now that $s_0, s_1, \ldots, s_K$ are safe terms and that t_j is either s_j or ηs_j for $j = 0, \ldots, K$, we must consider the two cases when S is either $\{t_0, \ldots, t_K\}$ or $\eta\{s_0, \ldots, s_K\}$. We first calculate $\overline{s}_0, \ldots, \overline{s}_K, \overline{t}_0, \ldots, \overline{t}_K$; then, by easy manipulations of these, calculate

$$\overline{\{t_0, \ldots, t_K\}} \quad \text{and} \quad \overline{\eta\{s_0, \ldots, s_K\}}.$$

Concerning in particular the latter, note that Lemma 4.10 enables us to find a subscript j in $\{0, \ldots, K\}$ for which $\overline{s}_j^X = min\{\overline{s}_0^X, \ldots, \overline{s}_K^X\}$. ∎

Corollary 4.10 *Knowing the characteristic tuple, it is possible to calculate, for any given* $m = 1, 2, 3, \ldots$ *, a normalized n-term* $\overline{S}_m$ *such that*

$$\overline{S}_m^X = B_m \setminus B_0.$$

Hence $\overline{S}_m$ *will have the form* $\{t_0, \ldots, t_K\}$, *where*

$$t_0^X < \cdots < t_K^X,$$

and

$$\{t_0^X, \ldots, t_K^X\} = B_m \setminus B_0.$$

Proof. In the preceding corollary, let S be the m-th term in the sequence $S_1, S_2, S_3, \ldots$ of safe terms in which

$$S_1 \text{ is } \{\emptyset\},$$
$$S_{i+1} \text{ is } ((S_i \cup y_1) \cup Pow(S_i \cup y_1)) \setminus y_1,$$

for $i = 1, 2, 3, \ldots$. Also let t_1 be the n-term $\{x_1 \ldots, x_n\}$, so that:

$$(y_1)_*^X = t_1^X = B_0,$$
$$(S_1)_*^X = S_1^X = \{\emptyset\} = B_1 \setminus B_0,$$
$$(S_i \cup y_1)_*^X = (B_i \setminus B_0) \cup B_0 = B_i,$$
$$(S_{i+1})_*^X = (B_i \cup Pow(B_i)) \setminus B_0 = B_{i+1} \setminus B_0,$$

for $i = 1, 2, 3, \ldots$. Thus we have

$$\overline{S}^X = \overline{S}_m^X = (S_m)_*^X = B_m \setminus B_0,$$

as desired.   ■

Corollary 4.11 *Let:*

(1) p *be a safe formula;*

(2) $y_1, \ldots, y_Q$ *be the distinct variables that appear in* p; *and*

(3) $t_1, \ldots, t_Q$ *be n-terms, other than variables.*

Also let p_* *be the formula that results from* p *after one replaces each variable* y_i *by the corresponding term* t_i, *all substitutions being performed simultaneously. The truth value* p_*^X *can be calculated, if the characteristic tuple is known.*

Proof. As explained in Corollary 4.9, we can find, for every term s that occurs in p, a normalized n-term $\overline{s}$ such that $\overline{s} \sim s_*$, i.e. s_* and $\overline{s}$ take the same value in the $\mathcal{B}$-interpretation. Moreover, if s is safe, s will have the form $\emptyset$ or the form $\{s_0, \ldots, s_K\}$.

It is then easy to evaluate the atomic sub-formulae of p_*. In fact, let $(s_* \otimes t_*)$, where $\otimes$ is $=$ or $\subseteq$ or $\in$, be one of these. After finding $\overline{s}$ and $\overline{t}$, we can apply, for the evaluation of $(\overline{s} \otimes \overline{t})$, the following criteria:

(1) $s^X = t^X$ holds if and only if $\bar{s}$ and $\bar{t}$ are the same term;

(2) $s^X \in t^X$ holds if and only if $\bar{t}$ has the form $\{ \cdots, \bar{s}, \cdots \}$;

(3) $s^X \subseteq t^X$ holds if and only if:
either $\bar{s}$ is $\emptyset$, or
$\bar{t}$ has the form $\{ t_0, t_1, \ldots, t_n \}$ and s has the form $\{ t_{i_0}, t_{i_1}, \ldots, t_{i_m} \}$,
with $0 \leq i_0 < i_1 < \cdots < i_m \leq n$.

Finally, taking into account the meaning of the propositional connectives, we complete the evaluation of p_*. ■

4.7 First semi-decision procedure

We are now ready to present a procedure which, given a safe formula p, will either:

- terminate discovering a finite model of p; or

- search for ever, in case p is not finitely satisfiable.

Procedure 4.1 *Let $y_1, \ldots, y_Q$ be the distinct variables appearing in the given safe formula p.*
 Step 1. *We generate all pairs*

$$(n, k_2, \ldots, k_n), m$$

consisting of a characteristic tuple (see Lemma 4.6) and of an $m \in \omega \setminus \{ 0 \}$. For each of these pairs, we perform the following two steps:
 Step 2. *Let us denote by B_0 and B_m the bottom and the m-th stage in the construction of a superstructure $\mathcal{B}$ whose characteristic tuple is*

$$(n, k_2, \ldots, k_n).$$

As in Corollary 4.10, we calculate the n-term $\{ t_0, \ldots, t_K \}$ that both denotes $B_m \setminus B_0$ in the $\mathcal{B}$-interpretation and is normalized with respect to $\mathcal{B}$.
 Step 3. *For every function*

$$y_i \mapsto t_{j_i} \quad (i = 1, \ldots, Q)$$

from the variables of p into the collection $t_0, \ldots, t_K$ of terms, we evaluate p_*^X, as in Corollary 4.11. Here p_* is the formula resulting from p when each variable y_i is replaced by the corresponding term t_{j_i}. If we obtain

$$p_*^X = \text{TRUE}$$

for one of these substitutions, we then stop, declaring that

$$y_i \overset{\varphi}{\mapsto} t_{j_i}^X \quad (i = 1, \ldots, Q)$$

is a finite model of p. Otherwise, we continue to search. □

The soundness of this test is entirely trivial, as it does not even attempt to prove p unsatisfiable. Completeness is also obvious, because, as we have noticed in the proof of Corollary 4.9,

$$p^\varphi = p_*^\chi$$

and hence

$$p^\varphi = \text{TRUE},$$

if φ corresponds to the replacement that led to the output 'p has a finite model'. A sort of strong completeness can also be proved, namely that

p is finitely satisfiable only if the above test is able to detect this fact.

As a matter of fact, if a finite model φ of p exists, then, by Corollary 4.5, there is a superstructure $\mathcal{B}$ such that

$$s^\varphi \in \mathcal{B} \text{ for every safe term } s \text{ that occurs in } p.$$

In particular $s^\varphi \in \mathcal{B}$ when s is a variable of p, and hence, for m large enough:

$$y^\varphi \in B_m \setminus B_0, \text{ for every variable } y \text{ in } p.$$

Our procedure will hence encounter the model φ while considering the pair $(n, k_2, \ldots, k_n)$, m, where $(n, k_2, \ldots, k_n)$ is the characteristic tuple of $\mathcal{B}$, in connection with the substitution $y \mapsto t_y$ that sends each variable y of p into the normalized n-term t_y for which $t_y^\chi = y^\varphi$.

4.8   An $\in$-$<$-isomorphism from B_m into $\mathcal{H}$

Let $X_1, X_2, \ldots, X_n$ be infinite sets arranged in increasing order with respect to $<$. Also, for technical reasons, we denote the empty set by X_0. Let $B_0 = \{X_1, \ldots, X_n\}$ and consider the m-th stage B_m, with $m > 0$, of the construction of the superstructure $\mathcal{B}$ over B_0.

In this section we show that there exists a one-to-one mapping $'$ from B_m into $\mathcal{H}$ which preserves both membership and the ordering relation $<$. This map will be used to show that if a safe formula has a model which maps variables into finite sets, then it has a hereditarily finite model.

Put

$$L = \lceil log_2 n \rceil, \quad N = 2^L, \tag{4.1}$$

so that L is the smallest integer with $0 < n \leq 2^L$. Let moreover ℓ be an integer such that

$$\ell > |B_m| + L + 2. \tag{4.2}$$

Note that since, by Lemma 4.1, $|B_m| \geq m$, (4.2) implies

$$\ell > m. \tag{4.3}$$

Next for each member x of B_m we define its *pseudo-rank*, $prk\,x$. Later on we will see that for all x in B_m, $rk\,x' = prk\,x$, where x' is the image (to be defined) of x via $'$.

Definition 4.9   *(a)* $prk\,X_0 = 0;$

$\quad$ *(b)* $prk\,X_j = (prk\,X_{j-1}) + min\{\ell, \alpha_{j-1}\},$
$\qquad$ *where* $rk\,X_j = (rk\,X_{j-1}) + \alpha_{j-1},$ *for* $j = 1,\ldots,n;$

$\quad$ *(c)* $prk\,x = (max_{z \in x}\,prk\,z) + 1,$ *for all x in* $B_m \setminus B_1.$ $\qquad\square$

Observe that the tuple $(prk\,X_1,\ldots, prk\,X_n)$ results from the replacement of every $k_j > \ell$ by ℓ in $(k_1,\ldots, k_n)$, where $k_1 = \omega$, and $(n, k_2,\ldots, k_n)$ is the characteristic tuple of $\mathcal{B}$. In particular $prk\,X_j = \ell$ when $k_j \notin \omega$, i.e. $k_j = \omega$.

Lemma 4.11   *(i) Let $1 \leq p \leq m$. For each y in B_p, there exist sets* $Z_0, Z_1, \ldots, Z_{k_y} \in B_p$ *such that*

$\quad$ *(i)$_1$ $0 \leq k_y < p$, and $k_y \geq 1$ if $y \notin B_1$;*

$\quad$ *(i)$_2$ $Z_0 \in Z_1 \in \cdots \in Z_{k_y}$;*

$\quad$ *(i)$_3$ $Z_0 \in B_1$ and $Z_{k_y} = y$;*

$\quad$ *(i)$_4$ $rk\,Z_{q+1} = (rk\,Z_q) + 1$, for $q = 0, 1, \ldots, k_y - 1$;*

$\quad$ *(i)$_5$ $prk\,Z_{q+1} = (prk\,Z_q) + 1$, for $q = 0, 1, \ldots, k_y - 1$;*

$\quad$ *(i)$_6$ $prk\,y = (prk\,Z_0) + k_y < n \cdot \ell + 1 + k_y < (n + 1) \cdot \ell.$*

(ii) $rk\,x < rk\,y$ iff $prk\,x < prk\,y$, for all x, y in B_m.

Proof. (i) and (ii) will be proved by way of interlocking induction. Note that the last inequality in (i)$_6$ is a trivial consequence of (i)$_1$ and (4.3), which imply $n \cdot \ell + 1 + k_y < n \cdot \ell + 1 + p \leq n \cdot \ell + 1 + m \leq n \cdot \ell + \ell.$

$\quad$ **Base case for** *(i)*. Let $y \in B_1$. In this case it is enough to put $k_y = 0$ and $Z_0 = y$. Concerning (i)$_6$, note that $y \in \{\emptyset, X_1, \ldots, X_n\}$; hence (a), (b) of Definition 4.9 imply $prk\,y \leq prk\,X_n \leq n \cdot \ell.$

$\quad$ **Base case for** *(ii)*. Let $x, y \in B_1$ with $rk\,x < rk\,y$. Then we have $x = X_i, y = X_j$ with $0 \leq i < j \leq n$. Let $rk\,X_{q+1} = (rk\,X_q) + \alpha_q$, for $q = i, i+1, \ldots, j-1$. Then $rk\,X_j = (rk\,X_i) + \alpha_i + \cdots + \alpha_{j-1}$, from which it follows easily that $\alpha_i + \ldots + \alpha_{j-1} \neq 0$, i.e. some α_q with $i \leq q < j$ is non-null. Since $prk\,X_j = (prk\,X_i) + min\{\ell, \alpha_i\} + \cdots + min\{\ell, \alpha_{j-1}\}$, where $min\{\ell, \alpha_q\} > 0$, we obtain $prk\,X_i < prk\,X_j$. The converse proof is similar to the one given, and we omit it.

$\quad$ Suppose now that (i) and (ii) hold for $1 \leq p < m$, and prove that they still hold for $p + 1$. Note that the inductive hypothesis (ii) says that $rk\,x < rk\,y$ iff $prk\,x < prk\,y$, for all x, y in B_p.

$\quad$ **Proof of the inductive step for** *(i)*. Let $y \in B_{p+1}$. By induction we can assume that $y \notin B_p$. As y is finite, it has a member z_0 of highest rank. Thus $rk\,y = (rk\,z_0) + 1$ and, moreover, $z_0 \in B_p$. Observe also that by inductive hypothesis (ii) if $z \in y$ then $prk\, z \leq prk\, z_0$. Therefore $prk\, y = (max_{z \in y}\, prk\, z) + 1 = (prk\, z_0) + 1$. Note in addition that since $z_0 \in B_p$, by inductive hypothesis (i) there exist sets $W_0, W_1, \ldots, W_{k_{z_0}} \in B_p$ such that $0 \leq k_{z_0} < p$, $W_0 \in W_1 \in \cdots \in W_{k_{z_0}}$, $W_0 \in B_1$ and $W_{k_{z_0}} = z_0$, $rk\, W_{q+1} = (rk\, W_q) + 1$ and $prk\, W_{q+1} = (prk\, W_q) + 1$ for $q = 0, 1 \ldots, k_{z_0} - 1$. By taking $k_y = k_{z_0} + 1$, $Z_k = W_k$ for $k = 0, 1, \ldots, k_{z_0}$, and $Z_{k_y} = y$, one easily checks that $(i)_1$–$(i)_6$ are all satisfied for $y \in B_{p+1}$. This concludes the proof of (i).

Proof of the inductive step for (ii). Let $x, y \in B_{p+1}$. By induction we can assume that $\{x, y\} \setminus B_p \neq \emptyset$. We distinguish the following three cases.

Case $x, y \notin B_1$. From (i) it follows that there exist $Z_x, Z_y \in B_p$ such that

$$prk\, x = (prk\, Z_x) + 1, \quad rk\, x = (rkZ_x) + 1, \tag{4.4}$$

and

$$prk\, y = (prk\, Z_y) + 1, \quad rk\, y = (rk\, Z_y) + 1. \tag{4.5}$$

Therefore

$$
\begin{aligned}
rk\, x &< rk\, y & &\text{iff (by } (4.4)_2 \text{ and } (4.5)_2) \\
rk\, Z_x &< rk\, Z_y & &\text{iff (by inductive hypothesis (ii))} \\
prk\, Z_x &< prk\, Z_y & &\text{iff (by } (4.4)_1 \text{ and } (4.5)_1) \\
prk\, x &< prk\, y,
\end{aligned}
$$

completing the proof of (ii) in this case.

Case $x \in B_1$, $y \in B_{p+1} \setminus B_p$. From (i) it follows that there exist k_y, $1 \leq k_y < p+1$ and sets $Z_0, Z_1, \ldots, Z_{k_y} \in B_{p+1}$ such that $(i)_2$–$(i)_6$ are satisfied.

We have

$$
\begin{aligned}
rk\, x &< rk\, y & &\text{iff (by } (i)_4 \text{ and } (i)_3) \\
rk\, x &< (rk\, Z_{k_y-1}) + 1 & &\text{iff} \\
rk\, x &\leq rk\, Z_{k_y-1} & &\text{iff (by inductive hypothesis (ii))} \\
prk\, x &\leq prk\, Z_{k_y-1} & &\text{iff} \\
prk\, x &< (prk\, Z_{k_y-1}) + 1 & &\text{iff (by } (i)_5 \text{ and } (i)_3) \\
prk\, x &< prk\, y,
\end{aligned}
$$

proving (ii) also in this case.

The final case to be considered is the following.

Case $x \in B_{p+1} \setminus B_p$, $y \in B_1$. We can choose k_x and $Z_0, Z_1, \ldots, Z_{k_x}$ so that the analogue of (i) holds, with x in place of y and $p + 1$ in place of p. Let $Z_0 = X_i$ and $y = X_j$, with $0 \leq i \leq n, 0 \leq j \leq n$. From $(i)_3$, $(i)_4$, and $(i)_5$ it follows that

$$rk\, x = (rk\, X_i) + k_x \tag{4.6}$$

and

$$prk\, x = (prk\, X_i) + k_x. \tag{4.7}$$

Assuming that $rk\,x < rk\,y$ we have by (4.6) that $(rk\,X_i) + k_x < rk\,X_j$, where we recall that $k_x < m$. Obviously $i < j$. Let q be such that $i \leq q < j$ and

$$rk\,X_q \leq (rk\,X_i) + k_x < rk\,X_{q+1}. \tag{4.8}$$

It follows immediately that putting $rk\,X_r = (rk\,X_{r-1}) + \alpha_{r-1}$ for $i < r \leq q$, one has $\alpha_{r-1} \leq k_x < m < \ell$ (see (4.3)). Hence by Definition 4.9(b) we have also $prk\,X_r = (prk\,X_{r-1}) + \alpha_{r-1}$, for $i < r \leq q$. Therefore we can write

$$rk\,X_q = (rk\,X_i) + (\alpha_i + \cdots + \alpha_{q-1}). \tag{4.9}$$

$$prk\,X_q = (prk\,X_i) + (\alpha_i + \cdots + \alpha_{q-1}). \tag{4.10}$$

Since by (4.8) and (4.9) $(rk\,X_q) + (k_x - (\alpha_i + \cdots + \alpha_{q-1})) < rk\,X_{q+1}$, and $k_x - (\alpha_i + \cdots + \alpha_{q-1}) < \ell$, Definition 4.9(b) implies

$$(prk\,X_q) + (k_x - (\alpha_i + \cdots + \alpha_{q-1})) < prk\,X_{q+1} \leq prk\,X_j. \tag{4.11}$$

Substituting (4.10) in (4.11) we obtain $(prk\,X_i) + k_x < prk\,X_j$, i.e., by (4.7), $prk\,x < prk\,y$. Conversely, if $prk\,x < prk\,y$, then by (4.7)

$$(prk\,X_i) + k_x < prk\,X_j.$$

Putting $rk\,X_r = (rk\,X_{r-1}) + \alpha_{r-1}$ for $r = i+1, \ldots, j$, we have $prk\,X_j = (prk\,X_i) + min\{\ell, \alpha_i\} + \cdots + min\{\ell, \alpha_{j-1}\}$. Hence $k_x < min\{\ell, \alpha_i\} + \cdots + min\{\ell, \alpha_{j-1}\}$. From this inequality it follows

$$(rk\,X_i) + k_x < (rk\,X_i) + \alpha_i + \cdots + \alpha_{j-1} = rk\,X_j,$$

i.e., by (4.6), $rk\,x < rk\,y$. This completes the analysis of the current case, and therefore the inductive proof of (ii) is concluded. ∎

Next we define the mapping $' : B_m \to \mathcal{H}$, by induction on $prk\,x$. In the rest of this section, when $\alpha \in \omega$ we indicate V_α by H_α to remind that $V_\alpha \subseteq \mathcal{H}$.

Definition 4.10 Let $x \in B_m$ and assume that y' has already been defined for all y such that $prk\,y < prk\,x$. We distinguish two cases.

(a) If $x \in B_0$, then $x = X_{i_0}$, for some $i_0 \in \{1, \ldots, n\}$. Let $c_x = |\{X \in B_0 : rk\,X = rk\,X_{i_0}$ and $X_{i_0} \leq X\}|$, and let d_x be the c_x-th last element in $H_{(prk\,x)-1}$. We put $x' = \{y' : y \in x \cap B_m\} \cup \{d_x\}$.

(b) If $x \notin B_0$, we put $x' = \{y' : y \in x\}$.

(Note that if $y \in x \cap B_m$, then certainly $rk\,y < rk\,x$, so that by (ii) of Lemma 4.11 $prk\,y < prk\,x$, and y' is already defined. We will show below that the element d_x introduced in (a) always exists.) □

Lemma 4.12 *Let* $r \geq \ell - 1$, *where* ℓ *is given by (4.2), and let* $S^* = \{s_0, s_1, \ldots, s_{L-1}\}$ *be the initial segment of the von Neumann universe* $\mathcal{V}$ *with* $s_0 < s_1 < \ldots < s_{L-1}$ *and* L *defined as in (4.1). Then*

(a) $S^* \subsetneq H_r$;

(b) $\{H_r \setminus S : S \subseteq S^*\}$ *is the (unordered) final segment of* H_{r+1} *of length* 2^L. *Moreover, for each* $S \subseteq S^*$,

    (b1) $|H_r \setminus S| > |B_m| + 1$;

    (b2) $rk(H_r \setminus S) = r$.

Proof. Since H_r itself is an initial segment of the von Neumann universe $\mathcal{V}$, to prove (a) it suffices to observe that $|H_r| \geq r \geq \ell - 1 > |B_m| + L + 1 > L$ by Lemma 4.1 and (4.2).

Proof of (b): Let $T \in H_{r+1}$ be such that $H_r \setminus S^* < T$. Then $H_r \setminus S^* \subseteq T$. Indeed, if by contradiction $H_r \setminus S^* \not\subseteq T$, then by Theorem 2.3 and Corollary 2.1 $max((H_r \setminus S^*) \setminus T) < max(T \setminus (H_r \setminus S^*))$, where $T \setminus (H_r \setminus S^*) = T \cap S^*$, since $T, S^* \subseteq H_r$. On the other hand, $T \not\subseteq H_r \setminus S^*$, hence $T \setminus (H_r \setminus S^*) \neq \emptyset$ and therefore

$$max(T \setminus (H_r \setminus S^*)) \in S^*,$$

which implies

$$max(T \setminus (H_r \setminus S^*)) < max((H_r \setminus S^*) \setminus T),$$

as S^* is an initial segment of H_r. This is a contradiction and therefore $H_r \setminus S^* \subseteq T$, i.e. $T \in \{H_r \setminus S : S \subseteq S^*\}$, showing that $\{H_r \setminus S : S \subseteq S^*\}$ is the final segment $\{T \in H_{r+1} : H_r \setminus S^* \leq T\}$ of H_{r+1}.

Moreover, if $S \subseteq S^*$, then $|H_r \setminus S| = |H_r| - |S| \geq r - L > |B_m| + 1$, by (a), Lemma 4.1, and (4.2), proving (b1). (b2) follows immediately observing that $rk(H_r \setminus S) = rk(max(H_r \setminus S)) + 1 = (rk\, max\, H_r) + 1 = rk\, H_r = r$. ∎

Lemma 4.13 *For all* x, y *in* B_0, w, z *in* B_m

(a) $d_x \neq z'$ *(cf. Definition 4.10)*;

(b) $rk\, d_x = (prk\, x) - 1$;

(c) *if* $x < y$ *then* $d_x < d_y$ *and if* $rk\, z' \leq rk\, d_x$ *then* $z' < d_x$;

(d) $rk\, w' = prk\, w$.

Proof. It is an obvious consequence of Definition 4.9 that $prk\, x \geq prk\, X_1 = \ell$; hence the preceding lemma applies, with $r = (prk\, x) - 1$. Since $c_x \leq n$, where c_x is as in Definition 4.10, it follows by (4.1) and Lemma 4.12(a) that $c_x \leq 2^L = |\{H_{(prk\, x)-1} \setminus S : S \subseteq S^*\}|$. Hence $d_x \in \{H_{(prk\, x)-1} \setminus S : S \subseteq S^*\}$, showing in particular that d_x, and therefore $'$, is well-defined. As

$|z'| \leq |\{v' : v \in z \cap B_m\}| + 1 \leq |B_m| + 1$, then by Lemma 4.12(b1) $z' \notin \{H_{(prk\,x)-1} \setminus S : S \subseteq S^*\}$ from which $d_x \neq z'$, proving (a).

(b) follows immediately from (b2) of Lemma 4.12.

Next, to prove (c) we assume $x < y$. Note that in this case Definition 4.9 implies $prk\,x \leq prk\,y$. If $prk\,x < prk\,y$, then by (b) $rk\,d_x < rk\,d_y$, which implies $d_x < d_y$. On the other hand, if $prk\,x = prk\,y$, then $c_x < c_y$. Therefore from the very definition of d_x and d_y it follows $d_x < d_y$. To complete the proof of (c) assume that $rk\,z' \leq rk\,d_x$. Then by (b) $z' \in H_{prk\,x}$. Since $\{H_{(prk\,x)-1} \setminus S : S \subseteq S^*\}$ is a final segment of $H_{prk\,x}$ and, as noticed in the proof of (a), $\{z', d_x\} \cap \{H_{(prk\,x)-1} \setminus S : S \subseteq S^*\} = \{d_x\}$, it follows that $z' < d_x$, establishing (c).

Only (d) remains to be proved. We do this by induction on $prk\,w$. If $prk\,w = 0$, then by Definition 4.9 $w = \emptyset$, from which $rk\,w' = 0$. Suppose that $rk\,w_1' = prk\,w_1$ for all $w_1 \in B_m$ with $prk\,w_1 \leq p$, and assume that $prk\,w = p+1$. If $w \in B_m \setminus B_0$, then $w' = \{w_1' : w_1 \in w\}$, and therefore by inductive hypothesis $rk\,w' = (max_{w_1 \in w}\,rk\,w_1') + 1 = (max_{w_1 \in w}\,prk\,w_1) + 1 = prk\,w$. On the other hand, if $w \in B_0$, then $w' = \{w_1' : w_1 \in w \cap B_m\} \cup \{d_w\}$, where $rk\,d_w = (prk\,w) - 1$ by (b). By inductive hypothesis and by Lemma 4.11(ii), $rk\,w_1' \leq rk\,d_w$ for all $w_1 \in w$, which shows $rk\,w' = (rk\,d_w) + 1 = prk\,w$, establishing (d) and concluding the proof of the lemma. ∎

The following proposition lists some useful reflection properties of the embedding '.

Corollary 4.12 *The function $v \mapsto v'$ defined on B_m is one-to-one. Moreover, assuming that $x, y \in B_m \setminus B_0$ and $w, z \in B_m$, one has*

(i) $w \in z$ iff $w' \in z'$;

(ii) $(x \cup y)' = x' \cup y'$;

(iii) $(x \cap w)' = x' \cap w'$;

(iv) $(x \setminus w)' = x' \setminus w'$;

(v) $a \subseteq x'$ iff there exists $t \in B_m$ such that $t \subseteq x$ and $t' = a$;

(vi) if $(Pow\,x) \in B_m$ then $(Pow\,x)' = Pow\,x'$;

(vii) if $(x \times y) \in B_m$ then $(x \times y)' = x' \times y'$;

(viii) $w < z$ iff $w' < z'$;

(ix) $(\eta\,x)' = \eta\,x'$. More generally, $(\eta\,z)' = \eta\,z'$ provided $\eta\,z \in B_m$.

Proof. Let us consider the WFG having nodes B_m and edges $\{(w, z) : w, z \in B_m \text{ and } w \in z\}$. We put $u_v = \{d_v\}$ and $\ddot{u}_v = \{v\}$ for all v in B_0 (note that the definition of $\ddot{u}_v$ is well-posed by Lemma 4.13(c)). It follows from Lemma 4.13(a)—also taking Definition 4.10 into account—that $(U,\,\ddot{})$, where $U = \{u_v : v \in B_0\}$, is a grafting which induces the function $v \mapsto v'$ on B_m (recall Definition 2.5).

It is easy to recognize that any two nodes w, z that belong to the same places (of the WFG or of the realization) necessarily coincide. In fact, if $w \in B_0$, then $z \in \ddot{u}_w$, i.e. $z = w$, as a consequence of $w \in \ddot{u}_w$. Symmetrically, $w = z$ follows from $z \in B_0$. If z, $w \notin B_0$, then—since z, $w \subseteq B_m$ by Lemma 4.2(1)—saying that z, w belong to the same places $\dot{v}$ with v in B_m is equivalent to saying that z, w have the same members, i.e. $z = w$. We conclude that $(U, \ddot{\ })$ is extensional, so that both the injectivity of $v \mapsto v'$ and the thesis (i) readily follow from Lemma 2.10. It is then easy to derive (ii) through (vii), by exploiting Lemma 4.2, Lemma 4.13(a) and the injectivity of $v \mapsto v'$.

Proof of $(viii)$: We begin by showing by induction on $p = max\{\, prk\, w\,,\, prk\, z\,\}$ that if $w < z$ then $w' < z'$. The base case, with $p = 0$, is vacuously true, since in this case $w = z = \emptyset$. Therefore suppose the result true for p, and assume that $w < z$ and $max\{\, prk\, w\,,\, prk\, z\,\} = p + 1$. We distinguish the following two cases.

Case A $rk\, w < rk\, z$. Then from Lemma 4.11(ii) $prk\, w < prk\, z$, which by (d) of Lemma 4.13 implies $rk\, w' < rk\, z'$. Hence $w' < z'$.

Case B $rk\, w = rk\, z$. (Hence from Lemma 4.11(ii) $prk\, w = prk\, z$.) We have the following subcases.

Subcase B1 Assume first w, $z \in B_0$. If $w_1 \in w \cap B_m$, then $rk\, w_1 < rk\, w$. Hence Lemma 4.11(ii) implies $prk\, w_1 < prk\, w$, i.e. by Lemma 4.13(b),(d), $rk\, w_1' \leq rk\, d_w$. Therefore Lemma 4.13(c) implies $w_1' < d_w < d_z$. Hence, by the anti-lexicographicity of $<$, $w' = \{\, w_1' : w_1 \in w \cap B_m\,\} \cup \{\, d_w\,\} < \{\, d_z\,\} \leq z'$, showing $w' < z'$ in Subcase B1.

Subcase B2 Now suppose that $w \in B_m \setminus B_0$, $z \in B_0$. Let $w_1 \in w$. Then $prk\, w_1 \leq prk\, w - 1 = prk\, z - 1 = rk\, d_z$ by Lemma 4.13(b). Therefore Lemma 4.13(c) implies $w_1' < d_z$ which yields $w' = \{\, w_1' : w_1 \in w\,\} < \{\, d_z\,\} \leq z'$, proving that $w' < z'$ in Subcase B2 also.

Subcase B3 Finally suppose $z \in B_m \setminus B_0$. Since z is finite, it follows that w must be finite, i.e. $w \in B_m \setminus B_0$. If $w \subseteq z$, then from (v) $w' \subseteq z'$. Moreover, since $w < z$, then the injectivity of $v \mapsto v'$ implies $w' \neq z'$, from which $w' \subsetneq z'$ and therefore $w' < z'$. So we can assume that $w \not\subseteq z$. In this case, by Theorem 2.3 and Corollary 2.1, $z \setminus w \neq \emptyset$, $w \setminus z \neq \emptyset$, and $max(\, w \setminus z\,) < max(\, z \setminus w\,)$. Therefore (i), (iv) and the inductive hypothesis imply $z' \setminus w' \neq \emptyset$, $w' \setminus z' \neq \emptyset$, and $max(\, w' \setminus z'\,) < max(\, z' \setminus w'\,)$, i.e. $w' < z'$, concluding the analysis of Case B.

So far we have proved that if $w < z$ then $w' < z'$. Conversely, if $w' < z'$ then one must have $w < z$. Indeed, if by contradiction $z \leq w$, then from the first half of the proof it would follow $z' \leq w'$, contradicting our initial assumption $w' < z'$. This completes the proof of (viii).

Proof of (ix): We have $z' = \{\, z_1' : z_1 \in z\,\} = \{\, z_1' : z_1 \in z \cap B_m\,\}$ or $z' = \{\, z_1' : z_1 \in z \cap B_m\,\} \cup \{\, d_z\,\}$, according to whether $z \in B_m \setminus B_0$ or $z \in B_0$. Moreover, in the latter case we have $z_1' < d_z$ for every $z_1' \in z \cap B_m$, by Lemma 4.13(b),(d),(c), and it follows from the assumption $\eta z \in B_m$ that at least one such z_1 exists. If $z = \emptyset$, then we have $(\eta z)' = \emptyset' = \emptyset = \eta \emptyset = \eta \emptyset' = \eta z'$. If $z \neq \emptyset$, then $\eta z'$ is the least z_1' with $z_1 \in z \cap B_m$. Hence $\eta z' \leq (\eta z)'$ because $\eta z \in z \cap B_m$. Conversely, by (viii), we have $(\eta z)' < z_1'$ for every $z_1 \in z \cap B_m$ with $z_1 \neq \eta z$, which implies $(\eta z)' \leq \eta z'$. The lemma is hence established. ∎

Corollary 4.13 *Let p be a safe formula. Then p is finitely satisfiable if and only if p is hereditarily finitely satisfiable.*

Proof. We will show that to each finite interpretation φ of p there corresponds an interpretation ψ, with $p^\psi = p^\varphi$, such that ψx is hereditarily finite for every variable x in p. In particular $p^\psi = TRUE$ if $p^\varphi = TRUE$, which proves our statement.

Let $\mathcal{B}$ be a superstructure as in Corollary 4.5, that is to say, with $t^\varphi \in \overline{\mathcal{B}}$ for every term t occurring in p. Since the number of terms in p is finite, a suitably large m will ensure $t^\varphi \in B_m$ for any term t in p, where B_m is the m-th stage in the construction of $\mathcal{B}$. If the bottom of $\mathcal{B}$ is $\emptyset$, it will suffice for our purposes to take $\psi = \varphi$. Otherwise, we consider the embedding $v \mapsto v'$ of B_m in $\mathcal{H}$ and define $\psi x = (\varphi x)'$ for every variable x in p. It follows from Corollary 4.12 that $t^\psi = (t^\varphi)'$ for every term t in p and that $q^\psi = q^\varphi$ for every atomic subformula q of p, so that $p^\psi = p^\varphi$ as desired. ∎

Thanks to this corollary, we can now simplify Procedure 4.1, as follows:

Procedure 4.2 *It is to be checked that a given safe formula p is finitely satisfiable. If p is not finitely satisfiable, the procedure will search for ever.*

Step 1. *For every m in $\omega \setminus \{0\}$, we determine the normalized 0-term $\{t_0, \ldots, t_K\}$ that denotes H_m, as in Corollary 4.10.*

Step 2. *For every function $y_i \mapsto t_{j_i}$ ($i = 1, \ldots, Q$) from the variables $y_1, \ldots, y_Q$ of p into the collection $t_0, \ldots, t_K$ of terms, we evaluate p_*^χ as in Corollary 4.11. Note that χ, which is the $\mathcal{H}$-interpretation here, is empty, because the bottom of $\mathcal{H}$ is empty. As in Procedure 4.1, p_* is the formula resulting from p when each variable y_i is replaced by the corresponding term t_{j_i}. If we obtain $p_*^\chi = TRUE$ for one of these substitutions, we then halt declaring that $y_i \overset{\varphi}{\mapsto} t_{i_j}^\chi$ ($i = 1, \ldots, Q$) is a hereditarily finite model of p. Otherwise, we continue to search with the next value of m.* □

5. ELEMENTARY SYLLOGISTICS

5.1 Introduction

In this chapter we develop satisfiability decision tests for three collections $\mathbf{F}_1$, $\mathbf{F}_2$, and $\mathbf{F}_3$ of unquantified formulae. In addition to set variables and to propositional connectives, the constructs involved by these collections of formulae are, respectively:

$$(\mathbf{F}_1) \quad \emptyset,\, \cap,\, \in,\, =,\, \subseteq,\, \textit{Finite};$$
$$(\mathbf{F}_2) \quad \emptyset,\, \eta,\, \in,\, =,\, <,\, \textit{Finite};\ \text{and}$$
$$(\mathbf{F}_3) \quad \emptyset,\, \eta,\, \in,\, =,\, <,\, \{\bullet, \cdots, \bullet\}.$$

The decidability results regarding $\mathbf{F}_1$ and $\mathbf{F}_2$ are based on a reflection lemma (see Section 5.2) which generalizes former conclusions, reached in Chapter 2, about the realizability of well-founded graphs. The decidability result regarding $\mathbf{F}_3$ is based on a somewhat different technique, due to Parlamento and Policriti (see [PP88b]); for this reason the last two sections are largely independent of the rest of the chapter.

The first decision methods that will be presented for $\mathbf{F}_1$ and $\mathbf{F}_2$ (see Section 5.3 and the brief discussion below) involve a non-deterministic choice from a usually large space of possibilities. We will see, however, that there is a specific choice that is better than any other, in the sense that it can reveal whether a given formula p is satisfiable or not, making it unnecessary to backtrack to other possible choices. In order to make our exposition simpler, in discussing this improvement upon the 'brute force' algorithm, we will not consider the full collection $\mathbf{F}_2$, but will only take into account the constructs

$$(\mathbf{F}_2) \quad \emptyset,\, \eta,\, \in,\, = .$$

To enter into more detail, let us denote by p a formula to be tested for satisfiability, belonging to $\mathbf{F}_1 \cup \mathbf{F}_2$, and by T the set of all terms that occur in $p\ \&\ \emptyset = \emptyset$. We will prove that unless the relator $\textit{Finite}$ appears in p, p can only be satisfiable if it has a model M with $rk\,M\,x < 2 \cdot |T| - 1$ for every variable x in p. This provides the following crude decision method:

'Generate all functions M : {variables of p} $\rightarrow V_{2 \cdot |T|-1}$ and, for each such function, evaluate p in M, until M is found to be a model of p'.

Note that the direct evaluation of p in each M is possible, as we know from Chapter 4, because T is finite.

If we do not forbid *Finite* to occur in p, some variables may have to be assigned infinite values in a model of p; therefore we can neither produce all potential models M of p, nor, in general, evaluate p in a given assignment of sets to the variables. We will therefore resort to 'surrogates' of the potential models of p, each surrogate being associated with many interpretations of p, which all yield the same truth value for p. The class of the interpretation surrogates for p must be finite, so that we can explore it systematically in a finite amount of time. It must be rich enough so that when p admits a model, p also has a model associated with one of its interpretation surrogates. Finally, we want a technique for calculating the common value of p in all interpretations associated with any given surrogate.

The interpretation surrogates proposed in this chapter for $\mathbf{F}_1$ and $\mathbf{F}_2$ will take the form of quintuples

$$T \overset{\sim}{\to} T, \ G = (V, \cdot), \ \prec, \ \lhd, \ I,$$

where:

> $\tilde{\tilde{r}} = \tilde{r}$ for every r in T, i.e., for every term r in p & $\emptyset = \emptyset$;
> $V = \{\tilde{r} : r \in T\}$;
> G is a well-founded graph;
> $\prec$ is a partial ordering of V;
> $\lhd$ is a total ordering of V which both extends $\prec$ and contains as a subset the set of all edges of G;
> $I \subseteq V$.

Each such quintuple represents the class of those interpretations M that satisfy, for all r, s in T, the conditions

> $r^M = s^M$ if and only if $\tilde{r} = \tilde{s}$;
> $r^M \in s^M$ if and only if $(\tilde{r}, \tilde{s})$ is an edge of G;
> $r^M \subsetneq s^M$ if and only if $\tilde{r} \prec \tilde{s}$;
> $r^M < s^M$ if and only if $\tilde{r} \lhd \tilde{s}$;
> r^M is infinite if and only if $\tilde{r} \in I$.

In dealing with $\mathbf{F}_1$, we will not choose $\lhd$ but will simply require that such an ordering exists. In connection with $\mathbf{F}_2$, we will instead see that it is not restrictive to always assume that $\prec = \{\tilde{\emptyset}\} \times (V \setminus \{\tilde{\emptyset}\})$.

The 'eurekas' for the improved decision algorithms that we have announced at the beginning are essentially criteria for fixing $\sim$, $\prec$, and I in the case of $\mathbf{F}_1$, and for fixing $\sim$ in the case of $\mathbf{F}'_2$ (in the latter case, $\prec$ is already fixed, as we have just noted, while I can be taken to be empty, because the relator *Finite* is not employed in $\mathbf{F}'_2$).

Let us now briefly outline the decision algorithms for $\mathbf{F}_3$. It will be seen in Section 5.7 that any satisfiable formula p in $\mathbf{F}_3$ is also finitely satisfiable. Since the normalization process introduced in Section 3.8 translates p into a safe formula q which also belongs to $\mathbf{F}_3$, and since we already have a semi-decision satisfiability test for safe formulae (namely Procedure 4.2), we immediately obtain a semi-decision satisfiability test for $\mathbf{F}_3$. To convert the latter into a *decision* test, it will be necessary to impose a finite bound to the infinite search process. How to do this without disrupting the correctness of the method, for a significantly broad collection of safe formulae, is one of the major issues discussed in this chapter (see Section 5.6).

5.2 A technique for realizing well-founded graphs whose nodes are ordered

The following reflection lemma is crucial for obtaining all results in the subsequent Sections 5.3 through 5.5.

Lemma 5.1 (Reflection) *Assume that:*

(a) $G = (V, \dot{\ })$ *is a WFG whose nodes form a set;*

(b) $\emptyset$ *is a member of* V;

(c) $\prec$ *is a partial ordering of* V *such that*

    (c1) $\emptyset \prec v$ *for every* v *in* $V \setminus \{\emptyset\}$;

    (c2) $w \in \dot{y}$, $w \prec v$ *implies* $v \in \dot{y}$ *for all* y, v, w *in* V;

    (c3) the pairs (v, w) *with* $w \in \dot{v}$ *or* $v \prec w$ *are among the edges of a WFG* $\Gamma = (V, \overset{*}{\ })$;

(d) I *is a subset of* $V \setminus \{\emptyset\}$, *with the property that*

    (d1) $w \in I$, $w \prec v$ *implies* $v \in I$, *for all* v, w *in* V.

Then there is a one-to-one realization $v \mapsto v'$ *of* G, *with* $\emptyset' = \emptyset$, *which satisfies, for any pair* v, w *of nodes, the conditions*

(A) $w' < v'$ *whenever* (w, v) *is an edge of* Γ,

(B) $w' \subsetneqq v'$ *if and only if* $w \prec v$,

(C) if the number of edges (y, x) *in* Γ *is finite for every node* x, *then* $|v'|$ *is infinite if and only if* $v \in I$.

Proof. Let $\lhd$ be a well-ordering of V such that $w \lhd v$ whenever (w, v) is an edge of Γ (the existence of such an ordering is insured by Corollary 2.6). By (c1), $\emptyset$ is the minimum of V with respect to $\lhd$. The minimum of $V \setminus \{\emptyset\}$ will be denoted $\mathfrak{l}$ in what follows. Without loss of generality, if $I \neq \emptyset$ we assume that $\mathfrak{l} \in I$ and that for any v in V, neither $v \in \dot{\mathfrak{l}}$ nor $\mathfrak{l} \prec v$ holds.

We define $R(\{s\}) = \{\{t\} : t \subseteq s \text{ and } rk\, t = rk\, s\}$ for any set s, so that $\{s\}$ belongs to $R(\{s\})$, and $\omega \leq |s|$ implies $\omega \leq |R(\{s\})|$ by Corollary 2.4. We also put

(1) $u_{\emptyset} = \emptyset' = \emptyset$,

(2a) $d_v = \begin{cases} \{\omega\} & \text{if both } v = \mathbf{1} \text{ and } I \neq \emptyset, \\ \{\bigcup_{y \lessdot v} y'\} & \text{otherwise}; \end{cases}$

(2b) $u_v = \begin{cases} \{d_v\} & \text{if } v \notin I, \\ R(d_v) & \text{if } v \in I; \end{cases}$

(3) $v' = u_v \cup \left(\bigcup_{w \prec v} u_w\right) \cup \{w' : v \in \dot{w}\}$,

for all v in $V \setminus \{\emptyset\}$. Since $\lhd$ is a well-ordering, clauses (1), (2a), (2b), (3) constitute a legal definition by transfinite recursion: u_v and v' are in fact defined in terms of values u_w, w' with $w \lhd v$. Note that every u_v trivially consists of singletons a_v with $rk\, a_v = rk\, d_v$, which is a successor ordinal.

We now prove that when $w \lhd v$, and in particular when $v \in \dot{w}$ or $w \prec v$, then $rk\, a_w < rk\, w' < rk\, a_v$ for any $a_w \in u_w$ and $a_v \in u_v$, so that $a_w < w' < a_v$. Hence it will readily follow from (3) that u_v is the final segment of v' for every v in $V \setminus \{\emptyset\}$. In the special case when $w = \emptyset$, there is no a_w in u_w and hence our local hypothesis simplifies into $rk\, w' < rk\, a_v$, which is obvious, since a_v has rank greater than 0, whereas w' is $\emptyset$. When $w \neq \emptyset$, then $w \lhd v$ implies $v \neq \mathbf{1}$, so that

$$w' \subseteq \bigcup_{y \lhd v} y' \in d_v, \text{ hence } rk\, w' \leq rk \bigcup_{y \lhd v} y' < rk\, d_v = rk\, a_v;$$

moreover $a_w \in w'$, and hence $rk\, a_w < rk\, w'$. The desired conclusion follows immediately.

It is also true that $w' < v'$ when $w \lhd v$. In fact, if $w = \emptyset$ then $w' = \emptyset < v'$ because $d_v \in u_v \subseteq v'$, which implies $v' \neq \emptyset$. If $w \neq \emptyset$ then

$$rk\, w' = rk\, d_w + 1 < rk\, d_v + 1 = rk\, v', \text{ hence } w' < v'.$$

Although the fact that $w \lhd v$ implies $w' < v'$ clearly suffices to prove (A), we incidentally observe that, trivially, the converse holds: if $w \ntriangleleft v$ then either $w = v$ and hence $w' = v'$, $w' \not< v'$; or $v \lhd w$, so that $v' < w'$ and again $w' \not< v'$.

We now want to prove that the function

$$u_x \longmapsto \{v : x \prec v\} \cup \{x\}$$

defined for all x in $V \setminus \{\emptyset\}$ is a grafting of V. In the first place we note that the sets u_x with $x \neq \emptyset$ are non-empty and pairwise disjoint, because, when

$x \neq y$, then $d_x \in u_x$ and $rk\, d_x = rk\, a_x \neq rk\, a_y = rk\, d_y$ hold for any $a_x \in u_x$, $a_y \in u_y$. Since ¨ clearly induces the function $v \mapsto v'$, and recalling Definition 2.5, for our current purposes it will suffice to show that no node v has a node x with $v' \in u_x$. Indeed, if there could be some u_x with $v' \in u_x$, then plainly $v \neq \emptyset$ and we would have

$$rk\, d_x = rk\, v' = rk\, d_v + 1.$$

As $rk\, d_v < rk\, d_x$, it follows $v \triangleleft x$, and in particular $v' \subseteq \bigcup_{y \triangleleft x} y'$. But

$$rk\, d_x = \left(rk \bigcup_{y \triangleleft x} y' \right) + 1 \geq rk\, v' + 1,$$

contradicting $rk\, d_x = rk\, v'$.

One easily recognizes that no two nodes v, w belong to the same places of the grafting ($\{ u_x : x \in V \setminus \{\emptyset\}\}$, ¨). Assuming for definiteness that $w \triangleleft v$, we have in fact that $w \notin \ddot{u}_v$ and $v \in \ddot{u}_v$. As a consequence the grafting is extensional (see Definition 2.5), and hence, by Lemma 2.10, the function $v \mapsto v'$ is a one-to-one realization of G.

We now prove (B); that is, we prove that

$$w' \subsetneq v' \text{ if and only if } w \prec v,$$

for every pair v, w of nodes. We first prove the 'if' part of this statement, assuming that $w \prec v$. We note that if the inclusion $w' \subseteq v'$ holds, then it is strict, because $d_v \in v' \setminus w'$. The inclusion holds for $w = \emptyset$, since $w' = \emptyset$ in this case. If $w \neq \emptyset$, then

$$v' = u_v \cup \left(\bigcup_{x \prec v} u_x \right) \cup \{y' : v \in \dot{y}\} \text{ and}$$

$$w' = u_w \cup \left(\bigcup_{x \prec w} u_x \right) \cup \{y' : w \in \dot{y}\},$$

by definition (3). In this case we note that $u_w \subseteq v'$ since $w \prec v$. Moreover, if $w \in \dot{y}$, then $v \in \dot{y}$ by (c2); hence

$$\{y' : w \in \dot{y}\} \subseteq \{y' : v \in \dot{y}\} \subseteq v'.$$

Finally, if $x \prec w$ then $x \prec v$ because $\prec$ is an ordering; hence

$$\{u_x : x \prec w\} \subseteq \{u_x : x \prec v\},$$

which implies

$$\bigcup_{x \prec w} u_x \subseteq \bigcup_{x \prec v} u_x \subseteq v',$$

and therefore $w' \subseteq v'$.

For the converse, assume that $w' \subsetneq v'$. If $w = \emptyset$, then $w \prec v$ by (c1). If $w \neq \emptyset$, then $u_w \subseteq w'$, hence $u_w \subseteq v'$, and therefore $w \prec v$ because the unions in (3) are disjoint, as we have already proved, and moreover $u_w \neq u_v$ as a consequence of $w \neq v$.

We now prove (C), making the assumption that the number of edges (w, v) in Γ is finite for every node v. It follows from this assumption that $\{w' : v \in \dot{w}\}$ and $\{u_w : w \prec v\}$ are both finite, for all v. Therefore, in view of (3), v' is infinite if and only if either u_v, or some u_w with $w \prec v$, is infinite. Let us consider first the case where $I = \emptyset$. In this case, every set u_v is singleton, and consequently no v' is infinite. To prove (C) also in the case $I \neq \emptyset$, we will show that, for every node x, u_x is infinite if and only if $x \in I$. Thus (C) will follow; in fact, if some u_w with $w \prec v$ is infinite, then (d1) will enable us to infer that $v \in I$, because $w \in I$.

The first node x such that u_x is infinite is $\mathbf{1}$, since $u_{\mathbf{1}} = R(\{\omega\})$, consisting of all singletons $\{X\}$ with X an infinite subset of ω. As a matter of fact $\mathbf{1} \in I$, as desired. If x is a node with an infinite u_x, then u_x must be of the form $R(d_x)$, and hence $x \in I$. Conversely, if $x \in I \setminus \{\mathbf{1}\}$, then $R(d_x)$ consists of those sets of the form $\{X\}$ where X is a subset of $Y = \bigcup_{y \lhd x} y'$ having the same rank as Y. But note that Y is infinite, because it includes $\mathbf{1}'$ as a subset. Therefore, by a remark we have made when we have introduced the notation $R(s)$, $u_x = R(d_x)$ will be infinite too, as we had to prove. ∎

Corollary 5.1 *In the realization $v \mapsto v'$ of the preceding lemma, if v_1, v_2 are nodes with $v_1' \not\subseteq v_2'$ and $v_2' \not\subseteq v_1'$, then $w' = v_1' \cap v_2'$ holds if and only if the following conditions are met:*

(i) $w \prec v_1$, $w \prec v_2$;

(ii) $w \in \dot{x}$ for every node x such that both $v_1 \in \dot{x}$ and $v_2 \in \dot{x}$;

(iii) $x \prec w$ for every node $x \notin \{w, \emptyset\}$ such that both $x \prec v_1$ and $x \prec v_2$.

Proof. We begin by assuming that $w' = v_1' \cap v_2'$, $v_1' \not\subseteq v_2'$, $v_2' \not\subseteq v_1'$, and proceed to show that (i), (ii), (iii) are satisfied. Our hypotheses readily yield that $w' \subsetneq v_1'$, $w' \subsetneq v_2'$, whence (i) follows by (B) of the lemma. When $v_1 \in \dot{x}$ and $v_2 \in \dot{x}$, we will have $x' \in v_1'$, $x' \in v_2'$ (by the very definition of a realization), so that $x' \in w'$ and hence (again by the definition of a realization) $w \in \dot{x}$. Thus (ii) is proved. When $x \neq \emptyset$ and $x \prec v_1$, $x \prec v_2$, then $u_x \subseteq v_1'$, $u_x \subseteq v_2'$; hence $u_x \subseteq w'$ and, therefore, either $u_x = u_w$ or $u_x = u_y$ for some $y \prec w$. Since $x \mapsto u_x$ is a one-to-one function, we will have $x = w$ or $x \prec w$. Thus (iii) is proved.

Let us now assume that (i), (ii), (iii), $v_1' \not\subseteq v_2'$, $v_2' \not\subseteq v_1'$ hold. We must prove that $w' = v_1' \cap v_2'$. From (i) and (B) of the lemma, we obtain $w' \subsetneq v_1'$, $w' \subsetneq v_2'$, and hence $w' \subseteq v_1' \cap v_2'$. We now consider a member x' of $v_1' \cap v_2'$.

Since $v \mapsto v'$ is a realization, $v_1 \in \dot{x}$ and $v_2 \in \dot{x}$ follow from $x' \in v_1' \cap v_2'$, so that $w \in \dot{x}$ by (ii), and therefore $x' \in w'$. To complete the proof, we must show that $a \in w'$ also holds for each element a in $v_1' \cap v_2'$ that differs from any x'. Indeed, such an a must belong to an u_x with $x \neq \emptyset$, $x \prec v_1$, $x \prec v_2$. If $x = w$, then $a \in w'$ holds trivially; otherwise $a \in w'$ holds because $x \prec w$ follows from $x \prec v_1$, $x \prec v_2$ in view of (iii).   ∎

Corollary 5.2 *Given distinct sets A_0, $A_1, \ldots, A_n$ with $n \in \omega$, $A_0 = \emptyset$, there exist*

- *a graph $G = (V, \cdot)$ with $n + 1$ nodes $v_0, \ldots, v_n$;*
- *orderings $\prec'$, $\prec''$ of V; and a*
- *subset I of V;*

such that the hypotheses of the lemma are satisfied with $\emptyset = v_0$ and with $\prec = \prec'$ or with $\prec = \prec''$. Correspondingly, one will obtain one-to-one realizations $v \mapsto v'$, $v \mapsto v''$, both satisfying the thesis of the lemma, such that

(0) v_i' (respectively, v_i'') is infinite if and only if A_i is infinite;

(1) $v_i' \in v_j'$ (respectively, $v_i'' \in v_j''$) if and only if $A_i \in A_j$;

(2a) $v_i' \subseteq v_j'$ if and only if $A_i \subseteq A_j$;

(2b) $v_i'' \subseteq v_j''$ if and only if $i = j$ or $i = 0$;

(3) $v_i' = v_j' \cap v_k'$ if $A_i = A_j \cap A_k$;

(4) $v_i'' < v_j''$ if and only if $A_i < A_j$;

(5) $v_i'' = \min v_j''$ if $A_i = \min A_j$;

for $i, j, k = 0, \ldots, n$. Here the minimum operation is referred to the ordering $<$. It is also possible to satisfy, instead of condition (0), the following condition:

(0′) the values v' (respectively, v'') of the nodes in either realization have consecutive even ranks ranging from 0 to $2 \cdot n$.

Proof. It will suffice to take as edges of G the pairs (v_i, v_j) with $A_i \in A_j$, putting also

$$
\begin{aligned}
\prec' &= \{(v_i, v_j) : A_i \subsetneq A_j\} \\
\prec'' &= \{(v_0, v_j) : j = 1, \ldots, n\} \\
I &= \begin{cases} \{v_i : \omega \leq |A_i|\} & \text{if condition (0) is to be satisfied,} \\ \emptyset & \text{if condition (0′) is to be satisfied.} \end{cases}
\end{aligned}
$$

The fact that G, so defined, is indeed a WFG was proved in Example 2.2. The hypotheses (a) through (d) of the reflection lemma are hence easily verified both for $\prec = \prec'$ and for $\prec = \prec''$. We limit ourselves to proving (c3) with

$\prec = \prec'$. For this purpose, it suffices to show that there exists no infinite chain $\ldots, A_{i_3}, A_{i_2}, A_{i_1}, A_{i_0}$ with $A_{i_{j+1}} \in A_{i_j}$ or $A_{i_{j+1}} \subsetneq A_{i_j}$ holding for each j. In view of the finiteness of n, one could in fact extract from such a chain a finite subchain $A_{i_k}, \ldots, A_{i_{j+1}}, A_{i_j}, \ldots, A_{i_{h+1}}, A_{i_h}$ with $A_{i_k} = A_{i_h}$ and $h < k$. No membership $A_{i_{j+1}} \in A_{i_j}$ can hold in this subchain, else one could easily infer $rk\, A_{i_k} < rk\, A_{i_h}$, conflicting with $A_{i_k} = A_{i_h}$. But then $A_{i_k} \subsetneq \cdots \subsetneq A_{i_{h+1}} \subsetneq A_{i_h}$, implying $A_{i_k} \subsetneq A_{i_h}$, which again conflicts with $A_{i_k} = A_{i_h}$.

The equivalences (0), (1), (2a), (2b), and (3) follow readily from the definitions of $\prec'$, $\prec''$, I, from the thesis of the lemma, and from the preceding corollary. The equivalence (4) follows easily if, in the the lemma, we take as edges of Γ the pairs (v_i, v_j) with $A_i < A_j$. Regarding the implication (5), let us observe that the general formula

$$v_j'' = u_{v_j} \cup \left(\bigcup_{v_k \prec'' v_j} u_{v_k} \right) \cup \{ v_k'' : v_j \in \dot{v}_k \},$$

holding for $j = 1, \ldots, n$, trivially simplifies—in view of the definition of $\prec''$—into $v_j'' = u_{v_j} \cup \{ v_k'' : v_j'' \in \dot{v}_k \}$. Here $v_k'' < a$ holds for any k with $v_j \in \dot{v}_k$ and for any a in u_{v_j}, as was observed in the proof of the lemma; moreover the sets v_k'' are ordered by $<$ like the corresponding A_ks.

If $(0')$ is to be satisfied instead of (0), then let σ be the permutation such that $v_{\sigma_0} \lhd v_{\sigma_1} \lhd \cdots \lhd v_{\sigma_n}$, where $\lhd$ is defined as in the proof of the lemma. Since we have taken $I = \emptyset$ in this case, the realization $v \mapsto v'$ is defined by the clauses

$$\begin{aligned}
v_{\sigma_0}' &= \emptyset, \\
d_{v_{\sigma_i}} &= \{ v_{\sigma_0}' \cup \cdots \cup v_{\sigma_{i-1}}' \}, \\
v_{\sigma_i}' &= \{ d_{v_{\sigma_i}} \} \cup \{ d_{v_{\sigma_j}} : v_{\sigma_j} \prec' v_{\sigma_i} \} \cup \{ v_{\sigma_j}' : v_{\sigma_i} \in \dot{v}_{\sigma_j} \},
\end{aligned}$$

where $i = 1, \ldots, n$ and where $v_{\sigma_j} \prec' v_{\sigma_i}$ or $v_{\sigma_i} \in \dot{v}_{\sigma_j}$ implies $j < i$. These clauses yield the values $0, 1, 2, 3, 4, \ldots, 2 \cdot n - 1, 2 \cdot n$ for the ranks of v_{σ_0}', $d_{v_{\sigma_1}}, v_{\sigma_1}', d_{v_{\sigma_2}}, v_{\sigma_2}', \ldots, d_{v_{\sigma_n}}, v_{\sigma_n}'$. The proof of $(0')$ for $v \mapsto v''$ is entirely analogous. ∎

5.3 Application of the reflection lemma to the decision problems of two elementary syllogistics

Let us consider a formula p belonging to the collection $\mathbf{F}_1$ defined in Section 5.1. From Section 5.1 we also borrow the notations:

- T for the set of all terms in $p \,\&\, \emptyset = \emptyset$;

- $\sim$ for a function $T \xrightarrow{\sim} T$ with $\tilde{\tilde{r}} = \tilde{r}$ for all r in T;

- V for the set $\{\tilde{r} \ : \ r \text{ in } T\}$.

Moreover, we put $\natural = \tilde{\emptyset}$. Note that there are finitely many possible values for $\sim$ and finitely many ways of making V into a WFG $G = (V, \cdot)$ and of choosing $\prec$, I so that the hypotheses (a) through (d) of Lemma 5.1 are satisfied. Moreover, the set of all possible choices for $\sim$, G, $\prec$, I is determinable by means of an algorithm.

A choice $\sim$, G, $\prec$, I is said to be *legal* iff $x \mapsto \tilde{x}'$ is an interpretation of p, where $'$ is constructed as in Lemma 5.1. Since we already know from Lemma 5.1 that $\natural' = \emptyset$, the choice is legal if and only if $\tilde{y}' = \tilde{x}_1' \cap \tilde{x}_2'$ for every term y of the form $x_1 \cap x_2$ in T. Let us put $w = \tilde{y}$, $v_1 = \tilde{x}_1$, $v_2 = \tilde{x}_2$. In view of Lemma 5.1 and of Corollary 5.1, the desired identity $w' = v_1' \cap v_2'$ holds if and only if either $w = v_1 = v_2$, or $w = v_1 \prec v_2$, or $w = v_2 \prec v_1$, or clauses (i),(ii),(iii) of Corollary 5.1 are satisfied. This remark clearly gives a method for testing whether a choice is legal or not.

For every legal choice, Lemma 5.1 enables us to calculate the truth value of p in the interpretation $x \mapsto \tilde{x}'$. E.g., a subformula $r = s$ of p holds iff $\tilde{r} = \tilde{s}$; a subformula $r \in s$ holds iff $\tilde{s} \in \tilde{r}$; etc.. We will declare that p is satisfiable if and only if there is a legal choice that yields the value *TRUE* for p.

To make sure that the satisfiability test described above is correct, we must prove that it will give the proper answer when p admits a model M. For this purpose, we select a representative member $\tilde{r}$ from every one of the classes $\{s \ : \ s^M = r^M\}$ with r in T, so that $\tilde{r}_1$ will coincide with $\tilde{r}_2$ if and only if $r_1^M = r_2^M$. We also take the pairs $(\tilde{r}, \tilde{s})$ with r^M in s^M as edges of G, and put: $\prec = \{(\tilde{r}, \tilde{s}) \ : \ r^M \subsetneq s^M\}$, $I = \{\tilde{r} : \omega \leq |r^M|\}$. Let us now identify in Corollary 5.2: the A_is with the distinct values r^M; $\prec$ with $\prec'$; the v_is with the representatives $\tilde{r}$, also requiring that $v_i^M = A_i$. Clause (3) of Corollary 5.2 then ensures that the choice $\sim$, G, $\prec$, I is legal, while clauses (0),(1),(2a) imply that p takes the same value, which is *TRUE*, in the interpretation $x \mapsto \tilde{x}'$ as in M. The correctness of the proposed decision test follows immediately.

An analogous satisfiability test can be designed for the collection $\mathbf{F}_2$ of formulae defined in Section 5.1. In this case, instead of choosing $\prec$, which will be invariably the relation $\{\natural\} \times (V \setminus \{\natural\})$, we directly choose the ordering $\lhd$ of V employed in the proof of Lemma 5.1. In defining the *legal* choices $\sim$, G, $\lhd$, I, we must now take into account the terms y of the form ηx in T. Every such term must satisfy either the condition $\tilde{y} = \tilde{x} = \natural$, or the condition $\tilde{x} \in \tilde{y}$ along with the requirement that $\tilde{y} \lhd z$ hold for every $z \neq \tilde{y}$ with $\tilde{x}$ in $\dot{z}$. In order to prove the correctness of the decision test one can proceed as in the preceding case, observing that the $\prec$ we have chosen is precisely the $\prec''$ of Corollary 5.2.

As we have anticipated in Section 5.1, if the relator *Finite* does not appear in p, and p belongs to either $\mathbf{F}_1$ or $\mathbf{F}_2$, then we can establish whether or not p

is satisfiable by simply evaluating p in every assignment $v \mapsto \bar{v}$ that sends each variable v of p into a set $\bar{v}$ with an even rank smaller than $2 \cdot |T| - 1$. Unless p is satisfied by one of these assignments, we will declare that p is unsatisfiable. The correctness proof regarding this simplified method is, both for $\mathbf{F}_1$ and for $\mathbf{F}_2$, analogous to the correctness proofs outlined above; with the difference, however, that in applying Corollary 5.2 we will require condition $(0')$, instead of condition (0), to be satisfied.

5.4   An improved decision algorithm for an elementary syllogistic involving η

In this section, we assume that p is an unquantified formula in which only the symbols $\emptyset$, variables, $=$, $\in$, η, and the propositional connectives, are allowed to occur. We present a decision algorithm that determines whether or not p is valid. In order to make our presentation simpler, here we avoid considering the constructs $<$, *Finite*. Except for this restriction, the algorithm that will be described is a refined variant of the decision method outlined in the preceding section.

Instead of directly testing a given formula p for validity, we bring its negation $\neg p$ to disjunctive normal form $q_0 \vee \ldots \vee q_m$ (see Definition 3.3 and Theorem 3.1), and then seek a model of some q_h. In other words, the strategy we follow is to systematically try to construct an interpretation

$$M : \{\text{variables}\} \rightarrow \{\text{sets}\}$$

in which some q_h is true. If all of our attempts fail, we conclude that p is valid; otherwise, M provides a counter-example to p.

Each q_h can be replaced by a set Q_h (here regarded as a conjunction) which is satisfiable if and only if q_h is satisfiable and which consists of literals of the forms

$$x = y, \, x \neq y, \, x \in y, \, x \notin y, \text{ and } y = \eta x,$$

where each x or y is either a variable or $\emptyset$. For this purpose, let us denote: by Ω a brand new variable; by $x_\emptyset$ the term $\emptyset$; by x_y, where y is a variable occurring in q_h, the variable y itself; and by x_t, where t is a term of the form ηs in q_h, a new variable, which is so chosen as to make the correspondence $t \mapsto x_t$ one-to-one. We first replace each outermost occurrence of a term t in q_h by the corresponding variable x_t. After this replacement, we conjoin with the resulting formula: all identities of the form $x_{\eta s} = \eta x_s$, where ηs is a term occurring in q_h; the identity $\emptyset = \eta \Omega$; all literals $x_t \in \Omega$ where t is $\emptyset$ or a term in q_h. The formula obtained in this manner is denoted Q_h.

For instance, if q_h is

$$\eta z = \eta\eta y \mathbin{\&} \emptyset \in \eta y,$$

with y, z variables, then Q_h will be

$$\{ x_{\eta z} = x_{\eta \eta y}, \; x_\emptyset \in x_{\eta y},$$

$$x_{\eta z} = \eta\, x_z, \; x_{\eta \eta y} = \eta\, x_{\eta y}, \; x_{\eta y} = \eta\, x_y,$$

$$\emptyset = \eta \Omega, \; x_\emptyset \in \Omega, \; x_y \in \Omega, \; x_z \in \Omega,$$

$$x_{\eta y} \in \Omega, \; x_{\eta z} \in \Omega, \; x_{\eta \eta y} \in \Omega \}.$$

Every model M of q_h can be extended to a model of Q_h, by simply putting $M x_t = t^M$ for every new variable x_t, and $M\Omega = \{ M x_t : t \text{ is } \emptyset \text{ or a term in } q_h \}$. Conversely, if M is a model of Q_h, then $t^M = M x_t$ for every term t in q_h; this is proved by an easy induction on $n \geq 0$, after having written t in the form $\eta^n u$, where η^n indicates n consecutive occurrences of η and u is $\emptyset$ or a variable. One concludes readily that M satisfies q_h. Hence q_h is satisfiable if and only if Q_h is satisfiable.

Let us denote by H_h the set of all those literals, in Q_h, which involve η. For each 'alternative' $A \subseteq H_h$, we indicate by Q_h^A the formula obtained from Q_h by adding, for every literal $y = \eta\, x$ in H_h:

$y \in x$, $if \; y = \eta\, x$ is in A; otherwise $x = \emptyset$ and $y = \emptyset$.

Since Q_h^A is a conjunction containing Q_h, every model of any Q_h^A is also a model of Q_h; conversely, every model M of Q_h is also a model for that Q_h^A for which

$$A = \{ y = \eta\, x \text{ in } H_h \; : \; M x \neq \emptyset \}.$$

Now we begin to describe an algorithm which will determine whether any given Q_h^A admits a model. It follows easily from the remarks just made that this can be exploited to determine whether Q_h (hence q_h, hence $\neg p$) is satisfiable. Having thus narrowed our problem, we focus on a single Q_h^A, which we denote by q; in particular, by 'variable' we mean 'variable occurring in q'.

Definition 5.1 *Let $\sim$ be an equivalence relation on the set*

$$W = \{ \text{variables} \} \cup \{ \emptyset \},$$

i.e. a binary relation on W such that for all x, y, z in W:

$$x \sim x;$$
$$x \sim y \text{ implies that } y \sim x;$$
$$x \sim y \text{ and } y \sim z \text{ implies that } x \sim z.$$

An equivalence class *(or '$\sim$-class') is a set of the form*

$$\{ x \in W \; : \; x \sim y \},$$

with y in W. A zipper *is a list*

$$
\left(
\begin{pmatrix} y_n & x_0 \\ y_0^* & x_0^* \end{pmatrix}
\ ;\
\begin{pmatrix} y_0 & x_1 \\ y_1^* & x_1^* \end{pmatrix}
,\right.
$$

$$
\left.
\begin{pmatrix} y_1 & x_2 \\ y_2^* & x_2^* \end{pmatrix}
,\ldots,\
\begin{pmatrix} y_{n-1} & x_n \\ y_n^* & x_n^* \end{pmatrix}
\right)
$$

of quadruples, such that:

$n \geq 0;$

$x_j , x_j^*, y_j , y_j^* \in W$ *and* $x_j \sim x_j^*, y_j \sim y_j^*$, *for* $j = 0, 1, \ldots, n;$

$y_n \in x_0$ *and* $y_j \in x_{j+1}$ *belong to* q, *for* $j = 0, 1, \ldots, n-1;$

$y_j^* = \eta\, x_j^*$ *belongs to* q, *for* $j = 0, 1, \ldots, n.$

A zipper is consistent *iff* $y_0 \sim y_j$ *for* $j = 1, \ldots, n$. *The equivalence relation* $\sim$ *is* consistent *iff the following two conditions are satisfied:*

(1) $u \sim v$ *whenever* $u = v$ *is a member of* $q;$

(2) every zipper is consistent.   □

A trivial example of a consistent zipper is one with $n = 0$, i.e. with only one quadruple. An equally trivial example of a consistent equivalence relation is the one in which all variables are equivalent to $\emptyset$. Less trivially:

Lemma 5.2 *Suppose that M is a model of q. Define* $u \equiv v$ *to mean that* $Mu = Mv$. *Then* $\equiv$ *is consistent.*

Proof. (1): Trivially $Mu = Mv$ when $u = v$ belongs to q. (2): In every zipper we have that $My_n \in Mx_0$, $My_0 = min\, Mx_0$, and that $My_j \in Mx_{j+1}$, $My_{j+1} = min\, Mx_{j+1}$, for $j = 0, \ldots, n-1$. It follows that

$My_0 \leq My_n$ and
$My_1 \leq My_0 , My_2 \leq My_1 , \ldots, My_n \leq My_{n-1},$

whence it follows that $My_0 = My_1 = \ldots = My_n$.   ■

Plainly, if $\sim_0$ and $\sim_1$ are consistent, so is their intersection $\sim$:

$$x \sim y \text{ iff both } x \sim_0 y \text{ and } x \sim_1 y.$$

There are finitely many variables in q, hence finitely many equivalence relations on W; if we intersect all the consistent equivalence relations together, we find the 'smallest' or the 'finest' of them all. This smallest consistent equivalence relation can, of course, be calculated if we have an algorithm that can detect whether or not an equivalence relation is consistent. The only difficulty in this connection arises from the apparent need to examine infinitely many zippers in order to establish whether an equivalence relation satisfies (2). The next two lemmas show that, as a matter of fact, it suffices to consider zippers consisting of at most $m + 1$ quadruples, where m is the total number of equivalence classes.

Lemma 5.3 *Consider a zipper*

$$
\left(\left(\begin{array}{cc} y_n & x_0 \\ y_0^* & x_0^* \end{array} \right) \; ; \; \left(\begin{array}{cc} y_0 & x_1 \\ y_1^* & x_1^* \end{array} \right) \; , \right.
$$

$$
\left. \left(\begin{array}{cc} y_1 & x_2 \\ y_2^* & x_2^* \end{array} \right) \; , \ldots, \; \left(\begin{array}{cc} y_{n-1} & x_n \\ y_n^* & x_n^* \end{array} \right) \right).
$$

If there exist h, k *with* $0 \leq h \leq k < n-1$ *for which*

$$
\left(\left(\begin{array}{cc} y_h & x_{h+1} \\ y_{h+1}^* & x_{h+1}^* \end{array} \right) \; , \ldots, \; \left(\begin{array}{cc} y_k & x_{k+1} \\ y_{k+1}^* & x_{k+1}^* \end{array} \right) \right)
$$

is a zipper, then, after removing this 'inner' zipper from the first zipper, we get an 'outer' zipper

$$
\left(\left(\begin{array}{cc} y_n & x_0 \\ y_0^* & x_0^* \end{array} \right) \; ; \; \left(\begin{array}{cc} y_0 & x_1 \\ y_1^* & x_1^* \end{array} \right) \; , \ldots, \; \left(\begin{array}{cc} y_{h-1} & x_h \\ y_h^* & x_h^* \end{array} \right) \; ; \right.
$$

$$
\left. \left(\begin{array}{cc} y_{k+1} & x_{k+2} \\ y_{k+2}^* & x_{k+2}^* \end{array} \right) \; , \ldots, \; \left(\begin{array}{cc} y_{n-1} & x_n \\ y_n^* & x_n^* \end{array} \right) \right).
$$

The original zipper is consistent if and only if both the inner and the outer zipper are consistent.

Proof. Concerning the first claim, that the third list of quadruples is a zipper, it suffices to show that $y_h^* \sim y_{k+1}$. Indeed, since the first two lists are zippers, we have $y_h^* \sim y_h \sim y_{k+1}^* \sim y_{k+1}$.

Trivially if the original zipper is consistent, so are the inner and the outer zipper. Conversely, if the inner zipper and the outer zipper are consistent, then

$$
y_h \sim \ldots \sim y_k \text{ and}
$$
$$
y_n \sim y_0 \sim \ldots \sim y_{h-1},
$$
$$
y_n \sim y_{k+1} \sim \ldots \sim y_{n-1}.
$$

Moreover, as we have already noticed,

$$
y_h \sim y_{k+1} ;
$$

this gives the necessary link to conclude that

$$
y_0 \sim \ldots \sim y_n ,
$$

i.e. the original zipper is consistent.   ∎

Lemma 5.4 *Every zipper whose length (i.e. number of quadruples) exceeds* $m + 1$*, where* m *is the total number of equivalence classes, contains an inner zipper.*

Proof. Consider a zipper

$$
\left(
\begin{pmatrix} y_n & x_0 \\ y_0^* & x_0^* \end{pmatrix} ;
\begin{pmatrix} y_0 & x_1 \\ y_1^* & x_1^* \end{pmatrix} , \ldots ,
\begin{pmatrix} y_{n-1} & x_n \\ y_n^* & x_n^* \end{pmatrix}
\right) ,
$$

with $m < n$. We assume that this zipper contains no inner zipper, i.e. that for $h = 0, 1, \ldots, n-2$ there exists no $i = 0, \ldots, n-2-h$ for which

$$
Z_{i,h} = \left(
\begin{pmatrix} y_i & x_{i+1} \\ y_{i+1}^* & x_{i+1}^* \end{pmatrix} , \ldots ,
\begin{pmatrix} y_{i+h} & x_{i+h+1} \\ y_{i+h+1}^* & x_{i+h+1}^* \end{pmatrix}
\right)
$$

is a zipper. By induction on h we will prove that in every list of this form the symbols y_i, $y_{i+1}^*, \ldots, y_{i+h+1}^*$ belong to distinct equivalence classes. However this leads to an absurd conclusion, because when $h = n-2$ (and hence $i = 0$) we get the sequence y_0, $y_1^*, \ldots, y_{n-1}^*$, which has more than m components, whereas we know that there are only m equivalence classes available.

For $h = 0$, the induction hypothesis follows from the remark that

$$
Z_{i,0} = \left(
\begin{pmatrix} y_i & x_{i+1} \\ y_{i+1}^* & x_{i+1}^* \end{pmatrix}
\right)
$$

is a zipper if and only if $y_i \sim y_{i+1}^*$; but we have assumed that this is never the case, so that $y_i \not\sim y_{i+1}^*$ for $i = 0, \ldots, n-2$.

For $h > 0$, on the one hand we are guaranteed by the induction hypothesis that the symbols y_i, $y_{i+1}^*, \ldots, y_{i+h}^*$ belong to distinct equivalence classes. On the other hand, we cannot have $y_{i+h+1}^* \sim y_i$, or the entire $Z_{i,h}$ would be a zipper; nor can we have $y_{i+h+1}^* \sim y_{i+j}^*$ with $0 < j \leq h$, because $y_{i+j}^* \sim y_{i+j}$ and hence $Z_{i+j,h-j}$ would be a zipper. ■

Corollary 5.3 *An equivalence relation is consistent if and only if*

   (1) $u \sim v$ *whenever* $u = v$ *is a member of* q;

   (2) every zipper whose length does not exceed $m + 1$, *where* m *is the total number of equivalence classes, is consistent.*

Proof. The 'only if' part of the thesis is trivial. Conversely, we prove that if (2) holds, then every zipper is consistent. In fact, if a zipper Z has length exceeding $m + 1$, then, by Lemma 5.4 and Lemma 5.3, it can be subdivided into an inner zipper Z_1 and an outer zipper Z_0, so that Z is consistent if and only if Z_0 and Z_1 are both consistent. If either Z_0 or Z_1 has length exceeding $m + 1$, we can split it again, and so on. At the conclusion of this process, we obtain a finite set $\mathcal{Z}$ of zippers, each having length less than or equal to $m+1$, such that Z is consistent if and only if each of the zippers in $\mathcal{Z}$ is consistent. Indeed these are consistent, by (2), and hence Z is consistent. ■

Algorithm 5.1 *This algorithm tests a given unquantified formula q for satisfiability: q is assumed to be a set (here identified with a conjunction) of literals of the forms*

$$x = y, \; x \neq y, \; x \in y, \; x \notin y, \; \text{and } y = \eta x,$$

where each x or y is either a variable or $\emptyset$. We moreover assume that for every literal $y = \eta x$ in q, either the literal $y \in x$ or the pair $x = \emptyset, y = \emptyset$ of literals belongs to q. Finally, we assume that there is a variable Ω such that the literals involving Ω, in q, are: $\emptyset \in \Omega$, $\emptyset = \eta \Omega$, and $x \in \Omega$, with x ranging over all variables distinct from Ω in q.

Step 1. *Calculate the intersection $\sim$ of all consistent equivalence relations on $W = \{$ variables appearing in $q \} \cup \{\emptyset\}$.*

Step 2. *Choose a representative member $\tilde{v}$ from within each equivalence class $\{u : u \sim v\}$; in particular let $\tilde{\emptyset}$ be $\emptyset$ and let $\tilde{\Omega}$ be Ω. Obtain a 'simplified' form $\tilde{q}$ of q, by replacing each variable v by $\tilde{v}$ in q.*

Step 3. *Check that $\tilde{q}$ contains no manifest contradiction of either the form*

(i)   *$u \neq u$; or the form*

(ii)   *$u \in v, u \notin v$.*

Step 4. *Check that there exists an ordering $\lhd$ of*

$$V = \{ \text{ variables appearing in } \tilde{q} \} \cup \{\emptyset\}$$

such that:

(a)   *$\emptyset$ is minimal in the ordering;*

(b)   *$v \lhd u$ whenever $v \in u$ is in $\tilde{q}$; and*

(c)   *$y \lhd u$ whenever:*

     y is distinct from u; and, for some x in V, the two literals

$$u \in x, \; y = \eta x$$

   belong to $\tilde{q}$.

Step 5. *If both **Step 3** and **Step 4** have been successful (in the sense that no manifest contradiction was found, but an ordering $\lhd$ was found), declare that q is satisfiable; otherwise declare that q is unsatisfiable.* $\square$

Note in the first place that this algorithm is total, because it always answers the question whether q is satisfiable or not: this is true, in part, because **Step 1** can actually be performed in a finite amount of time, as a consequence of Corollary 5.3. The following two lemmas prove the correctness of the algorithm.

Lemma 5.5 (Soundness) *Algorithm 5.1 is sound.*

Proof. Concerning **Step 2**, we note that every model M of q is also a model of $\tilde{q}$. In fact, since the equivalence relation:

$$u \equiv v \text{ iff } Mu = Mv,$$

defined for u, v in W, is consistent by Lemma 5.2, we have that

$$Mu = Mv \text{ whenever } u \sim v,$$

hence that

$$Mv = M\tilde{v} \text{ for all } v \text{ in } W,$$

and therefore that $\tilde{q}^M = q^M = TRUE$. Since the models of q are models of $\tilde{q}$ too, q cannot have any model if **Step 3** is unsuccessful, i.e. if $\tilde{q}$ is manifestly contradictory.

Concerning **Step 4**, assuming that q has a model M, we must prove that there exists an ordering $\lhd$ of V which satisfies (a), (b), and (c). For v, u in V, we define $v \lhd u$ to mean that for some n in ω there are $y_1, \ldots, y_n$ in V such that

$$v \lhd_0 y_1 \lhd_0 \ldots \lhd_0 y_n \lhd_0 u,$$

where $y \lhd_0 z$ means that either $My < Mz$, or :

$My = Mz$, but y and z are distinct and, for some x in V, both $z \in x$ and $y = \eta x$ belong to $\tilde{q}$.

Here $<$ is the usual well-ordering of all sets.

Trivially $\lhd$ is transitive; in order to prove that $\lhd$ is a (partial) ordering it will hence suffice to prove that there is no list $y_0, y_1, \ldots, y_n$ with $0 < n$ such that:

$$y_0 \lhd_0 y_1 \lhd_0 \ldots \lhd_0 y_n \text{ and } y_0 \text{ is } y_n.$$

Assuming, to the contrary, that such a list exists, we note that

$$My_j \leq My_{j+1} \quad for \ j = 0, \ldots, n-1,$$

but $My_j < My_{j+1}$ can never be the case, or it would be impossible to have $My_0 = My_n$. Thus, for $j = 0, \ldots, n-1$:

$My_j = My_{j+1}$ but y_j and y_{j+1} are distinct (hence, $1 < n$) and, for some x_j in V, both $y_{j+1} \in x_j$ and $y_j = \eta x_j$ belong to $\tilde{q}$.

Since y_n is the same as y_0, the list

$$\left(\begin{pmatrix} y_n & x_{n-1} \\ y_{n-1} & x_{n-1} \end{pmatrix}, \begin{pmatrix} y_{n-1} & x_{n-2} \\ y_{n-2} & x_{n-2} \end{pmatrix}, \ldots, \begin{pmatrix} y_1 & x_0 \\ y_0 & x_0 \end{pmatrix} \right)$$

originates from a zipper, via the substitution $z \mapsto \tilde{z}$. Since $\sim$ is consistent, it follows that $y_0, \ldots, y_{n-1}$ are all the same, which contradicts the fact that y_0 and y_1 must be distinct.

Now we show that:

(a) holds. In fact, $\emptyset \leq My$ for every y in V; hence we can only have $y \lhd_0 \emptyset$ if, for some x in V, the two literals $\emptyset \in x$, $y = \eta x$ belong to $\tilde{q}$. However, if this is the case, then the list

$$\left(\begin{pmatrix} \emptyset & x \\ y & x \end{pmatrix} , \begin{pmatrix} y & \Omega \\ \emptyset & \Omega \end{pmatrix} \right)$$

originates from a zipper via the substitution $z \mapsto \tilde{z}$; this implies that y and $\emptyset$ are the same, because $\sim$ is consistent. Hence we cannot have $y \lhd_0 \emptyset$ nor, as a consequence, can we have $y \lhd \emptyset$.

(b) holds, since M is a model of $\tilde{q}$ and hence $v \in u$ in $\tilde{q}$ implies $Mv \in Mu$, so that $Mv < Mw$, $v \lhd_0 w$, and $v \lhd w$.

(c) holds, because if y and u are distinct, and $u \in x$, $y = \eta x$ are in $\tilde{q}$, then $Mu \in Mx$ and $My = \min Mx$, and hence $My \leq Mu$. But, either $My < Mu$ or $My = Mu$, imply that $y \lhd_0 u$, by definition; hence $y \lhd u$. ∎

Lemma 5.6 (Completeness) *Algorithm 5.1 is complete. More specifically, when the algorithm declares that q is satisfiable, there exists a model M of q in which non-equivalent variables take distinct values, and in which*

$$rk\, Mx < 2 \cdot |V| - 1$$

for every variable x present in q.

Proof. Assume that the algorithm declares that q is satisfiable. Without loss of generality, we can assume that the ordering found by the algorithm is total, because every partial ordering of a finite set can be extended to a total ordering of the same set, using, e.g. a 'topological sort' algorithm [Knu75]. Moreover, the nature of the conditions (b), (c) imposed on $\lhd$ is such that their truth is automatically preserved in any extension. We can also easily manage to preserve (a), so that $\emptyset$ will be the first member of V. We can therefore make a single list

$$u_0 \lhd u_1 \lhd \ldots \lhd u_k$$

of all members of V, where u_0 is $\emptyset$ and $u_1, \ldots, u_k$ are variables. Recall that $k + 1 = |V|$ is the total number of $\sim$-classes.

We define G to be the well-founded graph $(V, \dot{\ })$ whose edges are the pairs (x, y) with $x \in y$ in $\tilde{q}$. Thus the hypotheses of Lemma 5.1 are satisfied, if we identify: $\mathbb{Q}$ with the symbol $\emptyset$, I with the empty set, $\prec$ with the set

$\{(u_0, u_j) : j = 1, \ldots, k\}$, the edges of Γ with the pairs (u_i, u_j) such that $i < j$. Reasoning in analogy with Corollary 5.2(0'),(1),(4),(5) one easily deduces that there is a function $v \mapsto v'$ from V to sets which satisfies the conditions:

$\emptyset' = \emptyset$, and, more generally, $rk\, u'_i = 2 \cdot i$;

$u'_i \in u'_j$ if and only if $u_i \in u_j$ belongs to $\tilde{q}$;

$u'_i < u'_j$ if and only if $i < j$;

$u'_i = min\, u'_j$ if and only if both $u'_i \in u'_j$ and $i \le h$
for every h such that $u'_h \in u'_j$;

for $i, j = 0, 1, \ldots, n$.

We now prove that by putting $Mx = u'_j$, where u_j is $\tilde{x}$, for every variable x of q, one obtains a model M of q. In the first place, M models $\emptyset$ correctly, since

$$M\emptyset = \emptyset' = \emptyset.$$

If $x = y$ is in q, then $x \sim y$, because $\sim$ is consistent. Assuming that $x \sim u_j$, we have $Mx = My = u'_j$. If $x \ne y$ is in q, then $x \not\sim y$, otherwise $\tilde{x} \ne \tilde{x}$ would be a manifest contradiction in $\tilde{q}$, and we have assumed that no such contradiction exists; as a consequence, $rk\, Mx \ne rk\, My$.

If $x \in y$ is in q, then $\tilde{x} \in \tilde{y}$ is in $\tilde{q}$; therefore $\tilde{x} \lhd \tilde{y}$, by hypothesis (b), and $Mx = u'_i \in u'_{j+1} = My$ for suitable i, j. If $x \not\in y$ is in q, then $\tilde{x} \not\in \tilde{y}$ is in $\tilde{q}$ and hence $\tilde{x} \in \tilde{y}$ is not in $\tilde{q}$, or we would have a manifest contradiction. As a consequence, $Mx = \tilde{x}' \not\in \tilde{y}' = My$.

Finally, let us suppose that $y = \eta x$ is in q. Then we can exploit the assumption that either the literal $y \in x$ or the pair $x = \emptyset$, $y = \emptyset$ of literals belongs to q. Accordingly, as we have already proved, $My \in Mx$ or $Mx = My = \emptyset$. We must show that in the first case $My = min\, Mx$. Assume that $y \sim u_i$, $x \sim u_{j+1}$, so that clearly $u_i \in u_{j+1}$, $u_i = \eta\, u_{j+1}$ belong to $\tilde{q}$, and $i \le j$, by hypothesis (b). Recalling that $My = u'_i$, $Mx = u'_{j+1}$, we must simply prove that $i < h$ for all $h \ne i$ such that $u_h \in u_{j+1}$ is in $\tilde{q}$. This is in fact an obvious consequence of hypothesis (c). ∎

5.5 An improved decision algorithm for the elementary syllogistic involving $\subseteq$, $\cap$, and $Finite$

In what follows, q^* is a set (here seen as a conjunction) of literals of the forms

$$x \subseteq y, \; x \not\subseteq y, \; x = y, \; x \ne y, \; x \in y, \; x \not\in y,$$
$$Finite\, x, \; \neg Finite\, x, \; z = x \cap y,$$

where each x, y, z is either the constant $\emptyset$ or a variable. By q_* we denote the largest subset of q^* that contains only inclusions and identities, which is the set (conjunction) consisting of all literals

$$x \subseteq y, \; x = y, \; z = x \cap y$$

that belong to q^*. For simplicity, we can assume that q^* and q_* involve the same variables (if needed, we could in fact add identities $x = x$ in both q^* and q_*): the set of symbols that consists of these variables together with the symbol $\emptyset$ is indicated by Z. Finally, we assume that the literal $Finite\,\emptyset$ belongs to q^*.

Definition 5.2 *The characteristic function of a subset W of Z is that function χ_W, defined for all x in Z, which sends into $\emptyset$ all members of $Z \setminus W$ and sends into $\{\,\emptyset\,\}$ all members of W. If χ_W is a model of q_*, then W is said to be a* place *(of q_*) : in particular we must have, for a place W,*

$$\chi_W\,\emptyset \ = \ \emptyset\,, \ \ i.e. \ \emptyset \ is \ not \ in \ W\,.$$

For x, y in Z we define:

$$
\begin{aligned}
x \preceq y \quad &iff \quad \chi_W\,x \subseteq \chi_W\,y \ in \ every \ place \ W\,, \ i.e.\\
&iff \quad y \ belongs \ to \ every \ place \ to \ which \ x \ belongs;\\
x \sim y \quad &iff \quad both \ x \preceq y \ and \ y \preceq x\,, \ i.e.\\
&iff \quad x \ and \ y \ belong \ to \ the \ same \ places.
\end{aligned}
$$

Suppose that u is an $\sim$-class. We say that W is a place *(of q^*) at u iff W is a place and, moreover, $y \in W$ (respectively, $y \notin W$) whenever there is a literal $x \in y$ (respectively, $x \notin y$) in q^*, with x in u.*

We say that W is an infinity placement *(of q^*) iff $y \in W$ (respectively, $y \notin W$) whenever the literal $\neg\,Finite\,y$ (respectively, $Finite\,y$) belongs to q^*, and moreover $y \in W$, $y \preceq z$ implies $z \in W$ for all y, z in Z.* $\qquad\square$

Remarks 5.1 *(A)* A trivial example of a place is the empty set of symbols. In fact, since $\chi_\emptyset\,v = \emptyset$ for every v in Z, one has readily that $\chi_\emptyset\,\emptyset = \emptyset$, and that $\chi_\emptyset\,x \subseteq \chi_\emptyset\,y$, $\chi_\emptyset\,x = \chi_\emptyset\,y$, $\chi_\emptyset\,z = \chi_\emptyset\,x \cap \chi_\emptyset\,y$, for all x, y, z in Z.

(B) The definition of *place at u* makes sense, because one immediately recognizes that $\sim$ is an equivalence relation, so that Z can be subdivided into pairwise disjoint non-empty equivalence classes.

(C) Although $\preceq$ is transitive and our notation seems to indicate that one can define a partial ordering $\prec$ by putting

$$x \prec y \ \text{iff} \ x \preceq y \ \text{and} \ x \neq y\,,$$

this is not the case. As a matter of fact, sometimes x and y belong to the same places (e.g., if the literal $x = y$ belongs to q_* —see Lemma 5.9(2) below), but they are distinct, so that $x \prec y \prec x$ while $x \not\prec x$. However it will follow from Lemma 5.7(4) below that $\preceq$ induces a partial ordering between $\sim$-classes, by the definition

$$
\begin{aligned}
u \prec v \ \text{iff:} \ &u \neq v \ \text{and there are an } x \text{ in } u \text{ and}\\
&\text{a } y \text{ in } v \text{ for which } x \preceq y,
\end{aligned}
$$

where u, v are $\sim$-classes. One easily recognizes that the $\sim$-class to which $\emptyset$ belongs precedes every other $\sim$-class in this induced ordering. $\qquad\square$

Lemma 5.7 *Suppose that M is a model of q_*, and that x, y, x^*, $y^* \in Z$. Then:*

(1) *for every set s, the subset $\ddot{s} = \{ z \in Z : s \in Mz \}$ of Z is a place;*

(2) *if $x \preceq y$, then $Mx \subseteq My$;*

(3) *if $x \sim y$, then $Mx = My$;*

(4) *if $x \sim x^*$, $y \sim y^*$, and $x \preceq y$, then $x^* \preceq y^*$;*

(5) *if X is the $\sim$-class of x, and W is a place or an infinity placement, then $x \in W$ if and only if $X \subseteq W$.*

Proof. Trivially, (3) follows from (2). Moreover, (2) follows from (1), because when $x \preceq y$ and $s \in Mx$, then $x \in \ddot{s}$ and hence $y \in \ddot{s}$, i.e. $s \in My$. Concerning (4): from the assumptions $x \sim x^*$, $y \sim y^*$, and $x \preceq y$, it follows that

$$\chi_W \, x^* = \chi_W \, x \subseteq \chi_W \, y = \chi_W \, y^*,$$

hence $\chi_W \, x^* \subseteq \chi_W \, y^*$, in every place W; i.e. $x^* \preceq y^*$.

Concerning (5), assume that $x \in W$. If $y \in X$, then $x \sim y$; therefore $\chi_W \, y = \chi_W \, x = \{ \emptyset \}$, hence $y \in W$. Since y is an *arbitrary* member of X, it follows that $X \subseteq W$. Conversely, if $X \subseteq W$, then $x \in W$, because $x \in X$.

Concerning (1), we must prove that

(a) if $x \subseteq y$ is a literal in q_* and $x \in \ddot{s}$, then $y \in \ddot{s}$;

(b) if either $x = y$ or $y = x$ belongs to q_*, and $x \in \ddot{s}$, then $y \in \ddot{s}$;

(c) if $z = x \cap y$ belongs to q_*, then

    (c′) if $x \in \ddot{s}$ and $y \in s$, then $z \in \ddot{s}$;

    (c″) if $x \notin \ddot{s}$ or $y \notin \ddot{s}$, then $z \notin \ddot{s}$.

Concerning (a): from the fact that $Mx \subseteq My$, it follows that $s \in Mx$ implies $s \in My$. Concerning (b): from the fact that $Mx = My$, it follows that $s \in Mx$ implies $s \in My$. Concerning (c): from the fact that $Mz = Mx \cap My$ it follows that

$$\text{if } s \in Mx \text{ and } s \in My, \text{ then } s \in Mz;$$
$$\text{if } s \notin Mx \text{ or } s \notin My, \text{ then } s \notin Mz.$$

$\blacksquare$

Corollary 5.4 *If M is a model of the entire q^*, u is an $\sim$-class, and s is the common image, via M, of all symbols that belong to u, then $\ddot{s}$ is a place at u. The set of all variables v for which Mv is infinite is an infinity placement.*

Proof. By (1), $\ddot{s}$ is a place. Moreover, trivially, if $x \in y$ is a literal in q^* with x in u, then we have $s = Mx \in My$, i.e. $y \in \ddot{s}$. Likewise, if $x \notin y$ is in q^* and x is in u, then $s = Mx \notin My$, i.e. $y \notin \ddot{s}$. $\blacksquare$

Lemma 5.8 *Let I be the intersection $\bigcap_{W \in F} W$ of a non-empty family F of subsets of Z. If every W in F is a place, then I is a place. If every W in F is a place at u, then I is a place at u. If every W in F is an infinity placement, then I is an infinity placement.*

Proof. The proof is entirely straightforward, and is left to the reader. Concerning in particular the first claim in the lemma, the proof can be modelled after the proof of Lemma 5.7(1). $\blacksquare$

Lemma 5.9 *Let x, y, $z \in Z$. Then:*

(1) if $x \subseteq y$ belongs to q^*, then $x \preceq y$;

(2) if $x = y$ belongs to q^*, then $x \sim y$;

(3) if $z = x \cap y$ belongs to q^*, then: $z \preceq x$, $z \preceq y$, and z belongs to every place to which both x and y belong.

Furthermore, if X, Y are the respective $\sim$-classes of x and y, and $\overline{X}$ is a place at X, then:

(4) if $x \in y$ belongs to q^*, then $Y \subseteq \overline{X}$;

(5) if $x \notin y$ belongs to q^*, then $Y \not\subseteq \overline{X}$.

Finally, if Y is the $\sim$-class of y, and W is an infinity placement, then

(6) if $\neg$ Finite y belongs to q^*, then $Y \subseteq W$;

(7) if Finite y belongs to q^*, then $Y \cap W \neq \emptyset$.

Proof. (1): If $x \subseteq y$ belongs to q^*, then $x \subseteq y$ also belongs to q_*, and hence $Mx \subseteq My$ in every model M of q_*. In particular $\chi_W x \subseteq \chi_W y$ in every place W, i.e. $x \preceq y$.   (2): If $x = y$ belongs to q^*, then $Mx = My$ in every model M of q_*. In particular $\chi_W x = \chi_W y$ in every place W, i.e. $x \sim y$.   (3): If $z = x \cap y$ belongs to q^*, then $Mz = Mx \cap My$ in every model M of q_*. In particular $\chi_W z = \chi_W x \cap \chi_W y$, hence $\chi_W z \subseteq \chi_W x$ and $\chi_W z \subseteq \chi_W y$ in every place W. Thus $z \preceq x$ and $z \preceq y$. If both x and y belong to the place W, then $\chi_W z = \chi_W x \cap \chi_W y = \{\emptyset\} \cap \{\emptyset\} = \{\emptyset\}$, so that $z \in W$.   (4), (5): If $x \in y$ (respectively, $x \notin y$) belongs to q^*, then, by the definition of place at X, we have $y \in \overline{X}$ (respectively, $y \notin \overline{X}$); this implies $Y \subseteq \overline{X}$ (respectively, $Y \not\subseteq \overline{X}$), by Lemma 5.7(5).   (6): If $\neg$ Finite y belongs to q^*, then $y \in W$, and therefore $Y \subseteq W$ by Lemma 5.7(5).   (7): If $z \in Y \cap W$, then $y \in W$, since $z \preceq y$; hence Finite y does not belong to q^*, by the definition of an infinity placement. $\blacksquare$

Lemma 5.10 *Suppose that there is a model M of q^*, so that there is at least one place at each $\sim$-class (see Corollary 5.4). For every $\sim$-class v, let $\overline{v}$ be the intersection of all places at v (which is itself a place at v, by Lemma 5.8); moreover, let $\dot{v} = \{\, u : u \subseteq \overline{v}\,\}$. Then there exists an ordering $\lhd$ of the $\sim$-classes, such that:*

(1) if $u \in \dot{v}$, then $v \lhd u$;

(2) if $v \prec u$, then $v \lhd u$;

(3) $\emptyset \!\!\!\backslash \lhd u$ or $\emptyset \!\!\!\backslash = u$;

for every pair u, v of $\sim$-classes. Here $\emptyset \!\!\!\backslash$ denotes the $\sim$-class to which the symbol $\emptyset$ belongs, and $\prec$ denotes the partial ordering among $\sim$-classes that is induced by $\preceq$ (see Remark 5.1(C) after Definition 5.1).

Proof. Throughout this proof, u and v indicate $\sim$-classes. For every $\sim$-class w, we denote by w_M the common image, via M, of all symbols in w (see Lemma 5.7(3)). We put

$$v \lhd u \text{ iff either } rk\, v_M < rk\, u_M \text{ or } v \prec u,$$

so that trivially $u \not\lhd u$.

In order to prove that $w \lhd v \lhd u$ implies $w \lhd u$, note first that $v \prec u$ implies $v_M \subseteq u_M$ by Lemma 5.7(2), hence $v \prec u$ implies that $rk\, v_M \leq rk\, u_M$, and $v \lhd u$ implies that $rk\, v_M \leq rk\, u_M$. Therefore, if we assume that $w \lhd v \lhd u$, then we have $rk\, w_M \leq rk\, v_M \leq rk\, u_M$. If either $rk\, w_M < rk\, v_M$ or $rk\, v_M < rk\, u_M$, then $rk\, w_M < rk\, u_M$, so that $w \lhd u$. If $rk\, w_M = rk\, v_M = rk\, u_M$, then $w \prec v \prec u$ follows from $w \lhd v \lhd u$; hence we have $w \prec u$ and $w \lhd u$. We conclude that $\lhd$ is a partial ordering.

Condition (2) follows trivially from the definition of $\lhd$; condition (3) readily follows from (2), because $\emptyset \!\!\!\backslash \prec u$ for every $\sim$-class $u \neq \emptyset \!\!\!\backslash$. Concerning (1), assume that $u \in \dot{v}$, i.e. $u \subseteq \overline{v}$. Since $\ddot{v}_M = \{\, z : v_M \in Mz \,\}$ is a place at v (see Corollary 5.4), and since $\overline{v}$ is the intersection of all places at v, we have that $\overline{v} \subseteq \ddot{v}_M$, hence $u \subseteq \ddot{v}_M$, i.e. $v_M \in u_M$. Therefore $rk\, v_M < rk\, u_M$, and hence $v \lhd u$. ∎

Algorithm 5.2 *A conjunction q^* of literals of the forms*

$$x \subseteq y,\ x \not\subseteq y,\ x = y,\ x \neq y,\ x \in y,\ x \notin y,$$
$$Finite\, x,\ \neg\, Finite\, x,\ z = x \cap y,$$

where each x, y, z is either $\emptyset$ or a variable, is to be tested for satisfiability. Without loss of generality, we assume that the literal $Finite\, \emptyset$ belongs to q^*.

Step 1. *Calculate the relations $\preceq$ and $\sim$, and the partial ordering $\prec$ among $\sim$-classes that is induced by $\preceq$ (see Definition 5.2 and the Remark 5.1(C)* *that follows it). For each $\sim$-class u, check that there exists a place at u, and, if this is the case, calculate the intersection $\overline{u}$ of all places at u and put $\dot{u} = \{ v : v \subseteq \overline{u} \}$. Check that there exists an infinity placement of q^*, and, if this is the case, calculate the intersection J of all infinity placements of q^* and put $I = \{ u \subseteq J : u$ is an $\sim$-class $\}$.*

Step 2. *Check that q^* contains no manifest contradiction of the form*

$$x \not\subseteq y \ \ with \ \ x \preceq y,$$

or of the form

$$x \neq y \ \ with \ \ x \sim y.$$

Step 3. *Check that there exists an ordering $\lhd$ of the $\sim$-classes such that $u \lhd v$ whenever either $u \prec v$ or $v \in \dot{u}$.*

Step 4. *According to whether or not **Steps 1, 2,** and **3** have been all successful, declare that q^* is satisfiable or unsatisfiable.* $\square$

The soundness of **Steps 1, 2,** and **3** follows from Corollary 5.4, Lemma 5.7(2),(3), and Lemma 5.10 respectively.

Lemma 5.11 (Completeness) *Algorithm 5.2 is complete. More specifically, if the algorithm declares that q^* is satisfiable, then q^* has a model M in which:*

$$Mx \subseteq My \ implies \ that \ x \preceq y, \ and \ hence$$
$$Mx = My \ implies \ that \ x \sim y; \ and \ moreover$$
$$\omega \leq |Mx| \ implies \ x \in J;$$

*for all x, y in Z. (Note: J is as in **Step 1** of Algorithm 5.2).*

Proof. We assume that the algorithm declares that q^* is satisfiable, and begin the proof by showing that a well-founded graph $G = (V, \ ^{\cdot})$ satisfying the hypotheses of Lemma 5.1 can be obtained by adopting:

- the $\sim$-classes as nodes of G,

- the pairs (u, v) with v in $\dot{u}$ (see **Step 1**) as edges of G,

and by also letting $\emptyset$ be the $\sim$-class of $\emptyset$. G is indeed a WFG, as an immediate consequence of the fact that **Step 3** has been successful; moreover, arguing as in the proof of Lemma 5.10(3), $\emptyset$ is the minimum of $\lhd$. Only conditions (c2) and (d) of Lemma 5.1 now remain to be verified.

By Lemma 5.8, every $\overline{u}$ is a place at u; therefore, if we assume $w \in \dot{u}$, i.e. $w \subseteq \overline{u}$, and also $w \prec v$, then we easily obtain $v \in \dot{u}$, which proves (c2) of Lemma 5.1. A similar reasoning proves (d): J is an infinity placement in view of Lemma 5.8, because J is obtained by intersecting together some infinity placements. Moreover $\emptyset \notin I$ by the definition of infinity placements and by Lemma 5.9(7), since we have put the literal $Finite\,\emptyset$ into q^*. Finally, taking Lemma 5.7(5) into account one easily derives (d1) from the definition of infinity placements.

Now define, for every x in Z: $Mx = [x]'$, where $[x]$ is the $\sim$-class to which x belongs. Since the hypotheses of Lemma 5.1 are satisfied by G, that lemma gives:

$$M\emptyset = \emptyset' = \emptyset;$$
$$Mx \subseteq My \text{ if and only if } x \preceq y;$$
$$Mx = My \text{ if and only if } x \sim y;$$
$$Mx \in My \text{ if and only if } [y] \subseteq \overline{[x]};$$
$$\omega \leq |Mx| \text{ if and only if } [x] \subseteq J;$$

for all x, y in Z. Moreover:

If $x \subseteq y$ belongs to q^*, then $x \preceq y$ by Lemma 5.9(1), and hence $Mx \subseteq My$;

if $x \not\subseteq y$ belongs to q^*, then $x \not\preceq y$ because **Step 2** has been successful, and hence $Mx \not\subseteq My$;

if $x = y$ belongs to q^*, then $x \sim y$ by Lemma 5.9(2), and hence $Mx = My$;

if $x \neq y$ belongs to q^*, then $x \not\sim y$ because **Step 2** has been successful, and hence $Mx \neq My$;

if $x \in y$ belongs to q^*, then $[y] \subseteq \overline{[x]}$ by Lemma 5.9(4), and hence $Mx \in My$;

if $x \notin y$ belongs to q^*, then $[y] \not\subseteq \overline{[x]}$ by Lemma 5.9(5), and hence $Mx \notin My$;

if $\neg\,Finite\,x$ belongs to q^*, then $[x] \subseteq J$ by Lemma 5.9(6), and hence Mx is infinite;

if $Finite\,x$ belongs to q^*, then $[x] \not\subseteq J$ by Lemma 5.9(7), and hence Mx is finite.

In order to complete the proof that M is a model of q^*, it remains to be shown that $Mz = Mx_1 \cap Mx_2$, assuming that $z = x_1 \cap x_2$ belongs to q^*. In view of Corollary 5.1, this amounts to proving that either $w = v_1 = v_2$ or $w = v_1 \prec v_2$ or $w = v_2 \prec v_1$ or the following three conditions are satisfied:

(i) $w \prec v_1,\, w \prec v_2$;

(ii) $w \subseteq \overline{u}$ for every $\sim$-class u such that both $v_1 \subseteq \overline{u}$ and $v_2 \subseteq \overline{u}$;

(iii) $u \prec w$ for every $\sim$-class $u \notin \{w, \emptyset\}$ such that both $u \prec v_1$ and $u \prec v_2$.

where $w = [z]$, $v_1 = [x_1]$, and $v_2 = [x_2]$. Recall that, by Lemma 5.9(3): $z \preceq x_1,\, z \preceq x_2$, and z belongs to every place of q^* to which both x_1 and x_2 belong. Hence $z \sim x_1$ implies $x_1 \preceq x_2$, and $z \sim x_2$ implies $x_2 \preceq x_1$; and therefore either $w = v_1 = v_2$ or $w = v_1 \prec v_2$ or $w = v_2 \prec v_1$ follows from the assumption $z \sim x_1$ or $z \sim x_2$. If we instead assume that $z \not\sim x_1$, $z \not\sim x_2$, we readily obtain (i). In this case, by Lemma 5.7(5), (ii) simplifies into

(ii') $z \subseteq \bar{u}$ for every place of the form $\bar{u}$ to which both x_1 and x_2 belong,

which is obvious. Moreover, by the definition of $\prec$ among $\sim$-classes, (iii) simplifies into

(iii') $y \preceq z$ for every y with $y \not\sim z$, $y \not\sim x_1$, $y \not\sim x_2$ such that both $y \preceq x_1$ and $y \preceq x_2$.

In order to prove this, suppose that W is a place to which y belongs. Under the premises of (iii'), we have $x_1, x_2 \in W$, and hence $z \in W$.   ∎

5.6   A collection of safe formulae for which finite satisfiability is decidable

Lemma 5.12 *Let $\mathcal{A}$, $\mathcal{B}$ be families such that*

$$|\mathcal{B} \cup Un\,\mathcal{A}| < \omega, \quad max\,\mathcal{B} \leq max\,\mathcal{A}.$$

Then there exists a function

$$^*: \mathcal{A} \cup \mathcal{B} \longrightarrow V_{2 \cdot |\mathcal{A}| + \alpha + \beta},$$

where

$$\alpha = 1 + 2 \cdot |\mathcal{A}| + |\mathcal{B}|, \quad \beta = \alpha + 2^{\cdot^{\cdot^2}}/^{\alpha'},$$

such that

$$b_0^* < b_1^* \text{ if and only if } b_0 < b_1,$$
$$b^* \in a^* \text{ if and only if } b \in a,$$
$$a^* \subseteq \{c^* : c \in \mathcal{A}\} \text{ if } a \subseteq \mathcal{A},$$

for arbitrary a in $\mathcal{A}$ and b, b_0, b_1 in $\mathcal{A} \cup \mathcal{B}$.

Proof. Let $\{\emptyset\} \cup \mathcal{A} = \{A_0, A_1, \ldots, A_n\}$ with $\emptyset = A_0 < A_1 < \ldots < A_n$, and let $f_i = max(A_i \setminus A_{i-1})$ for $i = 1, \ldots, n$. Also let $\mathcal{F} = \{f_1, \ldots, f_n\}$ and

$$\{\emptyset\} \cup \mathcal{A} \cup \mathcal{F} \cup \mathcal{B} = \{A_0\} \cup \bigcup_{j=1}^{n} \{d_{j1}, \ldots, d_{jm_j}, A_j\}$$

with $A_{j-1} < d_{j1} < \cdots < d_{jm_j} < A_j$ for $j = 1, \ldots, n$. Trivially, $|\{\emptyset\} \cup \mathcal{A} \cup \mathcal{F} \cup \mathcal{B}| \leq \alpha$. We will define a function

$$\{A_0\} \cup \bigcup_{j=1}^{i} \{d_{j1}, \ldots, d_{jm_j}, A_j\} \xrightarrow{*_i} V_{2 \cdot i + \alpha + \beta}$$

for each $i = 0, \ldots, n$, so that $*_{i+1} \supseteq *_i$. Every $*_i$ will fulfil suitable conditions, that are adequate to guarantee at the end that $*_{n|_{\mathcal{A} \cup \mathcal{B}}}$ is the sought $*$. Such conditions are:

(i) $b_0^{*_i} < b_1^{*_i}$ if and only if $b_0 < b_1$,
for all b_0, b_1 in $\{A_0\} \cup \bigcup_{j=1}^{i} \{d_{j1}, \ldots, d_{jm_j}, A_j\}$;

(ii) $A_j^{*_i} = \{x \in A_j : x \leq A_I\} \cup \{x^{*_i} : x \in A_j \cap (\{\emptyset\} \cup \mathcal{A} \cup \mathcal{F} \cup \mathcal{B}) : A_I < x\}$,

for $j = 0, \ldots, i$, where I is the smallest h in $\{0, \ldots, n-1\}$ for which $A_h < f_{h+1}$, if any such A_h exists, and $I = n$ otherwise. We will also have

(iii) $rk\, d_{I+1,1}^{*_i} = 2 \cdot I + \alpha + \beta - 2$, if $I < i$,

(iv) $x^{*_i} = x$ for all $x \in \{A_0\} \cup \bigcup_{j=1}^{min\{i,I\}} \{d_{j1}, \ldots, d_{jm_j}, A_j\}$.

Before carrying out the construction of the $*_i$s, let us prove and add to our induction hypotheses the following straightforward consequence of (ii):

(v) $|A_j^{*_i}| < \beta$.

An easy induction shows in fact that $rk\, A_h \leq h$ for $h = 0, 1, \ldots, I$. This is obvious for $h = 0$. For $h > 0$, $f_h = max\,(A_h \setminus A_{h-1}) \leq A_{h-1}$ implies that $rk\, A_h = rk\,(A_h \cap A_{h-1}) \cup rk\,(A_h \setminus A_{h-1}) \leq rk\, A_{h-1} \cup (rk\, f_h + 1) \leq rk\, A_{h-1} + 1$, where $rk\, A_{h-1} \leq h - 1$ by inductive hypothesis. It follows from the inequality $rk\, A_I \leq I$ that

$$|\{x : x \leq A_I\}| \leq |V_{I+1}| = 2^{\cdot^{\cdot^{\cdot^2}}} \big\} I',$$

hence $|\{x : x \leq A_I\}| < 2^{\cdot^{\cdot^{\cdot^2}}} \big\} \alpha'$, because $I < n + 1 \leq |\mathcal{A}| < \alpha$. This entails that, as long as (ii) is satisfied, $|A_j^{*_i}| < 2^{\cdot^{\cdot^{\cdot^2}}} \big\} \alpha' + |\{\emptyset\} \cup \mathcal{A} \cup \mathcal{F} \cup \mathcal{B}| \leq \beta$ will hold too, implying (v).

For $i = 0$, we are forced by (ii) to put $A_0^{*_i} = \emptyset$; then (i), (iii), and (iv) are obvious.

In defining $*_i$ for $i > 0$, we will distinguish the following three cases:

(1) $f_i \leq A_{i-1}, f_i \leq A_I$;

(2) $A_{i-1} < f_i$;

(3) $A_I < f_i \leq A_{i-1}$;

of which (1) applies for $i = 1, \ldots, I$, while (2) applies for $i = I + 1$. As a preparatory remark we observe that in cases (1) and (3), assuming that $*_{i-1}$ has been defined already, and since $*_i$ is to extend $*_{i-1}$ which is defined for all x in $(A_i \setminus A_{i-1}) \cap (\{\emptyset\} \cup \mathcal{A} \cup \mathcal{F} \cup \mathcal{B})$—hence for all x in $A_i \cap (\{\emptyset\} \cup \mathcal{A} \cup \mathcal{F} \cup \mathcal{B})$—condition (i) forces us to put

$$A_i^{*_i} = \{x \in A_i : x \leq A_I\} \cup \{x^{*_{i-1}} : x \in A_i \cap (\{\emptyset\} \cup \mathcal{A} \cup \mathcal{F} \cup \mathcal{B}) : A_I < x\}.$$

Case (1) We begin by putting $x^{*i} = x^{*i-1}$ for all x in the domain of $*_{i-1}$, and by defining A_i^{*i} as anticipated in the preceding paragraph. Then we proceed to define $d_{i\,1}^{*i}, \ldots, d_{i\,m_i}^{*i}$, as follows. Since A_{i-1} precedes A_i anti-lexicographically, it is plain that these two sets can be decomposed as

$$A_{i-1} = C_{i-1} \cup B,$$
$$A_i = C_i \cup \{f_i\} \cup B,$$

with $c < f_i$ for every c in $C_{i-1} \cup C_i$ and $f_i < b$ for every b in B. It follows from the induction hypotheses that

$$A_{i-1}^{*i} = C_{i-1} \cup B_*,$$
$$A_i^{*i} = C_i \cup \{f_i\} \cup B_*,$$

where every member b of B_* either belongs to A_I or has the form x^{*i} for some $x > f_i$, and, in either case, satisfies the inequality $b > f_i = f_i^{*i}$. It is easily recognized that the sets x with $A_{i-1} < x < A_i$ (respectively, with $A_{i-1}^{*i} < x < A_i^{*i}$) are those of the form $X \cup B$ (respectively, $X \cup B_*$), where $C_{i-1} < X < C_i \cup \{f_i\}$. It will therefore suffice to take $x_* = (x \setminus B) \cup B_*$ in order to establish a one-to-one correspondence $x \mapsto x_*$, preserving $<$, between $\{x : A_{i-1} \le x \le A_i\}$ and $\{x : A_{i-1}^{*i} \le x \le A_i^{*i}\}$. We now put $d_{ij}^{*i} = (d_{ij})_*$ for $j = 1, \ldots, m_i$, thus finishing the definition of $*_i$ in such a manner that (i) through (v) are fulfilled.

Case (2) Since $m_i < \alpha < \beta - 1$, it is possible to choose sets $D_1 < \cdots < D_{m_i}$ of rank $2 \cdot i + \alpha + \beta - 2$ (these D_ks will have rank exceeding I, because $i > I$ and $\beta > i + 2$; furthermore any set of the form x^{*i-1} has rank smaller than $2 \cdot i + \alpha + \beta - 2$, by inductive hypothesis). We put $b_{ij}^{*i} = D_j$ for $j = 1, \ldots, m_i$; in this manner f_i^{*i} will be defined among others. We finally add A_i to the domain of $*_i$, on the basis of (ii). It will turn out in this manner that A_i^{*i} has rank $2 \cdot i + \alpha + \beta - 1$, and that (iii), along with all other inductive hypotheses, holds for i.

Case (3) Since $*_i$ is to extend $*_{i-1}$, the set f_i^{*i} is determined already, and (by (i) and (iii)) $rk\, f_i^{*i} \ge 2 \cdot I + \alpha + \beta - 2$, so that f_i^{*i} has at least $2 \cdot I + \alpha + \beta - 3$ predecessors in the ordering $<$. This number exceeds by at least m_i units the largest cardinality possible for A_{i-1}^{*i} and for A_i^{*i}, which is $\beta - 1$ by (v). This implies that if $p_{\beta-1+m_i} < \cdots < p_\beta < p_{\beta-1} < \cdots < p_2 < p_1$ are the immediate predecessors of f_i, then the sets

$$D_j = \{x \in A_i^{*i} : f_i^{*i} < x\} \cup \{p_1, \ldots, p_{\beta-1}\} \cup \{p_{\beta-1+j}\},$$

with $j = 1, \ldots, m_i$, satisfy the inequalities $A_{i-1}^{*i} < D_j < A_i^{*i}$. We will therefore put $b_{ij}^{*i} = D_j$, for $j = 1, \ldots, m_i$, and in this manner (i) through (v) will be fulfilled. ∎

Corollary 5.5 *Given a family $\mathcal{A}$ with $\emptyset \in \mathcal{A}$ and $|Un\,\mathcal{A}| < \omega$, and putting*

$$\alpha = 1 + 2 \cdot |\mathcal{A}| + |\mathcal{A}|^3, \quad \beta = \alpha + 2^{\cdot^{\cdot^2}} \diagup^{\alpha'},$$

there exists a function

$$^*: \mathcal{A} \cup \{\eta A : A \in \mathcal{A}\} \longrightarrow V_{2 \cdot |\mathcal{A}| + \alpha + \beta}$$

such that

$$
\begin{array}{llllll}
b_0^* & < & b_1^* & \text{if and only if} & b_0 & < & b_1, \\
b^* & \in & a^* & \text{if and only if} & b & \in & a, \\
a_0^* & \subseteq & a_1^* & \text{if and only if} & a_0 & \subseteq & a_1, \\
\eta\, a^* & = & (\eta\, a)^*, & & & & \\
a_0^* & = & \eta\, a_1^* & \text{if and only if} & a_0 & = & \eta\, a_1, \\
a_0^* & = & a_1^* \cap a_2^* & \text{if and only if} & a_0 & = & a_1 \cap a_2, \\
a_0^* & = & a_1^* \setminus a_2^* & \text{if and only if} & a_0 & = & a_1 \setminus a_2, \\
a_0^* & = & a_1^* \cup a_2^* & \text{if and only if} & a_0 & = & a_1 \cup a_2, \\
a_0^* & = & \{a_1^*, \ldots, a_m^*\} & \text{if and only if} & a_0 & = & \{a_1, \ldots, a_m\},
\end{array}
$$

for all m in ω, $a, a_0, a_1, \ldots, a_m$ in $\mathcal{A}$, and b, b_0, b_1 in $\mathcal{A} \cup \{\eta A : A \in \mathcal{A}\}$.

Proof. Put

$$
e_{ABC} = \begin{cases}
\eta(A \setminus (B \cup C)) & \text{if } A \not\subseteq B \cup C, \\
\eta((B \cap C) \setminus A) & \text{if } A \subsetneq B \cap C, \\
\eta((A \cap B \cap C) \setminus A) & \text{otherwise,}
\end{cases}
$$

for all A, B, C in $\mathcal{A}$, and

$$\mathcal{B} = \{e_{ABC} : A, B, C \in \mathcal{A}\} \setminus \{\emptyset\},$$

so that $|\mathcal{B}| \leq |\mathcal{A}|^3$.

Note that $\eta A \in \mathcal{A} \cup \mathcal{B}$ for all A in $\mathcal{A}$. In fact, either $\eta A \in \mathcal{A}$ or $\eta A = e_{AAA}$.

If $A, B \in \mathcal{A}$ and $A \not\subseteq B$, then $e_{ABB} \in (A \setminus B)$.

If $A, B, C \in \mathcal{A}$ and $A \neq B \cap C$, then either $A \not\subseteq B \cup C$, or $A \subseteq B \cup C$ but $A \not\subseteq C$, or $A \subseteq C$ but $A \not\subseteq B$, or $A \subseteq B \cap C$ but $B \cap C \not\subseteq A$. In the four cases we define i_{ABC} to be $e_{ABC}, e_{ACC}, e_{ABB}$, and e_{ABC} respectively, so that $i_{ABC} \in (A \setminus (B \cap C)) \cup ((B \cap C) \setminus A)$.

If $A, B, C \in \mathcal{A}$ and $A \neq B \setminus C$, then either $A \not\subseteq B$, or $A \subseteq B$ but $B \setminus C \not\subseteq A$, or $A \subseteq B$, $B \setminus C = \emptyset$ and $\eta A \in A \cap B \cap C$, or $A \subseteq B$, $\emptyset \neq B \setminus C \subseteq A$. In the four cases we define d_{ABC} to be $e_{ABB}, e_{BAC}, \eta A$, and e_{ABC} respectively, so that $d_{ABC} \in (A \setminus (B \setminus C)) \cup ((B \setminus C) \setminus A)$.

If $A, B, C \in \mathcal{A}$ and $A \neq B \cup C$, then either $A \not\subseteq B \cup C$, or $A \subseteq B \cup C$ but $B \not\subseteq A$, or $A \subseteq B \cup C$ and $B \subseteq A$ but $C \not\subseteq A$. In the three cases we define u_{ABC} to be e_{ABC}, e_{BAA}, and e_{CAA} respectively, so that $u_{ABC} \in (A \setminus (B \cup C)) \cup ((B \cup C) \setminus A)$.

We now define $\mathcal{A} \cup \mathcal{B} \xrightarrow{*} \mathcal{H}$ as in the preceding lemma (note that now $\emptyset \in \mathcal{A}$ and $\mathcal{F} = \{e_{A_{i+1}A_{i+1}A_i} : i = 0,\ldots,n-1\} \subseteq \mathcal{B}$). In this manner the equality $\eta\, a^* = (\eta\, a)^*$ and the equivalences $b_0^* < b_1^*$ iff $b_0 < b_1$, $b^* \in a^*$ iff $b \in a$, $a_0^* = \eta\, a_1^*$ iff $a_0 = \eta\, a_1$ in the thesis become obvious, as well as the implications: $a_0^* \subseteq a_1^*$ if $a_0 \subseteq a_1$, $a_0^* = a_1^* \cap a_2^*$ if $a_0 = a_1 \cap a_2$, etc.. In order to prove the reverse implications, it suffices to take into account the elements $e_{ABC}, i_{ABC}, d_{ABC}$, and u_{ABC} introduced above. For example, if $a_0 \neq a_1 \cap a_2$, then $i_{a_0 a_1 a_2}^* \in (a_0^* \setminus (a_1^* \cap a_2^*)) \cup ((a_1^* \cap a_2^*) \setminus a_0^*)$, so that $a_0^* \neq a_1^* \cap a_2^*$. The operators $\setminus$ and $\cup$, and the relator $\subseteq$, are treated similarly.

If $A, a_1, \ldots, a_m \in \mathcal{A}$ and $A \neq \{a_1, \ldots, a_m\}$, then either $a_i \notin A$ for some i in $\{1,\ldots,m\}$—in which case we readily have $a_i^* \notin A^*$ and therefore $A^* \neq \{a_1^*, \ldots, a_m^*\}$—or $\{a_1, \ldots, a_m\} \subsetneq A$. In the latter case, if $A \subseteq \mathcal{A}$ then $|A^*| = |A| > m$ and hence $A^* \subsetneq \{a_1^*, \ldots, a_m^*\}$; otherwise $e_{AAA} = \eta(A \setminus A) \in A$, hence $e_{AAA}^* \in A^* \setminus \{c^* : c \in \mathcal{A}\} \subseteq A^* \setminus \{a_1^*, \ldots, a_m^*\}$, so that again $A^* \neq \{a_1^*, \ldots, a_m^*\}$. ∎

Consider the family $\mathcal{C}$ of those safe formulae p (see Definition 4.2) that do not involve the constructs $\times$ and Pow and such that any non-safe term $\eta\, s$ in p occurs as a side of an equality of p. An easy consequence of the preceding corollary is the following:

Corollary 5.6 (Decidability) *Let p be a formula in $\mathcal{C}$, $\mathcal{T}$ be the set of all terms that occur in $p \,\&\, \emptyset = \emptyset$, and $\alpha = 1 + 2 \cdot |\mathcal{T}| + |\mathcal{T}|^3$, $\beta = \alpha + 2^{\cdots^{2}} \Big/ {}^{\alpha^{\diagup}}$. Then p is finitely satisfiable only if p admits a model M with $rk\, t^M < 2 \cdot |\mathcal{T}| + \alpha + \beta$ for all t in $\mathcal{T}$.*

Proof. Let φ be a model of p with $|\varphi x| < \omega$ for every variable x occurring in p. By Corollary 4.3, s^φ is finite for any safe term s in $\mathcal{T}$ and, in particular, for any term $\eta\, s$ that occurs in p. We take $\mathcal{A} = \{s^\varphi : s$ is a safe term in $\mathcal{T}\}$ in the preceding corollary, so that

$$^*: \{t^\varphi : t \in \mathcal{T}\} \longrightarrow V_{2 \cdot |\mathcal{A}| + \alpha + \beta}\,,$$

where $V_{2 \cdot |\mathcal{A}| + \alpha + \beta} \subseteq V_{2 \cdot |\mathcal{T}| + \alpha + \beta}$ because $|\mathcal{A}| \leq |\mathcal{T}|$. It is a straightforward consequence of the thesis of the preceding corollary that the function $t \mapsto (t^\varphi)^*$, defined for all t in $\mathcal{T}$, is induced by the interpretation $x \xmapsto{M} (\varphi x)^*$ and that the latter assigns to p the same truth value as φ, which is *TRUE*. ∎

Corollary 5.7 *The finite satisfiability problem is decidable for $\mathcal{C}$.*

Proof. Given p in $\mathcal{C}$, it is possible to generate all assignments

$$M : \{\text{variables of } p\} \longrightarrow V_{2 \cdot |\mathcal{T}| + \alpha + \beta}\,.$$

By the preceding corollary, p is finitely satisfiable if and only if p^M (which can be directly calculated) is *TRUE* for some such M. ∎

5.7   A decision algorithm for an elementary syllogistic involving $\{\bullet, \ldots, \bullet\}$ and η

In this section we explain how to establish a one-to-one correspondence $x \mapsto x'$ between any given finite family $\mathcal{X}$ and a collection $\mathcal{F}$ of *finite* sets, so that membership, ordering, and equalities of the two kinds $y = \eta x$, $z = \{x_1, \ldots, x_n\}$ are preserved by this correspondence. As a consequence, if we define $\mathcal{C}_*$ to consist of those formulae in $\mathcal{C}$ (see preceding section) that do not involve $\cap$, $\backslash$, or $\cup$, then any p in $\mathcal{C}_*$ is satisfiable if and only if p is finitely satisfiable. This yields, by Corollary 5.7, a decision algorithm for satisfiability in $\mathcal{C}_*$. From Section 3.8 we know that any unquantified formula involving only set variables, propositional connectives, and the constructs $=$, $\in$, η and $\{\bullet, \ldots, \bullet\}$, can be translated, in a satisfiability-preserving fashion, into a formula belonging to $\mathcal{C}_*$. This enables us to immediately extend our decidability result to the satisfiability problem of this larger class of formulae, in which, unlike in $\mathcal{C}_*$, no restrictions are placed on the way η can occur.

Starting from $\mathcal{X}$, we consider the smallest finitely transitive set $\mathcal{Y}$ to which $\mathcal{X} \cup \{\emptyset\} \cup \{\eta X : X \in \mathcal{X}\}$ belongs—recall that by Corollary 4.4, which exploits König's lemma, such a $\mathcal{Y}$ exists and is finite. Thus $\emptyset \in \mathcal{Y}, \mathcal{X} \subseteq \mathcal{Y}, \{\eta X : X \in \mathcal{X}\} \subseteq \mathcal{Y}, |\mathcal{Y}| < \omega$; and $F \subseteq \mathcal{Y}$ when $F \in \mathcal{Y}$ and $|F| < \omega$.

Let $\alpha_0 = 0, \alpha_1, \ldots, \alpha_K$ be all ordinals α such that $\mathcal{Y}$ has members of rank α. Assume furthermore that $\alpha_0 < \alpha_1 < \cdots < \alpha_K$. We indicate by $C_{h1}, \ldots, C_{hf_h}$ all members of $\mathcal{Y}$ that have *finite* cardinality and rank α_h, and by $C_{h,f_h+1}, \ldots, C_{h,f_h+i_h}$ the *infinite* sets of rank α_h in $\mathcal{Y}$, for $h = 0, 1, \ldots, K$. It is plain that: $0 < f_h + i_h < \omega$ for all h; $C_{hf_h} < C_{h,f_h+1}$ when both $f_h > 0$ and $i_h > 0$; $C_{h,f_h+i_h} < C_{h+1,1}$ if $h < K$. We also assume that $C_{h1} < \cdots < C_{hf_h}$ and $C_{h,f_h+1} < \cdots < C_{h,f_h+i_h}$.

Lemma 5.13 *There is a number I with $0 < I \leq K+1$ such that $\alpha_h < \omega$, $f_h \neq 0$, $i_h = 0$ for all $h < I$, whereas $\alpha_I \geq \omega$, $f_I = 0$, and $i_I \neq 0$ if α_I, f_I, and i_I are defined, that is to say, if $I \neq K+1$.*

Proof. Observe that $\alpha_0 = 0$, $f_0 = 1$, and $i_0 = 0$. For $h > 0$, assume that $\alpha_k < \omega$, $f_k \neq 0$, and $i_k = 0$ for all $k < h$. If $f_h \neq 0$, then—by the finite transitivity of $\mathcal{Y}$—$C_{h1} \subseteq \mathcal{Y}$ and hence $C_{h1} \subseteq \{x \in \mathcal{Y} : rk\, x < \alpha_h\} = \{x \in \mathcal{Y} : rk\, x \leq \alpha_{h-1}\}$; therefore $rk\, C_{h1} \leq \alpha_{h-1} + 1$, because $\mathcal{Y}$ is finite, which implies $\alpha_h = \alpha_{h-1} + 1 < \omega$ and $i_h = 0$. If $f_h = 0$ instead, then we must have $i_h \neq 0$ and therefore $\alpha_h > \omega$. ∎

We now construct a DAG with nodes C_{hj} ($h = 0, 1, \ldots, K$ and $j = 1, \ldots, f_h + i_h$) and additional nodes ι_{hq} ($h = I, \ldots, K$ and $q = 1, \ldots, i_h$), where I is as in the lemma. As edges we take all pairs $(C_{h_0 j_0}, C_{h_1 j_1})$ with $C_{h_0 j_0} \in C_{h_1 j_1}$ and all pairs $(\iota_{hq}, C_{h,f_h+q})$. Then we define a 'pseudo-rank' function ϱ on the set of nodes, as follows: for every node of the form C_{hj}, we put

$$\varrho C_{hj} = \bigcup_{(N, C_{hj}) \text{ is an edge}} A(\varrho N + 1),$$

and, for every node of the form ι_{hq}, we put

$$\varrho \iota_{hq} = \begin{cases} \text{first successor ordinal that exceeds both} \\ \omega \text{ and } \varrho C_{h-1,1}, \text{ if } f_h = 0; \\ \\ (\varrho C_{h1}) - 1, \text{ if } f_h > 0. \end{cases}$$

The last branch of this definition makes sense because, since there are finitely many edges, ϱC_{hj} is a successor ordinal for every node $C_{hj} \neq C_{01}$.

Lemma 5.14 $\varrho C_{h_0 j_0} < \varrho C_{h_1 j_1}$ *if and only if* $rk\, C_{h_0 j_0} < rk\, C_{h_1 j_1}$, *for* h_0, h_1 *in* $\{0, 1, \ldots, K\}$ *and* j_b *in* $\{1, \ldots, f_{h_b} + i_{h_b}\}$ $(b = 0, 1)$. *Moreover,* $\varrho \iota_{hq}$ *is a successor ordinal such that* $\varrho C_{h-1,1} \leq \varrho \iota_{hq} < \varrho C_{h1}$ *for* $h = I, \ldots, K$ *and* $q = 1, \ldots, i_h$. *In particular* $\varrho C_{I1} = \omega + 2$.

Proof. For $h < I$ and $j = 1, \ldots, f_h$, plainly $\varrho C_{hj} = rk\, C_{hj} < \omega$. Since, moreover, it is obvious that $\varrho \iota_{I1} = \cdots = \varrho \iota_{I,i_I} = \omega + 1 > \omega > \varrho C_{I-1,1}$ —so that $\varrho C_{I1} = \varrho \iota_{I1} + 1 = \omega + 2 > \varrho C_{I-1,1}$—it only remains to prove the lemma for $h_0 \geq I, h_1 \geq I$.

Without loss of generality, we can assume that either $h_1 = h_0$ or $h_1 = h_0 + 1$. Thus, indicating h_1 by h and j_1 by j, in order to take both cases into account we must prove that $\varrho C_{h j_0} = \varrho C_{h j_1}$, that $\varrho \iota_{h1}$ is a successor, and that $\varrho C_{h-1,1} \leq \varrho \iota_{h1} < \varrho C_{hj}$.

Assume first that $f_h = 0$. In this case, since $\varrho \iota_{h, j_b} = \varrho \iota_{h1}$ is a successor ordinal bigger than $\varrho C_{h-1,1}$ by definition, we have $\varrho C_{h, j_b} = \varrho \iota_{h1} + 1$ for $b = 0, 1$, which readily yields the thesis. Assume then $f_h \neq 0$, and consider first a C_{hj} with $j \leq f_h$. We recall that the members of C_{hj} have themselves the form $C_{h_* j_*}$, and that the biggest among these must satisfy the identities $rk\, C_{h_* j_*} = \alpha_h - 1 = \alpha_{h-1}$; therefore $h_* = h - 1$ and, by inductive hypothesis, $\varrho C_{hj} = \varrho C_{h-1,1} + 1$. Consider then a C_{hj} with $j > f_h$. Every member $C_{h_* j_*}$ in C_{hj} has $\varrho C_{h_* j_*} \leq \varrho C_{h-1,1} = \varrho C_{h1} - 1$, and therefore $\varrho C_{hj} = \varrho \iota_{h1} + 1 = \varrho C_{h1}$. Thus ϱ sends $C_{h1}, \ldots, C_{h, f_h + i_h}$ all to the same image. The equalities $\varrho C_{h-1,1} = \varrho \iota_{h1} = \varrho C_{hj} - 1$—the first of which ensures that $\varrho \iota_{h1}$ is a successor ordinal—are also immediate. ∎

For $h = 0, \ldots, K$ we put

$$C'_{hj} = \{ x' : x \in C_{hj} \}, \text{ for } j = 1, \ldots, f_i;$$

moreover, for $h = I, \ldots, K$ and $q = 1, \ldots, i_h$, we put

$$d_{hq} = q\text{-th set of rank } \varrho \iota_{hq}$$
$$\text{that follows } C'_{h-1, f_{h-1} + i_{h-1}} \text{ with respect to } <;$$
$$C'_{h, f_h + q} = \{ x' : x \in C_{h, f_h + q} \cap \mathcal{Y} \} \cup \{ d_{hq} \}.$$

It is obvious that $C'_{01} = \emptyset$ according to this definition; that every x' intervening in the definition of a C'_{hj} is defined before C'_{hj}; that every C'_{hj} defined by the above equalities has finite cardinality (indeed, $|C'_{hj}| \leq |\mathcal{Y}| + 1$), so that selecting the d_{hq}s—which in turn will be finite—is possible by Corollary 2.4. It is easy to check that $rk\, C'_{hj} = \varrho\, C_{hj}$ for $h = 0, \ldots, K$ and $j = 1, \ldots, f_h + i_h$. It is also plain that d_{hq} is always the maximum of C'_{h,f_h+q} and that, thanks to its presence, C'_{h,f_h+q} is bigger than C'_{h,f_h} if $f_h \neq 0$. Thus, by anti-lexicographicity, $x \mapsto x'$ is an order-preserving mapping of $\mathcal{Y}$ into itself, with $x' \neq d_{hq}$ for every x in $\mathcal{Y}$ and for every 'scion' d_{hq}. It is straightforward, in view of these facts, to prove that

Lemma 5.15 *The function $x \mapsto x'$ is a realization of the graph $(\mathcal{Y}, \cdot)$, whose edges are the pairs (x, y) with x, y in $\mathcal{Y}$ such that $x \in y$. This realization, induced by the grafting $\{\, d_{hq} \} \mapsto \{ C_{hq} \}$, fulfills the following conditions:*

$$|x'| < \omega,$$
$$x' < y' \qquad \text{if and only if } x < y,$$
$$x' = \{\, y'_1, \ldots, y'_n \,\} \quad \text{if and only if } x = \{\, y_1, \ldots, y_n \,\},$$
$$e' = \eta\, a' \qquad \text{if and only if } e = \eta\, a,$$

for all $x, y, y_1, \ldots, y_n$ in $\mathcal{Y}$ and a, e in $\mathcal{X}$. ☐

As we have briefly discussed at the beginning of this section, this lemma has important consequences, namely

Corollary 5.8 (Decidability) *Let p be a formula in $\mathcal{C}$, without occurrences of $\cap$, $\backslash$, or $\cup$. Then p is satisfiable if and only if p admits a model M with $rk\, t^M < 2 \cdot |\mathcal{T}| + \alpha + \beta$ for all t in $\mathcal{T}$, where $\mathcal{T}$, α, and β have the same meaning as in Corollary 5.6.*

Proof. Given a model φ of p, we take $\mathcal{X} = \{\, s^\varphi : s \text{ is a term occurring in } p \,\}$ and define the function $x \mapsto x'$ as before. The preceding lemma yields that the function $t \mapsto (t^\varphi)'$ defined for all t in $\mathcal{T}$ is induced by the interpretation $x \overset{M}{\mapsto} (\varphi x)'$ and assigns to p the same truth value as φ, which is *TRUE*. Moreover Mx is finite for every variable x in p, which primes Corollary 5.6. ∎

Corollary 5.9 (Decidability) *The satisfiability problem is decidable for the unquantified formulae that involve, in addition to set variables and to propositional connectives, only the constructs $=$, $\in$, η, and $\{\, \bullet, \cdots, \bullet \,\}$.*

Proof. By exploiting the normalization process described in Section 3.8, one can translate any formula q in the class under consideration into a formula p, as in the preceding corollary, which is satisfiable—indeed, hereditarily finitely satisfiable—if and only if q is satisfiable. To test p for satisfiability, we proceed as in Corollary 5.7. ∎

6. MULTILEVEL SYLLOGISTIC

6.1 Introduction

The study of satisfiability decision tests for the collection of unquantified set-theoretic formulae that involve, in addition to set variables and to propositional connectives, the constructs

$$\emptyset, \cap, \setminus, \cup, \in, =, \subseteq,$$

was initiated in [FOS80a] and is known in the literature as *multilevel syllogistic* or just MLS. A theory slightly more general than MLS is developed in this chapter. Moreover, it is shown in Section 6.7 that MLS subsumes ordinary two-level syllogistic—see below.

Stating a decision algorithm for MLS so that it can also treat the relator *Finite* requires only a little additional effort. Furthermore, as long as one is *not* concerned with the computational complexity of the decision algorithm, it is not problematic to tackle the construct $\{\bullet, \ldots, \bullet\}$ along with the constructs of MLS. This is why in the first part of this chapter we take into account the collection F of unquantified formulae that involves the constructs of MLS with the addition of $\{\bullet, \ldots, \bullet\}$ and *Finite*. In the second part of the chapter, when we set the ground for more effective decision tests, we drop $\{\bullet, \ldots, \bullet\}$ but still retain *Finite*.

In the same manner as in [CGO88], we want to prove that F enjoys the following property, stronger than decidability: every p in F can be put in a suitable canonical form p', which will coincide with the propositional constant *FALSE* if and only if p is unsatisfiable. Having this goal in mind, from among the formulae of F we single out, for any given finite family X of set variables, a certain finite collection $\Sigma(X)$ of formulae, to be called the *syllogistic schemes over X*. Then we show that any formula p in F can be decomposed as a disjunction $p' = \bigvee_{\sigma \in \Sigma_p} \sigma$, where $\Sigma_p \subseteq \Sigma(X)$, X being the family of all variables of p. This reformulation of p, that will be called the *syllogistic canonical form of p*, can be regarded as a non-redundant and exhaustive description of all possible models of p. We will see, indeed, that every syllogistic scheme is satisfiable, while no two syllogistic schemes in $\Sigma(X)$ have any model in common.

A naive algorithm for determining whether or not a given p in F is satisfiable is to calculate the set Σ_p of all disjuncts of p' and then check whether this set is non-empty. As a matter of fact, a technique is established in this chapter for extracting the schemes that form Σ_p from the collection $\Sigma(X)$ of all syllogistic schemes over the variables X of p. In other words, this is a technique for evaluating p in any $\sigma \in \Sigma(X)$; in fact, σ will belong to Σ_p if and only if the result of the evaluation is *TRUE*.

By restricting our attention to those formulae p in F that do not involve $\{\bullet, \ldots, \bullet\}$, we will be able to characterize the schemes σ in Σ_p in such a way that Σ_p can be generated *directly*, instead of being obtained by filtering out the schemes in $\Sigma(X)$ that do not satisfy p. We also prove that if Σ_p is non-empty, then it contains 'minimum effort' schemes (in a sense to be explained in the next paragraph—and, in more detail, in Section 6.4). Thus in order to check p for satisfiability only such schemes are to be sought for. The most significant details of the decision algorithm resulting from this improvement are presented in Section 6.5—cf. also [Omo84, GO85].

To be more specific, even at this informal level of discussion, we anticipate that every syllogistic scheme σ in $\Sigma(X)$ is identified by a quadruple $(\sim, G, F, Z)$, where $\sim$ is an equivalence relation over X and G is a directed acyclic graph ('DAG') whose nodes are the $\sim$-classes. We will see that any σ in Σ_p corresponds to an equivalence relation $\sim$ of a particular kind, which will be called a *p-compatible relation*, and a DAG of a particular kind, to be called a *p-$\sim$-compatible DAG*. If Σ_p is non-empty, then a finest p-compatible relation $\sim_p$ exists (this would no longer be the case if the construct $\{\bullet, \ldots, \bullet\}$ were admitted). Moreover this relation, which can be determined by methods of propositional calculus (see [End72]), provably has some σ in Σ_p associated to it. Any σ associated to $\sim_p$ is what we have called above a minimum effort syllogistic scheme; establishing the existence of one such scheme amounts to searching for a p-$\sim_p$-compatible DAG. This search, although it involves backtracking, can be piloted quite effectively.

This chapter ends with two appendices establishing links between what we are calling 'syllogistic' and the meaning traditionally attached to this word. In assertoric Aristotelian syllogistic (cf., [LJ52, McC63, Qui61, She56, FO78]), as well as in its recasting due to the logicians of the second half of the 19th century (cf. [Boc72a, Boc72b, Gar68, Car77]), the validity problem is referred to a 'flat universe' where no hierarchy exists among sets (indeed, every set consists of member-less entities often called *individuals*, or *urelemente*). The situation is essentially the same with monadic calculus (see [Ack62]), whose greater richness with respect to earlier syllogistics is not due to the presence of a hierarchy, but to the possibility of quantifying individual variables as well as set variables. In contrast, our sets are organized as a hierarchy; however, since we are dealing with no individuals (indeed our sets are ultimately based on the empty set), it is *a priori* not obvious that a connection between our syllogistics and classical extensional syllogistics should exist. Nevertheless, after outlining in Section 6.7 a dialect of quantifier-free two-level syllogistic, we will be successful in proving that its validity problem reduces to the validity problem for a restricted class of formulae in MLS.

The second appendix (Section 6.8) shows the not surprising fact that a weak connection exists between syllogistic schemes and classic Venn diagrams. Historically, the latter were devised (see [Ven34]) with the aim of depicting boolean relations holding inside a family of sets. In addition to relations of that kind, our schemes reflect membership chains too.

The iconic appeal of Venn diagrams is undoubtedly a reason why they have become so popular (see [Gar68]). However, we intentionally fly over issues about graphics. This makes it possible to take also *infinite* Venn diagrams into account in our discourse.

6.2   Syllogistic schemes over a family of set variables

This section shows that a specific procedure can be used, given a family X of variables, to partition the collection of all possible assignments from X to sets into a finite number of equivalence classes. Each equivalence class is characterized by a certain formula σ of set theory which is satisfied by all assignments in the class, and only by these. Formulae having the syntactic structure exemplified by σ will be called *syllogistic schemes*. Then, in the next section, we will show that for every conjunction p of literals of the types $x = y$, $x \neq y$, $x = y \cup z$, $x = y \setminus z$, $x \in y$, $x \notin y$, $x = \{y\}$, $Finite\, x$, $\neg\, Finite\, x$, having X as its set of free variables, the class of all assignments that satisfy p is a union of classes of the partition above. Moreover, it will be shown that there is an algorithm to test whether the assignments in a given class of the partition satisfy p. This will provide at once a satisfiability procedure for conjunctions of this particular kind, thus leading us to the rediscovery of a decidability result that was proved in [FOS80b] for the first time.

We begin by considering sets $a_0, \ldots, a_I$, with I finite. The relations among these sets may, by a first approximation, be described by a DAG $(I, \dot{\,})$ defined by

$$i \in \dot{j} \text{ if and only if } a_j \in a_i,$$

for i, j in I. This is but a first crude approximation, but it allows each a_i to be written as

$$a_i = b_i \cup \{a_j : i \in \dot{j}\}$$

for appropriate sets b_i, none of which intersects $\{a_0, \ldots, a_I\}$. To characterize the relations among the sets a_i more accurately, it remains to determine the boolean algebra B generated by the b_is (see [Sik64]). This is done by forming the atoms of B. Since $b_0 \cup \cdots \cup b_I$ is the maximum of B, the complement $\bar{b}$ of each member b of B is $\bar{b} = (b_0 \cup \cdots \cup b_I) \setminus b$, and the atoms of B are the non-empty sets c_h of the form

$$c_h = \left(\bigcap_{i \in h} b_i\right) \cap \left(\bigcap_{j \in I \setminus h} \bar{b}_j\right) = \left(\bigcap_{i \in h} b_i\right) \setminus \left(\bigcup_{j \in I \setminus h} b_j\right),$$

with $h \subseteq I$, $h \neq \emptyset$. Thus we have

$$b_i = \bigcup_{i \in h} c_h \, ,$$

and hence

$$a_i = \left(\bigcup_{i \in h} c_h \right) \cup \left(\bigcup_{i \in j} \{a_j\} \right) ,$$

for all i in I. It is plain from the last identity that a_i is infinite if and only if there is an infinite set c_h with i in h.

Next, we consider a collection $X = \{x_0, x_1, \ldots, x_I\}$ of set variables. We introduce a collection $Y = \{y_0, y_1, \ldots, y_J\}$ of auxiliary set variables such that

(i) X and Y have no variables in common;

(ii) there is a bijection $^\circ : Y \rightarrow Pow(X) \setminus \{\emptyset\}$ (so that $|Y| = 2^{|X|} - 1$).

Any assignment $x_i \mapsto a_i$ of set values to the variables in X determines a boolean algebra B, as explained above. Roughly speaking, the variables y_j that we have just introduced are 'names' for the atoms generating B. Notice, though, that the atoms may fail to be in a one-to-one correspondence with the non-empty subsets of X, either because the a_is are *not* pairwise different, or because some of the above-defined sets c_h are empty. Therefore, in order to fully characterize the assignment $x_i \mapsto a_i$, one has to indicate which subset Z of Y consists of 'official names' of the atoms of B. Moreover, if one wants to specify which ones, among the atoms c_h, are finite sets, one must indicate the subset F of Z formed by the names of such atoms.

Let us now consider an equivalence relation $\sim$ over X. Referring to a pre-established ordering $\prec$ of all variables of set theory in a denumerable sequence $v_0 \prec v_1 \prec v_2 \prec \cdots$, we can partition X into equivalence classes

$$\{x_{00}, x_{01}, \ldots, x_{0L_0}\}, \{x_{10}, x_{11}, \ldots, x_{1L_1}\}, \ldots, \{x_{N0}, x_{N1}, \ldots, x_{NL_N}\},$$

with $x_{n0} \prec x_{n1} \prec \cdots \prec x_{nL_n}$ for $n = 0, 1, \ldots, N$, and with $x_{00} \prec x_{10} \prec \cdots \prec x_{N0}$. We put $s_n = x_{n0}$ for $n = 0, 1, \ldots, N$, and choose $S = \{s_0, s_1, \ldots, s_N\}$ as the set of representatives of the equivalence classes of $\sim$. Finally, we consider a DAG structure $(S, \cdot)$ (see Definition 2.4) superimposed to S.

Definition 6.1 *Let* $X, Y, \sim, (S, \cdot)$ *be as above. Let* $F \subseteq Z \subseteq Y$ *be such that*

(a) $\overset{\circ}{z} \subseteq S$, *for all* z *in* Z;

(b) *there are no distinct* v, w *in* S *such that*

(b1) $v \in \dot{s}$ if and only if $w \in \dot{s}$, for all s in S, and

(b2) $v \in \overset{\circ}{z}$ if and only if $w \in \overset{\circ}{z}$, for all z in Z

hold together.

Let δ, ϱ_{ZF} denote the formulae

$$\delta \quad =_{\mathrm{Def}} \quad \&_{j \leq J} \quad \left(y_j \neq \emptyset \,\&\, \&_{i \leq I}\, x_i \notin y_j \,\&\, \&_{j < k \leq J}\, y_j \cap y_k = \emptyset \right),$$

$$\varrho_{ZF} \quad =_{\mathrm{Def}} \quad \&_{n \leq N} \left(s_n = x_{n1} = \cdots = x_{nL_n} = \bigcup_{\substack{z \in Z \\ s_n \in \overset{\circ}{z}}} z \,\cup\, \right.$$

$$\left. \bigcup_{\substack{r \in S \\ s_n \in \dot{r}}} \{r\} \right) \,\&\,$$

$$\left(\&_{f \in F}\, Finite(f) \right) \,\&\, \left(\&_{z \in Z \setminus F}\, \neg\, Finite(z) \right).$$

We put $\sigma_{ZF} =_{\mathrm{Def}} \delta \,\&\, \varrho_{ZF}$ and call σ_{ZF} a syllogistic scheme *over X (relative to X, $\sim$, $\dot{}$).* □

Remark 6.1 It is clear that there are only a finite number of possible syllogistic schemes over a given set X of variables and that they can be determined by an algorithm. Also, each set X admits some syllogistic scheme. □

We now prove some useful properties of the class of models of a given syllogistic scheme.

Lemma 6.1 *Let M be a model of a syllogistic scheme σ_{ZF}. Then M assigns different values to distinct variables in S. Moreover, $v^M \in w^M$ holds if and only if $w \in \dot{v}$, for every v, w in S.*

Proof. By inspection of σ_{ZF} one easily sees that the function $z^M \mapsto \overset{\circ}{z}$ is well-defined on $U = \{ z^M : z \in Z \}$, and that, furthermore, $(U, \ddot{})$ is an extensional grafting of $(S, \dot{})$. The function $s \mapsto s'$ induced on S by this grafting coincides with the restriction of M to S. The thesis hence follows from Lemma 2.10. ■

Lemma 6.2 *Let M be a model of a syllogistic scheme σ_{ZF}. Then for every s in S one has that*

$$s^M \text{ is infinite if and only if } s \in \overset{\circ}{z} \text{ for some } z \text{ in } Z \setminus F.$$

Proof. If $s \in \overset{\circ}{z}$ for some z in $Z \setminus F$, then $\neg\, Finite(z)$ occurs in σ_{ZF}, thus implying that z^M is infinite. But $z^M \subseteq s^M$, hence s^M is infinite too.

Conversely, if s^M is infinite, z^M must be infinite for some z in Z for which $s \in \overset{\circ}{z}$. Therefore $z \in Z \setminus F$, proving the lemma in both directions. ∎

Lemma 6.3 *Let M be a model of the syllogistic scheme σ_{ZF}. Then, putting*

$$\bar{z} =_{\mathrm{Def}} \left(\bigcap_{v \in \overset{\circ}{z}} v^M \right) \setminus \left(\bigcup_{v \in S \setminus \overset{\circ}{z}} v^M \right),$$

for all z in Z, we have

$$z^M \subseteq \bar{z} \subseteq z^M \cup \{ s^M : s \text{ in } S \},$$

(so that z^M is infinite if and only if $\bar{z}$ is infinite).

Proof. It is clear that $z^M \subseteq \bar{z}$. On the other hand, since for each v in S, v^M is a finite union of sets z_0^M and $\{ s^M \}$, with z_0 in Z and s in S, and the sets of type z_0^M are pairwise disjoint, it follows that $\bar{z}$ is contained in the union of $\{ s^M : s \text{ in } S \}$ with a finite union of disjoint sets of type z_0^M. As a matter of fact, if $\bar{z} \subseteq z_0^M$, then $z_0^M \subseteq v^M$ for all v in $\overset{\circ}{z}$, and $z_0^M \cap v^M = \emptyset$, for all v in $S \setminus \overset{\circ}{z}$, which in turn imply $v \in \overset{\circ}{z}_0$ if and only if $v \in \overset{\circ}{z}$, i.e. $z_0 = z$. Thus $\bar{z} \subseteq z^M \cup \{ s^M : s \text{ in } S \}$, and the lemma is proved. ∎

We are now in a good position to prove that the syllogistic schemes over X determine a partition of all possible assignments of set values to the variables in X.

Theorem 6.1 (Mutual exclusion) *Let σ_0 and σ_1 be two distinct syllogistic schemes over X and let M_b be a model of σ_b, for $b = 0, 1$. Then $M_{0|X} \neq M_{1|X}$, i.e. there is some x in X such that $M_0 x \neq M_1 x$.*

Proof. Let σ_0 and σ_1 be relative to $(\sim_0, \cdot, Z_0, F_0)$ and $(\sim_1, {}^*, Z_1, F_1)$ respectively. We consider the following cases.

Case A If $\sim_0 \neq \sim_1$, then there exist v, w in X such that $v \sim_b w$ and $v \not\sim_{1-b} w$. By Lemma 6.1 $v^{M_{1-b}} \neq w^{M_{1-b}}$, whereas obviously $v^{M_b} = w^{M_b}$, so that $M_{0|X} \neq M_{1|X}$.

Case B Suppose now that $\sim_0 = \sim_1$, but $(S, \cdot) \neq (S, {}^*)$. This means that there exist v, w in S such that $w \in \dot{v}$ if and only if $w \notin \overset{*}{v}$. Assume for definiteness that $w \in \dot{v}$ and $w \notin \overset{*}{v}$. Therefore, again by Lemma 6.1, $v^{M_0} \in w^{M_0}$ and $v^{M_1} \notin w^{M_1}$, which proves the lemma in the present case.

Case C Consider now the case in which $\sim_0 = \sim_1$, $(S, \cdot) = (S, {}^*)$, but $Z_0 \neq Z_1$. Let $z \in Z_{b^*} \setminus Z_{1-b^*}$, $b^* \in \{0, 1\}$.

Observe that as an immediate consequence of Lemma 6.3 one proves

$$z^{M_{b^*}} = \left(\left(\bigcap_{v \in \overset{\circ}{z}} v^{M_{b^*}} \right) \setminus \left(\bigcup_{v \in S \setminus \overset{\circ}{z}} v^{M_{b^*}} \right) \right) \setminus \{ s^{M_{b^*}} : s \text{ in } S \}.$$

Therefore, if by contradiction $M_{0|x} = M_{1|x}$, then we would have

$$\left(\bigcap_{v \in \overset{\circ}{z}} v^{M_{1-b^*}} \right) \setminus \left(\bigcup_{v \in S \setminus \overset{\circ}{z}} v^{M_{1-b^*}} \right) \neq \emptyset,$$

which implies $z \in Z_{1-b^*}$, a clear contradiction.

Case D The last case to be considered is

$$\sim_0 = \sim_1, \ (S, \cdot) = (S, \overset{*}{\ }), \ Z_0 = Z_1, \text{ but } F_0 \neq F_1.$$

In such a case there exists an f in $F_{b^*} \setminus F_{1-b^*}$, for some b^* in $\{0\ 1\}$. Lemma 6.3 therefore implies that

$$\left(\bigcap_{v \in \overset{\circ}{f}} v^{M_{b^*}} \right) \setminus \left(\bigcup_{v \in S \setminus \overset{\circ}{f}} v^{M_{b^*}} \right) \text{ is finite}$$

and

$$\left(\bigcap_{v \in \overset{\circ}{f}} v^{M_{1-b^*}} \right) \setminus \left(\bigcup_{v \in S \setminus \overset{\circ}{f}} v^{M_{1-b^*}} \right) \text{ is infinite},$$

showing that even in this case the lemma holds. ∎

Next, we show that the class of all models of a syllogistic scheme is non-empty.

Theorem 6.2 (Irredundancy) *Every syllogistic scheme is satisfiable.*

Proof. Let σ_{ZF} be a syllogistic scheme relative to X, $\sim$, $\cdot$ and let S be the set of those variables that are representatives of the equivalence classes of $\sim$ (see paragraphs preceding Definition 6.1). Put

$$V = S \cup (Y \setminus (Z \setminus F)) \cup \bigcup_{z \in Z \setminus F} I_z,$$

where the I_zs are pairwise disjoint infinite sets of variables which are disjoint from $X \cup Y$ too. Extend $\cdot$ from S to V by putting

$$\begin{aligned}
\overset{\cdot}{f} &= \overset{\circ}{f} && \text{if } f \in F; \\
\overset{\cdot}{t} &= \overset{\circ}{z} && \text{if } t \in I_z \text{ for some } z \text{ in } Z \setminus F; \\
\overset{\cdot}{y} &= \emptyset && \text{if } y \in Y \setminus Z.
\end{aligned}$$

Observe that each newly introduced node has no predecessors in $(V, \cdot)$, therefore $(V, \cdot)$ is a DAG.

Let us show, now, that S is an extensional set of nodes for $(V, \dot{\ })$ (cf. Definition 2.6). Indeed, if x, y are variables in S such that $\{v \in V : x \text{ in } \dot{v}\} = \{v \in V : y \text{ in } \dot{v}\}$, then $\{s \in S : x \text{ in } \dot{s}\} = \{s \in S : y \text{ in } \dot{s}\}$ and $\{z \in Z : x \text{ in } \overset{\circ}{z}\} = \{z \in Z : y \text{ in } \overset{\circ}{z}\}$, which by the very definition of a syllogistic scheme (cf. Definition 6.1) implies $x = y$ and in particular $\dot{x} = \dot{y}$.

From the fact that S is an extensional set of nodes for $(V, \dot{\ })$, it follows in view of Lemma 2.11 that $(V, \dot{\ })$ has an extensional grafting $(U, \ddot{\ })$, with $\ddot{\ }$ a bijection from U onto $\{\{v\} : v \text{ in } V \setminus S\}$, which induces a function $'$ on V defined by:

$$w' = \{v' : v \text{ in } V \text{ and } w \text{ in } \dot{v}\} \cup \{u \in U : \{w\} = \ddot{u}\}.$$

Moreover, such a function has the property that if $v' = w'$ then $v, w \in S$ and $\dot{v} = \dot{w}$. Putting:

$$\begin{cases} x_{n\ell}^M = x_{n0}' & \text{for all } x_{n\ell} \text{ in } X \\ z^M = \{t' : t \text{ in } I_z\} & \text{if } z \in Z \setminus F \\ y^M = \{y'\} & \text{if } y \in Y \setminus (Z \setminus F), \end{cases}$$

it is an easy matter to verify that M so defined is a model of σ_{ZF}. Indeed the very definition of M implies at once that $y^M \neq \emptyset$ for all y in Y. Also, if $x^M \in y^M$ for some x in X and y in Y, then there exist s in S such that $s^M \in y^M$ which, in turn, implies $s' = t'$ for some t in $V \setminus X$, a contradiction. This shows that M satisfies $\underset{\substack{x \in X \\ y \in Y}}{\&}\ x \notin y$. Much in the same way it can be seen that M also satisfies the remaining conjuncts of δ (see Definition 6.1).

Furthermore $s_n^M = x_{n1}^M = \cdots = x_{nL_n}^M$, $n = 0, 1, \ldots, N$, is immediate. In addition, for each s in S we have

$$\begin{aligned} s^M = s' &= \{v' : v \text{ in } V \text{ and } s \text{ in } \dot{v}\} \cup \{u \in U : \ddot{u} = \{s\}\} \\ &= \{t' : t \text{ in } F \cup (\textstyle\bigcup_{z \in Z \setminus F} I_z) \text{ and } s \text{ in } \dot{t}\} \cup \\ &\quad \{r' : r \text{ in } S \text{ and } s \text{ in } \dot{r}\} \\ &= \{f' : f \text{ in } F \text{ and } s \text{ in } \overset{\circ}{f}\} \cup \\ &\quad \{t' : t \text{ in } I_z \text{ with } z \text{ in } Z \setminus F \text{ and } s \text{ in } \overset{\circ}{z}\} \cup \\ &\quad \{r^M : r \text{ in } S \text{ and } s \text{ in } \dot{r}\} \\ &= \bigcup_{\substack{z \in Z \\ s \in \overset{\circ}{z}}} z^M \cup \bigcup_{\substack{r \in S \\ s \in \dot{r}}} \{r^M\}. \end{aligned}$$

Finally, if $z \in F$, then $z^M = \{z'\}$ showing that z^M is finite, whereas if $z \in Z \setminus F$ then $z^M = \{t' : t \text{ in } I_z\}$, which is infinite. The last paragraph proves that M satisfies ϱ_{ZF} too, hence M is a model of σ_{ZF}, proving the lemma. ∎

We end this section by proving that every assignment on X is a model of some syllogistic scheme over X.

Theorem 6.3 (Exhaustivity) *Let Σ be the (finite) family of all syllogistic schemes over X. Then $(\exists_{y \in Y} y)\left(\bigvee_{\sigma \in \Sigma} \sigma\right)$ is valid.*

Proof. Given an assignment $g : X \to V$, put $x \sim_g y$ if and only if $x^g = y^g$, for x, y in X. Let S_g be defined as usual with respect to $\sim_g$. For v, w in S_g we put

$$w \in \dot{v} \text{ if and only if } v^g \in w^g .$$

Note that $(S_g, \dot{})$ so defined is a DAG. Moreover we define Z to be the set of all z in Y such that $\overset{\circ}{z} \subseteq S_g$ and $\hat{z} \neq \emptyset$, where

$$\hat{z} =_{\text{Def}} \left(\left(\bigcap_{v \in \overset{\circ}{z}} v^g\right) \setminus \left(\bigcup_{w \in S \setminus \overset{\circ}{z}} w^g\right)\right) \setminus \{x^g : x \text{ in } S_g\} .$$

Finally, we put

$$F =_{\text{Def}} \{f \in Z : \hat{f} \text{ is a finite set}\} .$$

This defines a syllogistic scheme σ_{ZF} over X.

Now it is an easy matter to extend g over $X \cup Y$ in such a way as to satisfy σ_{ZF}. As a matter of fact it suffices to put $z^g = \hat{z}$ for all z in Z, and to put

$$w_1^g = \{\{x^g : x \text{ in } S_g\}\}, w_2^g = \{w_1^g\}, \dots, w_R^g = \{w_{R-1}^g\},$$

where $w_1 \prec \dots \prec w_R$ are the elements of $Y \setminus Z$. Details are left to the reader. ∎

6.3   Syllogistic schemes as generators of the models of formulae

Let p be a conjunction consisting of literals of the following types:

$$
\begin{array}{ll}
(=) & x = y, x = y \cup z, x = y \setminus z, \\
(\neq) & x \neq y, \\
(\in, \notin) & x \in y, x \notin y, \\
(\{\bullet\}) & x = \{y\}, \\
(F, \not{F}) & Finite\, x, \neg\, Finite\, x,
\end{array}
$$

where each x, y, z is a set variable.

The next lemma shows that if a model M of p is also a model for some syllogistic scheme σ_{ZF} then every model of σ_{ZF} satisfies p too.

Lemma 6.4 *Let p be a conjunction of literals of the types listed above and let σ_{ZF} be a syllogistic scheme over the set X of variables occurring in p, relative to $\sim$, $(S, \dot{})$. Then either the implication $\sigma_{ZF} \to p$ is valid or $\sigma_{ZF} \to \neg p$ is valid. Moreover it is possible to establish algorithmically which of the two cases holds.*

Proof. It is enough to show that for every literal λ in p, either $\sigma_{ZF} \to \lambda$ is valid or $\sigma_{ZF} \to \neg \lambda$ is valid and, moreover, there is an algorithm to recognize which is the case. Observe that the following statements hold for all v, w, w_0, w_1 in X and s_v, s_w, s_{w_0}, s_{w_1} in S:

(a) As a consequence of Lemma 6.1 we have:

    (a1) $v \sim w$ implies that $\sigma_{ZF} \to v = w$ is valid;

    (a2) $v \not\sim w$ implies that $\sigma_{ZF} \to v \neq w$ is valid.

(b) Again from Lemma 6.1 we obtain:

    (b1) $v \sim s_v \& w \sim s_w \& s_w \in \dot{s}_v$ implies that $\sigma_{ZF} \to v \in w$ is valid;

    (b2) $v \sim s_v \& w \sim s_w \& s_w \notin \dot{s}_v$ implies that $\sigma_{ZF} \to v \notin w$ is valid.

(c1) $v \sim s_v \& w_0 \sim s_{w_0} \& w_1 \sim s_{w_1} \&$
    $\{ z \in Z : s_v \in \overset{\circ}{z} \} = \{ z \in Z : s_{w_0} \in \overset{\circ}{z} \vee s_{w_1} \in \overset{\circ}{z} \} \&$
    $\{ x \in S : s_v \in \dot{x} \} = \{ x \in S : s_{w_0} \in \dot{x} \vee s_{w_1} \in \dot{x} \}$
    implies that $\sigma_{ZF} \to v = w_0 \cup w_1$ is valid;

(c2) $v \sim s_v \& w_0 \sim s_{w_0} \& w_1 \sim s_{w_1} \&$
    $(\{ z \in Z : s_v \in \overset{\circ}{z} \} \neq \{ z \in Z : s_{w_0} \in \overset{\circ}{z} \vee s_{w_1} \in \overset{\circ}{z} \} \vee$
    $\{ x \in S : s_v \in \dot{x} \} \neq \{ x \in S : s_{w_0} \in \dot{x} \vee s_{w_1} \in \dot{x} \})$
    implies that $\sigma_{ZF} \to v \neq w_0 \cup w_1$ is valid;

(d) Changing $\vee$ into $\& \neg$ and $\cup$ into $\setminus$ in (c1) and (c2), we obtain the analogous statements for the literal $v = w_0 \setminus w_1$.

(e1) $v \sim s_v \& w \sim s_w \& \{ z \in Z : s_v \in \overset{\circ}{z} \} = \emptyset \& \{ x \in S : s_v \in \dot{x} \} = \{ s_w \}$
    implies that $\sigma_{ZF} \to v = \{ w \}$ is valid;

(e2) $v \sim s_v \& w \sim s_w \& (\{ z \in Z : s_v \in \overset{\circ}{z} \} \neq \emptyset \vee \{ x \in S : s_v \in \dot{x} \} \neq \{ s_w \})$
    implies that $\sigma_{ZF} \to v \neq \{ w \}$ is valid;

(f) From Lemma 6.2 it follows that

    (f1) $v \sim s_v \& s_v \in \bigcup_{z \in Z \setminus F} \overset{\circ}{z}$ implies that $\sigma_{ZF} \to \neg \, Finite \, v$ is valid;

    (f2) $v \sim s_v \& s_v \notin \bigcup_{z \in Z \setminus F} \overset{\circ}{z}$ implies that $\sigma_{ZF} \to Finite \, v$ is valid.

                                                               ■

An immediate consequence of Lemma 6.4 is contained in the following corollary.

Corollary 6.1 (Exhaustivity) *Let p be a conjunction as above and let* Σ *be the set of all syllogistic schemes over the set X of variables occurring in p. Then it is possible to determine a subset* Σ_p *of* Σ *such that*

$$(\star) \qquad p \leftrightarrow (\exists_{y \in Y}\, y) \left(\bigvee_{\sigma \in \Sigma_p} \sigma \right)$$

is valid.

Proof. Put

$$\Sigma_p = \{\, \sigma \in \Sigma : \sigma \to p \text{ is valid}\,\}.$$

The preceding lemma implies that Σ_p can be effectively constructed. It only remains to show that $(\star)$ holds true for such a choice of Σ_p.

Since $\sigma \to p$ is valid for all σ in Σ_p, it follows that

$$(\exists_{y \in Y}\, y) \left(\bigvee_{\sigma \in \Sigma_p} \sigma \right) \to p$$

is valid.

On the other hand Theorem 6.3 shows that $(\exists_{y \in Y}\, y) (\bigvee_{\sigma \in \Sigma} \sigma)$ is valid, and therefore *a fortiori* $p \to (\exists_{y \in Y}\, y) (\bigvee_{\sigma \in \Sigma} \sigma)$ is valid too. However, the preceding lemma implies that $p \to (\forall_{y \in Y}\, y) (\&_{\sigma \in \Sigma \setminus \Sigma_p} \neg\sigma)$ is valid; hence $p \to (\exists_{y \in Y}\, y)(\bigvee_{\sigma \in \Sigma_p} \neg\sigma)$ is valid too. This concludes the proof of the corollary. ∎

Let MLSSF (MultiLevel Syllogistic extended by the Singleton operator and the predicate *Finite*) denote the class of unquantified formulae whose operators are $\emptyset, \cap, \setminus, \cup, \{\bullet, \dots, \bullet\}$ and whose relators are $\in, \subseteq, =, Finite$. The previous results give a *canonization method* for MLSSF. This is stated in the following theorem:

Theorem 6.4 (Canonization) *Let q be a formula of MLSSF. Then there is an algorithm for determining certain sets* $X_1, \dots, X_D, Y_0$ *of variables, and, for each* $d = 1, \dots, D$, *a family* Σ_d *of syllogistic schemes over* X_d, *so that*

$$q \leftrightarrow (\exists_{y \in Y_0 \cup Y_1 \cup \cdots \cup Y_D}\, y) \left(\bigvee_{d=1}^{D} \bigvee_{\sigma \in \Sigma_d} \sigma \right)$$

is valid, where each Y_d *is related to* X_d *in the same manner Y was related to X in Section 6.2.* □

An immediate consequence of Theorem 6.4 is

Corollary 6.2 (Decidability) *The class MLSSF of formulae has a solvable satisfiability problem.*

Proof. Just observe that the preceding theorem implies that q can be put in the (syllogistic canonical) form

$$(\exists_{y \in Y_0 \cup Y_1 \cup \cdots \cup Y_D}\, y)\ \left(\bigvee_{d=1}^{D} \bigvee_{\sigma \in \Sigma_d} \sigma \right).$$

Then Theorem 6.2 ensures that q is solvable if and only if $\Sigma_d \neq \emptyset$ for some d in $\{1,\ldots,D\}$. (Any decision algorithm of practical value will of course avoid the detour through the syllogistic canonical form. How to do this will be explained in the next few sections.)   ■

Proof of Theorem 6.4. Let q be a formula of MLSSF. Compound terms in q can be easily 'simplified' by adding suitable new literals. For example $x \cap (y \setminus (z \cup \{v, w\})) = u$ can be simplified into

$$v_1 = \{v\}\ \&\ w_1 = \{w\}\ \&\ v_2 = v_1 \cup w_1\ \&\ z_1 = z \cup v_2$$
$$\&\ y_1 = y \setminus z_1\ \&\ x_1 = x \setminus y_1\ \&\ u = x \setminus x_1.$$

Note that such a simplification is both effective and satisfiability-preserving.

Assume that q^* is the result of so eliminating compound terms from q and that Y_0 is the collection of all variables introduced by the elimination process, so that $q \leftrightarrow (\exists_{y \in Y_0})\, q^*$ is valid. By putting q^* in disjunctive normal form we obtain a disjunction $\bigvee_{d=1}^{D} q_d$, where q_d is a conjunction of literals of the forms $(=),\ldots,(\not{F})$ shown at the beginning of this section. Now let X_d be formed by the variables of q_d, and apply Corollary 6.1.   ■

Thanks to Lemma 3.4, MLSSF can be replaced in this theorem and in its corollary by the family of all left prenex formulae with matrix belonging to MLSSF (cf. Definition 3.10).

6.4   A decision algorithm for MLSF

We now restrict our attention to those formulae of MLSSF (see preceding section) that do not involve the construct $\{\bullet,\ldots,\bullet\}$. The resulting collection of formulae is indicated MLSF. By a simple normalization process (see Section 3.8), the satisfiability problem for MLSF can be reduced to the problem of testing for satisfiability conjunctions of literals of the types $(=)$, $(\neq)$, $(\in, \notin)$, $(F, \not{F})$ introduced at the beginning of Section 6.3.

Let p be such a conjunction, and let X be the set of all variables appearing in p. We denote by $p_=$ the conjunction of the literals of type $(=)$ belonging to p. Moreover, p_* will stand for the propositional formula obtained from $p_=$ by replacing the symbols $=$, $\cup$, and $\setminus$ by $\leftrightarrow$, $\vee$, and $\&\neg$ respectively, accordingly regarding each set variable as a propositional variable. (Notice that in our use of the word 'propositional' we are conforming here to a well-established tradition —see, e.g., [Men64]. In particular, by 'propositional variable' we mean a variable that ranges over the truth values *FALSE, TRUE*.)

In the following definition of a *p-compatible* equivalence relation $\sim$ over X, our aim is to capture the properties that $\sim$ must enjoy in order it can be 'induced' by a model M of p, in the sense that $x \sim y$ if and only if $x = y$ is true in M, for all x, y in X. It will be obvious that there exists an algorithm for establishing whether any given $\sim$ is p-compatible or not.

Definition 6.2 *Let $\sim$ be an equivalence relation over the set X of all variables that occur in a given conjunction p of literals $(=)$, $(\neq)$, $(\in, \notin)$, and $(F, \boldsymbol{F})$; moreover, let $p_\sim$ denote the propositional formula $p_* \,\&\, \&_{x\sim y} (x \leftrightarrow y)$.*

An acceptable place *of the pair p, $\sim$ is any set $P \subseteq X$ whose characteristic function χ_P is a model of $p_\sim$, where*

$$\chi_P(x) = \begin{cases} \text{TRUE} & \text{if } x \in P \\ \text{FALSE} & \text{if } x \in X \setminus P. \end{cases}$$

By abuse of language, we will often say that a place satisfies a given propositional formula to mean that its characteristic function satisfies it.

An equivalence relation $\sim$ over X is said to be p-compatible iff the following conditions are met:

(a) $v \sim w$ *if (and only if) $v \leftrightarrow w$ is satisfied by every acceptable place of p, $\sim$;*

(b) *if $v \neq w$ belongs to p, then $v \not\sim w$;*

(c) *if $v \in z_0$, $w \notin z_1$ belong to p and $z_0 \sim z_1$, then $v \not\sim w$;*

(d) *if $\neg$ Finite v belongs to p then there must exist an acceptable place P_v such that $v \in P_v$ and P_v does not contain any variable w for which Finite w belongs to p.* □

The following lemma shows that any satisfiable conjunction p admits p-compatible equivalence relations, induced by the models of p:

Lemma 6.5 *Let M be a model of p, and let $x \sim_M y$ if and only if $x^M = y^M$. Then*

(I) $P_\xi = \{x \in X : \xi \in x^M\}$ *is an acceptable place of p, $\sim_M$, where $\xi \in \bigcup_{x \in X} x^M$;*

(II) $\sim_M$ *is p-compatible.*

Proof. Let $\xi \in \bigcup_{x \in X} x^M$. Clearly if $x \sim_M y$, then $\chi_{P_\xi}(x) = \chi_{P_\xi}(y)$, i.e. $(x \leftrightarrow y)^{\chi_{P_\xi}} = TRUE$. Moreover if $x \leftrightarrow (y \vee z)$ is in p_*, then $x = y \cup z$ is in p and $x^M = y^M \cup z^M$. Hence $\xi \in x^M$ if and only if either $\xi \in y^M$ or $\xi \in z^M$, i.e. $\chi_{P_\xi}(x) = \chi_{P_\xi}(y) \vee \chi_{P_\xi}(z)$, proving that χ_{P_ξ} is a model for $x \leftrightarrow y \vee z$. Analogously it can be shown that χ_{P_ξ} satisfies all remaining conjuncts in

$p_\sim$, which proves that χ_{P_ξ} is an acceptable place of p, $\sim_M$, establishing (I). Note in addition that χ_{P_ξ} also satisfies all formulae $\neg(v \leftrightarrow w)$ for which $\xi \in (v^M \setminus w^M) \cup (w^M \setminus v^M)$.

Next we show that $\sim_M$ is p-compatible. First of all, it is clear that $\sim_M$ is an equivalence relation. If $v \not\sim_M w$, for any two variables v, w occurring in p, then $v^M \neq w^M$, so that there exists $\xi \in (v^M \setminus w^M) \cup (w^M \setminus v^M)$. Therefore $(v \leftrightarrow w)^{\chi_{P_\xi}} = FALSE$, which proves that (a) of Definition 6.2 holds.

(b) is an immediate consequence of the very definition of $\sim_M$, in view of the fact that M satisfies p.

If $v \in z_0$, $w \notin z_1$ belong to p and $z_0 \sim_M z_1$, then $v^M \in z_0^M$, $w^M \notin z_1^M$, and $z_0^M = z_1^M$, which imply $v^M \neq w^M$ and in turn $v \not\sim_M w$. Having thus shown that (c) of Definition 6.2 holds too, we now proceed to prove the only remaining condition, (d). Let us assume that $\neg\, Finite\ v$ belongs to p. Hence v^M is infinite. By (I), P_ξ is an acceptable place of p, $\sim_M$ for every ξ in v^M; moreover, since obviously there are only finitely many acceptable places of p, $\sim_M$, it follows that there must exist an acceptable place P such that $\{\xi \in v^M : P_\xi = P\}$ is infinite. So, if $w \in P$, then $\{\xi \in v^M : P_\xi = P\} \subseteq w^M$, i.e. w^M is infinite, and hence $Finite\ w$ cannot belong to p. This completes the proof of the p-compatibility of $\sim_M$, establishing the lemma. ∎

An important equivalence relation is introduced by the following definition.

Definition 6.3 *Given p as above, for any x and y occurring in p we put*

$$x \sim_p y \text{ if and only if } p_* \rightarrow (x \leftrightarrow y) \text{ is tautological,}$$

where p_* is obtained from p as in Definition 6.2. □

Lemma 6.6 $\sim_p$ *is an equivalence relation.*

Proof. The lemma follows at once from the reflexivity, symmetry and transitivity of $\leftrightarrow$. ∎

An important property of the equivalence relation $\sim_p$ is stated in the following lemma.

Lemma 6.7 *Let $\sim$ be a p-compatible equivalence relation. Then*

(i) if $x \sim_p y$ then $x \sim y$, for x, y in X;

(ii) $\sim_p$ is p-compatible.

(In other words, $\sim_p$ is the finest p-compatible equivalence relation over the set X of variables occurring in p).

Proof. Let $x \sim_p y$. Then $p_* \to (x \leftrightarrow y)$ is a tautology. If P is an acceptable place of p, $\sim$, then χ_P satisfies $p_\sim$, and in particular satisfies p_*. Hence $(x \leftrightarrow y)^{\chi_P} = TRUE$. This shows that $x \leftrightarrow y$ is satisfied by all acceptable places of p, $\sim$. Therefore the p-compatibility of $\sim$ implies $x \sim y$, establishing (i).

In order to prove that $\sim_p$ is p-compatible, we show that conditions (a)-(d) of Definition 6.2 are met for $\sim = \sim_p$. Observe that the acceptable places of p, $\sim_p$ are those sets P such that χ_P satisfies p_*. Therefore, if $v \leftrightarrow w$ is satisfied by every acceptable place of p, $\sim_p$, it follows that $p_* \to (v \leftrightarrow w)$ is a tautology, i.e. $v \sim_p w$, proving (a).

To establish (b), assume that $v \neq w$ occurs in p. If $v \sim_p w$, then by (i) above $v \sim w$, contradicting the p-compatibility of $\sim$. Thus $v \not\sim_p w$, and (b) is proved.

Next suppose that $z_0 \sim_p z_1$, and assume also that $v \in z_0$ and $w \notin z_1$ occur in p. Then again by (i) $z_0 \sim z_1$, implying $v \not\sim w$. Thus, as above, $v \not\sim_p w$, concluding the verification of (c).

Finally, let $\neg Finite\, v$ occur in p. From the p-compatibility of $\sim$ it follows that there exists an acceptable place P_v of p, $\sim$ such that $v \in P_v$ and P_v does not contain any variable w for which $Finite\, w$ occurs in p. Then it is enough to observe that P_v is also an acceptable place of p, $\sim_p$. This shows that (d) of Definition 6.2 is also fulfilled, which in turn establishes (ii) of the present lemma.   ∎

An immediate consequence of the preceding lemma is given in the following corollary.

Corollary 6.3 *If p is satisfiable, then $\sim_p$ is p-compatible.*

Proof. Let M be a model of p. Lemma 6.5 shows that $\sim_M$ is p-compatible, and therefore Lemma 6.7 entails the p-compatibility of $\sim_p$.   ∎

The preceding corollary says that the p-compatibility of $\sim_p$ is a necessary (algorithmically verifiable) condition for p to be satisfiable. In what follows, we will bring to light another necessary condition for the satisfiability of p. Taken together with the p-compatibility of $\sim_p$, the latter condition will suffice to imply that p admits a model M_p such that $\sim_p$ coincides with $\sim_{M_p}$ (cf. Lemma 6.5).

In view of Lemmas 6.5(II), 6.7(i), this will entail that when p is satisfiable then any two variables x, y of p are $\sim_{M_p}$-equivalent if and only if $x \sim_M y$ (i.e. the set-values of x and y coincide) in every model M of p.

We incidentally note that the existence of a finest equivalence $\sim_M$, with M a model of p, is no longer insured if one adds new operators to those admitted in MLSF. For example, the models of the formula

$$z = z \setminus z \,\&\, s = \{x\} \,\&\, y \setminus s = z$$

of MLSSF can be grouped into two classes, with

$$y \sim_{M_0} z, \quad y \not\sim_{M_0} s,$$

$$y \not\sim_{M_1} z, \quad y \sim_{M_1} s,$$

(and hence

$$y \sim_{M_0} z \text{ not implying } y \sim_{M_1} z,$$

$$y \sim_{M_1} s \text{ not implying } y \sim_{M_0} s)$$

whenever M_0 belongs to the first class and M_1 belongs to the second. Other formulae with similar pathology are $y \in \{a, b\} \,\&\, a \neq b$ and $x, y, z \in \{a, b\}$.

The above discussion indicates that $\sim_p$ is mainly related to the equalities holding in a sought model of p. To fully take into account the membership literals as well, we introduce the p-$\sim$-compatible directed acyclic graphs.

Definition 6.4 *Let $\sim$ be a p-compatible equivalence relation and let S be the collection of representatives of the variables occurring in p (for example, if we assume that all variables of set theory are arranged in a denumerable sequence, then we can choose the first variable in each equivalence class C as the representative of C). A DAG $(S, \dot{\ })$ (cf. Definition 2.4) is said to be p-$\sim$-compatible iff*

(a) *for each v in S, there is an acceptable place P_v of p, $\sim$ such that $\dot{v} = P_v \cap S$;*

(b) *if $v \in w$ (respectively, $v \notin w$) occurs in p, then $s_w \in \dot{s_v}$, (respectively, $s_w \notin \dot{s_v}$) where $s_v, s_w \in S$, and $s_v \sim v$, $s_w \sim w$.*   □

Later in this section we will show that a normalized conjunction p of MLSF has a model if and only if $\sim_p$ is p-compatible and there exists a p-$\sim_p$-compatible DAG $(S_p, \dot{\ })$, where S_p is the set of all $\sim_p$-representatives. This provides at once a satisfiability algorithm (alternative to the one described in Corollary 6.2) for MLSF: in fact, the number of possible DAGs $(S_p, \dot{\ })$ is clearly finite and, moreover, there is an algorithm to test whether a given DAG is p-$\sim_p$-compatible.

A first step towards proving the completeness of the above test is the following lemma.

Lemma 6.8 *If p is satisfiable, then there exists a p-compatible equivalence relation $\sim$ which admits a p-$\sim$-compatible DAG.*

Proof. Let M be a model of p. Consider the relation $\sim_M$ defined by putting $x \sim_M y$ if and only if $x^M = y^M$. (II) of Lemma 6.5 shows that $\sim_M$ is p-compatible. It only remains to be shown that the set S_M of $\sim_M$-representatives of the variables of p can be given the structure of a p-$\sim_M$-compatible DAG. To this end we put

$$\dot{v} = \{ w \in S_M : v^M \in w^M \}, \text{ for every } v \text{ in } S_M,$$

so that $\dot{v} = \{\, x \in X : v^M \in x^M \,\} \cap S_M$, where X denotes the collection of all variables occurring in p. Since by Lemma 6.5(I) the set $\{\, x \in X : v^M \in x^M \,\}$ is an acceptable place of p, $\sim_M$, (a) of Definition 6.4 is satisfied. Observing that (b) of the same definition is an immediate consequence of the fact that M is a model of p, it follows that $(S_M ,\ \cdot\ )$ is a p-$\sim_M$-compatible DAG, which proves the lemma. ∎

Lemma 6.9 *Let p be a normalized conjunction of MLSF. If there exists a p-compatible equivalence relation $\sim$ admitting a p-$\sim$-compatible DAG, then $\sim_p$ (cf. Definition 6.3) is also p-compatible and admits a p-$\sim_p$-compatible DAG.*

Proof. Lemma 6.7 ensures the p-compatibility of $\sim_p$. Next, let $(S_\sim ,\ \cdot\ )$ be a p-$\sim$-compatible DAG. We will show how the set S_p of $\sim_p$-representatives of the variables occurring in p can be given a structure of p-$\sim_p$-compatible DAG.

For each v in S_p let $v_* \in S_\sim$ be such that $v \sim v_*$ and put $\overset{*}{v} = \{\, w \in S_p : w_* \in \dot{v}_* \,\}$. We must simply show that $(S_p ,\ \overset{*}{\ }\ )$ is a p-$\sim_p$-compatible DAG. Since $(S_\sim ,\ \cdot\ )$ is a DAG, so is $(S_p ,\ \overset{*}{\ }\ )$ (cf. Definition 2.4). The p-$\sim_p$-compatibility of $(S_p ,\ \overset{*}{\ }\ )$ can be proved as follows. Let $v \in S_p$. As $(S_\sim ,\ \cdot\ )$ is p-$\sim$-compatible, $\dot{v}_* = P \cap S_\sim$, for some acceptable place P of p, $\sim$. We have $\overset{*}{v} = P \cap S_p$. Indeed, if $w \in \overset{*}{v}$ then $w_* \in \dot{v}_* = P \cap S_\sim$. In particular $w_* \in P$, and since $w \sim w_*$ we have $w \in P$. Therefore $w \in P \cap S_p$. Conversely, if $w \in P \cap S_p$, from the fact that P is an acceptable place of p, $\sim$ it follows $w_* \in P \cap S_\sim = \dot{v}_*$, so that $w \in \overset{*}{v}$. Hence (a) of Definition 6.4 is proved for $(S_p ,\ \overset{*}{\ }\ )$.

Assume now that $S_\sim \ni s_v \sim v \sim_p g_v \in S_p$ and that $S_\sim \ni s_w \sim w \sim_p g_w \in S_p$. Lemma 6.7(i) implies that $g_v \sim s_v$ and $g_w \sim s_w$, so that $\overset{*}{g}_v = \{\, x \in S_p : x_* \in \dot{s}_v \,\}$. If $v \in w$ occurs in p, then by the p-$\sim$-compatibility of $(S_\sim ,\ \cdot\ )$ we have $s_w \in \dot{s}_v$ showing $g_w \in \overset{*}{g}_v$. If instead $v \notin w$ is a literal of p, then $g_w \notin \overset{*}{g}_v$, because otherwise $(g_w)_* \in \dot{s}_v$, i.e. $s_w \in \dot{s}_v$, which contradicts the hypothesis of p-$\sim$-compatibility of $(S_\sim ,\ \cdot\ )$. This completes the proof that $(S_p ,\ \overset{*}{\ }\ )$ is p-$\sim_p$-compatible, and the lemma is established. ∎

From the preceding two lemmas the following corollary is immediately derived.

Corollary 6.4 (Completeness) *Let p be a normalized conjunction of MLSF. If p is satisfiable then $\sim_p$ is p-compatible and it admits a p-$\sim_p$-compatible DAG. Moreover, there is an algorithm to test whether $\sim_p$ is p-compatible and it admits a p-$\sim_p$-compatible DAG.* □

The following lemma gives the soundness of the satisfiability test whose completeness has just been established.

Lemma 6.10 (Soundness) *Let p be a normalized conjunction of MLSF and let X be the collection of all variables occurring in p. If there exists a p-compatible equivalence relation $\sim$ which admits a p-$\sim$-compatible DAG $(S, \cdot)$, then p is satisfiable.*

Proof. We will show that from the hypothesis it follows that there exists a syllogistic scheme σ_{ZF} relative to X, $\sim$, $\cdot$ whose models correctly model p. This, in view of Theorem 6.2, gives the result. We construct such a syllogistic scheme as follows.

For each pair s_0, s_1 of distinct variables in S for which $\{x : s_0 \in \dot{x}\} = \{x : s_1 \in \dot{x}\}$, we consider an acceptable place $P_{s_0 s_1}$ of p, $\sim$ such that

$$s_0 \in P_{s_0 s_1} \text{ iff } s_1 \notin P_{s_0 s_1}.$$

Moreover, for each s in S for which $\neg Finite\ v$ occurs in p with $s \sim v$, we consider an acceptable place P_s such that $s \in P_s$ and P_s does not contain any variable w with $Finite\ w$ occurring in p. Note that the existence of such places $P_{s_0 s_1}$ and P_s is ensured by the p-compatibility of $\sim$ (cf. Definition 6.2). Let Y be a set of variables, disjoint from X, and let $\overset{\circ}{}$ be a bijection from Y onto $Pow(X) \setminus \{\emptyset\}$. We denote by Z the set of all y in Y with $\overset{\circ}{y} = P \cap S$, where P is a place of the type $P_{s_0 s_1}$ or of the type P_s. Moreover we denote by F the set of all z in Z with $\overset{\circ}{z} = P \cap S$ where P is *not* of the form P_s. It is an easy matter to verify that σ_{ZF} is a syllogistic scheme over X, $\sim$, $\cdot$.

In order to complete the proof of the lemma it only remains to show that every model of σ_{ZF} is a model of p. Let, therefore, M be a model of σ_{ZF}. Lemmas 6.1 and 6.2 imply at once that M correctly models all literals in p of type $x = y$, $x \neq y$, $x \in y$, $x \notin y$, $Finite\ x$, and $\neg Finite\ x$. If $v = w_1 \cup w_2$ occurs in p, then supposing $v \sim s_v$, $w_1 \sim s_{w_1}$, and $w_2 \sim s_{w_2}$, we have (cf. Definition 6.1)

$$v^M = s_v^M = \bigcup_{\substack{z \in Z \\ s_v \in \overset{\circ}{z}}} z^M \cup \bigcup_{\substack{r \in S \\ s_v \in \dot{r}}} \{r^M\}$$

$$= \left(\bigcup_{\substack{z \in Z \\ s_{w_1} \in \overset{\circ}{z}}} z^M \cup \bigcup_{\substack{r \in S \\ s_{w_1} \in \dot{r}}} \{r^M\} \right) \cup \left(\bigcup_{\substack{z \in Z \\ s_{w_2} \in \overset{\circ}{z}}} z^M \cup \bigcup_{\substack{r \in S \\ s_{w_2} \in \dot{r}}} \{r^M\} \right)$$

$$= s_{w_1}^M \cup s_{w_2}^M = w_1^M \cup w_2^M,$$

since, by the p-$\sim$-compatibility of $(S, \cdot)$, sets $\dot{r}$ and $\overset{\circ}{z}$ are of the form $P \cap S$ for some acceptable place P of p, $\sim$, so that $s_v \in P$ if and only if $s_{w_1} \in P$

or $s_{w_2} \in P$. This shows that M is a model of the literals in p of type $v = w_1 \cup w_2$. Analogously it can be shown that the remaining clauses of type $v = w_1 \setminus w_2$ are also modelled correctly, thus showing that M is model of p. This completes the proof of the lemma.   ∎

Combining together Corollary 6.4 and Lemma 6.10 we rediscover that the class MLSF of formulae has a decidable satisfiability problem:

Corollary 6.5 (Decidability) *An algorithm for testing a normalized conjunction p of MLSF for satisfiability is:*

(1) construct the equivalence relation $\sim_p$;

(2) check whether $\sim_p$ is p-compatible;

(3) check whether $\sim_p$ admits a p-$\sim_p$-compatible DAG.   □

6.5 An improved decision algorithm for MLSF

In this section we refine the decision algorithm for MLSF introduced in Corollary 6.5, by developing the details of a possible implementation thereof.

6.5.1 Determination of $\sim_p$

The construction of the equivalence relation $\sim_p$ (see Definition 6.3) starts with the generation of a model of p_*. Models can be returned by any propositional decider—say the Davis-Putnam procedure [CL73, Lov78]. Note that at least one model exists for p_*: this is the trivial model in which all variables are *false*. The generated model splits the variables into two disjoint sets according to their truth value. Variables lying on different sides of this partition can *not* be $\sim_p$-equivalent. Hence, we need to bind together pairs x, y of variables such that either both x and y or both $\neg x$ and $\neg y$ are true in the model. To do this, we look for a model of $p_* \ \& \ \neg (x \leftrightarrow y)$. Two situations may arise: either no model exists, in which case $x \sim_p y$, or the new model once again partitions the variables. The same line of reasoning followed in connection with the previous model can now be repeated: pairs of variables such that they, or their negations, do not both hold in the model are eliminated from the set of pairs of potentially equivalent variables.

As the algorithm proceeds in this 'divide-and-conquer' fashion, the set of potential equivalents narrows down to the actual equivalents. The rate at which this happens depends on the propositional decider generating models. In order to achieve the best possible performance, it would be necessary to design a model generator providing nearly equally-sized partitions of variables, thus ensuring a massive pruning during each iteration.

Using a 'pidgin' variant of SETL (a set-theoretic programming language described in [SDDS86] which should be rather self-explanatory in our context), and assuming that *clauset* is a conjunction of disjunctions of propositional literals that is satisfiable if and only if p_* is satisfiable, the above algorithm can be specified as follows:

Algorithm 6.1 *procedure entailingEq(clauset);*

This routine determines all equivalences $x \leftrightarrow y$ (where x and y are distinct variables) which are logical consequences of *clauset*.

Method: we are assuming that *clauset* is a set of disjunctive propositional clauses which admits a model (usually the model that maps every variable to *FALSE*). We begin by calculating a model and remark that $x \leftrightarrow y$ can be a consequence of *clauset* only if both x and y or both $\neg x$ and $\neg y$ hold in the model. For each of the potential equalities so determined, we look for a model M of *clauset* $\& \neg (x \leftrightarrow y)$. If no such model exists, then $x \leftrightarrow y$ is a consequence of *clauset*; otherwise, we take advantage of the model M that we have just found and remove from the potential equivalences every $xx \leftrightarrow yy$ such that neither both xx, yy nor both $\neg xx$, $\neg yy$ hold in M.

$$
\begin{array}{lll}
poteq & \leftarrow \emptyset; & \text{potential equalities} \\
neceq & \leftarrow \emptyset; & \text{necessary equalities} \\
posmod & \leftarrow \emptyset; & \text{positive part of propositional model} \\
negmod & \leftarrow \emptyset; & \text{negative part of propositional model} \\
model & \leftarrow propositionalDecider(clauset); &
\end{array}
$$

Propositional variables are encoded by positive integers.

$+(ABS\, x) \in model$ indicates: x is true in the model;

$-(ABS\, x) \in model$ indicates: x is false in model;

a variable x of *clauset* may have no value assigned to it.

Ideally (in order to achieve best performance), *propositionalDecider* should find a model subdividing variables (according to their truth-value) into two classes of as equal size as possible.

($\forall x \in model$)
    if $x > 0$ then
        $poteq \leftarrow poteq \cup \{ [x, y] : y \in posmod \};$
        $posmod \leftarrow posmod \cup \{ x \};$
    else
        $poteq \leftarrow poteq \cup \{ [-x, y] : y \in negmod \};$
        $negmod \leftarrow negmod \cup \{ -x \};$
    end if;
  end $\forall$;
(while $\exists z \in poteq$)
    $poteq \leftarrow poteq \setminus \{ z \};$
    $[x, y] \leftarrow z;$ the components of z are thus extracted
    if thereExists(model $\leftarrow$ propositionalDecider(clauset $\cup$
                $\{ \{ x, y \}, \{ -x, -y \} \}))$
    then

a propositional model exists: equality of x and y is not entailed; drop potential equalities that are false in this model

$(\forall w \in poteq)$
    if $(w(1) \notin model \vee w(2) \notin model)$ &
        $(-w(1) \notin model \vee -w(2) \notin model)$
    then
        $poteq \leftarrow poteq \setminus \{w\};$
    end if;
    end $\forall;$
  else
    $neceq \leftarrow neceq \cup \{z\};$
  end if;
end while;
return neceq;
end procedure.

                                                            □

6.5.2 p-compatibility test

Phase (2) of the decision algorithm consists in checking whether $\sim_p$ is p-compatible. This step can be directly based on Definition 6.2, whose condition (a), however, is automatically satisfied when $\sim$ coincides with $\sim_p$. Checking clause (c) of Definition 6.2 is unnecessary too; in fact this condition will hold automatically, provided phase (3) of the decision algorithm (see discussion below) is successful.

6.5.3 Generation of a p-$\sim_p$-compatible DAG

Phase (3) of the decision algorithm consists in checking that a p-$\sim_p$-compatible DAG exists. We will solve the problem in the following more general form:

Problem 6.1 *Given a finite set V and, for each v in V, a family $C_v \subseteq Pow(V)$; find a directed acyclic graph $(V, \,\dot{} \,)$ such that $\dot{v} \in C_v$ for all v in V.* □

The procedure to be described will also be able to diagnose when there is no solution $(V, \,\dot{} \,)$.

To show the correspondence between our former problem and Problem 6.1, we identify V with the set S of all representative variables associated with $\sim = \sim_p$ (see Definition 6.4). We obtain $\tilde{p}$ from p by replacing each variable x by the representative variable $\tilde{x}$ for which $x \sim \tilde{x}$. For every v in V, we indicate by p_v the formula obtained by conjoining all literals of the forms $x = y, x = y \cup z, x = y \setminus z$ in $\tilde{p}$ with:

- the literal $v = \emptyset$,

- all literals $w = \emptyset$ for which $v \notin w$ is in $\tilde{p}$,

- all literals $w \neq \emptyset$ for which $v \in w$ is in $\tilde{p}$.

Finally, for every v in V, we identify C_v with the family of those subsets W of V whose characteristic function (sending to $\emptyset$ the members of $V \setminus W$ and to $\{\emptyset\}$ the members of W) is a model of p_v.

Before we begin to describe an algorithm for the solution of Problem 6.1, let us introduce a convenient terminology. For our current purposes, we can restrict the sense of the word 'graph' to mean a DAG whose set W of nodes is contained in the given set V, and whose edges satisfy the condition

$$\dot{w} \in C_w \text{ for all } w \text{ in } W.$$

Moreover, we will call:

- *nodes*, the members of V;

- *contours of v*, where v is a node, the members of C_v (metaphorically, we could call each of these an 'admissible progeny' of v);

- *graphs* the functions
$$W \overset{\cdot}{\to} Pow(V)$$
  for which

  − $W \subseteq V$,
  − $C_w \ni \dot{w} \subseteq W$ for all w in W, and
  − there exist no *cycles*, i.e. no $(n + 2)$-tuples of the form
  $$(w_0, \dots, w_{n+1})$$
  with
  $$n \in \omega, \quad w_0 = w_{n+1} \in W,$$
  and $w_{i+1} \in \dot{w}_i$ for $i = 0, 1, \dots, n$.

Conforming with the Bourbakian tradition, we will peacefully identify a graph
$$G : W \overset{\cdot}{\to} Pow(V)$$
with the set—belonging to $\prod_{w \in W} C_w$—$\{(w, \dot{w}) : w \text{ in } W\}$ of ordered pairs. The *domain of G*, which is W, will be written $dom\,G$; the ordered pairs (w, z) with $z \in \dot{w} = G(w)$ will be called the *edges of G*, the members $w \in W$ the *nodes of G*. Every $(n + 1)$-tuple $(w_0, w_1, \dots, w_n)$, with $n \in \omega$, such that (w_{i-1}, w_i) is an edge for $i = 1, \dots, n$, is called a *path of G*. If such a path exists, we say that w_0 is an *ancestor of w_n* in G. If a node has no ancestors but itself, we say that it is a *root of* the graph.

We need to introduce a few more notions:

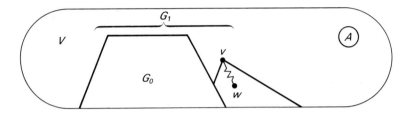

Fig. 6.1. G_1 is a v-extension of G_0 away from A

Definition 6.5 *A graph*

$$G_1 : W \dashrightarrow Pow(V)$$

is said to be an extension of *another graph*

$$G_0 : Z \dashrightarrow Pow(V)$$

iff it is the case that $Z \subseteq W$ and

$$\ddot{z} = \dot{z} \text{ for all } z \text{ in } Z.$$

If we have, in addition,

$$A \cap W = A \cap Z = \emptyset$$

(with $A \subseteq V$), then we say that G_1 is an extension of G_0 away from A. If G_1 is an extension of G_0 with

$$v \in (\, dom\, G_1 \setminus dom\, G_0\,),$$

and every node w such that

$$w \in (\, dom\, G_1 \setminus dom\, G_0\,)$$

has v among its ancestors, then G_1 is said to be a v-extension of G_0 (see Figure 6.1). □

Let us give a rather trivial example: the function $\emptyset \rightarrow Pow(V)$ is the empty graph, ◊, of which every other graph is an extension. Saying that a graph $G : W \dashrightarrow Pow(V)$ is an extension of ◊ away from A is therefore equivalent to asserting that $A \cap W = \emptyset$. Saying that G is a v-extension of ◊ amounts to saying that v is *the* root of G.

6.5.4   Algorithm

The algorithm that we will describe is animated by a self-calling routine, written
$$expand(\,v\,,\,A\,),$$
which modifies a global variable G in such a manner as to satisfy the following proposition:

Claim 6.1 *Assume that the initial value G_0 of G, when*

$$expand(\,v\,,\,A\,)$$

is called, is an extension of $\emptyset$ away from $A \cup \{\,v\,\}$, with

$$v \notin A \subseteq V.$$

Then the execution of this routine will eventually terminate, and the final value G^∞ of G will be an extension of G_0 away from A; moreover one will have

$$v \in dom\,G^\infty,$$

provided that G_0 admits an extension away from A, defined in v.   □

The procedure *expand* almost reaches our goal, which is finding a graph whose domain is the entire set V. Without loss of generality, we can in fact assume that there is a node r, not belonging to any contour, such that

$$C_r = \{\,V \setminus \{\,r\,\}\,\}.$$

In case such a node r should be missing, we could create one *ad hoc*: thus the solutions to the original problem would be in a one-to-one correspondence with the solutions to the modified problem. The simple presence of r in a graph G would suffice to insure that G exhausts all available nodes, and our algorithm could hence be very easily stated as follows:

'initialize G to $\emptyset$, and then call *expand*$(\,r\,,\,\emptyset\,)$'.

If r has been introduced artfully, then it must be deleted, together with all edges that leave r, from the final value of G.

Here we express in pidgin SETL the first version of the procedure *expand*; this will soon be improved. The correctness proof (i.e. the proof of Claim 6.1) will be developed in Section 6.6.

Algorithm 6.2

     1.   procedure expand(v, A);
     2.   ($\forall K \in C_v \mid K \cap (A \cup \{v\}) = \emptyset$)
     3.   (while $\exists w \in K \setminus dom\,G$)

4.   *expand(w, A ∪ {v});*
5.   *if w ∉ dom G then*
6.   *continue ∀ K;*
7.   *end if;*
8.   *end while ∃ w;*
9.   *G ← G ∪ {(v, K)};*
10.   *return;*
11.   *end ∀ K;*
12.   *return;*
13. end procedure.

□

It is apparent that this algorithm proceeds by trial and error in seeking a value K for the progeny $\dot{v}$ of v. A failure situation arises when the inner call

$$expand(w, A ∪ \{v\})$$

(see line 4) does not succeed in adding w to the domain of G. The algorithm escapes from the innermost loop (also bypassing lines 9 and 10) as soon as it detects such a failure (see lines 5-7), and then makes its next attempt with a brand new value K (if any exists). Of course K is to be discarded *a priori* if either $v ∈ K$ (an edge (v, v) would in fact constitute a cycle by itself), or K intersects A (the expansion of v is in fact subject to the prohibition to introduce as new nodes of G any members of A—these latter are, in a sense, 'taboo' nodes).

Let us now briefly explain the sense of the inner call

$$expand(w, A ∪ \{v\}).$$

This call is made under temporary assumption that K will become the progeny of v. Since $w ∈ K$, following the classical 'depth first' strategy (see, e.g., [AHU76]), the algorithm immediately begins to expand w, while it suspends the expansion of v. However, since the expansion of v is subject to the taboos A, w in turn inherits this set of prohibitions. As v is expected to become the father of w, it is a new taboo: as a matter of fact, by introducing v during the expansion of w one would predispose a cycle, which would be closed by the execution of line 9.

Note a good quality of the backtracking technique that we are proposing: the evolution of G is purely incremental. That is,

> after it has been added to the graph under construction, an edge is never removed; not even if, subsequently, one or more failure situations arise.

The procedure *expand(v, A)* can be speeded up—without any large additional burden in the proof of Claim 6.1—if we introduce in it a new local variable F, which keeps a record of all nodes w that have obstructed the expansion of v away from A. Also the evolution of F is, with respect to each pair (v, A), purely incremental; moreover, once w has been recorded on F, no contour K of v that includes w among its members will ever be considered a possible progeny of v in connection with the same set A of taboo nodes. Metaphorically, we can say that there is no obstacle that is not insurmountable (if the prohibitions remain the same). Each failed attempt to expand v away from A thus contributes to drastically reduce the search space. Here we present the improved version of *expand*:

Algorithm 6.3

> *1a. procedure expand(v, A);*
> *1b. $F \leftarrow \emptyset$;*
> *2.   (while $\exists K \in C_v \mid K \cap (A \cup \{v\} \cup F) = \emptyset$)*
> *3.   (while $\exists w \in K \setminus \text{dom } G$)*
> *4.   expand$(w, A \cup \{v\})$;*
> *5.   if $w \notin \text{dom } G$ then*
> *6a. $F \leftarrow F \cup \{w\}$;*
> *6b.   continue while $\exists K$;*
> *7.   end if;*
> *8.   end while $\exists w$;*
> *9. $G \leftarrow G \cup \{(v, K)\}$;*
> *10.   return;*
> *11.   end while $\exists K$;*
> *12.   return;*
> *13. end procedure.*

$\square$

6.6   Proof of correctness for the improved decision algorithm

The following lemma is crucial for our proof of Claim 6.1.

Lemma 6.11 *Let both G_1 and G_2 be extensions of G_0 away from A, and*

$$y \in (\text{dom } G_2 \setminus \text{dom } G_1).$$

We define:

(a) $\text{dom } G = \text{dom } G_1 \cup \{x : \text{there is a path } (y, \ldots, x) \text{ in } G_2 \text{ that does not touch any } u \text{ in } \text{dom } G_1\}$;

(b) $G(z) = \begin{cases} G_1(z) & \text{if } z \in \text{dom } G_1, \\ G_2(z) & \text{otherwise,} \end{cases}$
for every node z in $\text{dom } G$.

Then G is a y-extension of G_1 away from A (see Figure 6.2). $\square$

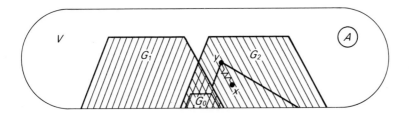

Fig. 6.2. Lemma 6.11

Before we supply a proof of this lemma, let us state two corollaries of it. The first is simply a restatement of the lemma:

Corollary 6.6 *If there exists an extension $\widetilde{G}$ of G' away from A, which is defined for a particular $v \notin dom\, G'$; and if, moreover, G'' is an extension of G' away from $A \cup \{v\}$; then also G'' admits an extension $\widehat{G}$ away from A such that $v \in dom\, \widehat{G}$.*

Proof. It suffices to establish the following correspondence between the names appearing in this corollary and those in Lemma 6.11:

$$\widetilde{G} \mapsto G_2,\, G' \mapsto G_0,\, G'' \mapsto G_1,\, \widehat{G} \mapsto G,\, v \mapsto y$$

(see Figures 6.2 and 6.3). ■

Corollary 6.7 *If $v,\, w \in V \setminus (A \cup dom\, G_0)$, and G_0 admits no extension away from $A \cup \{v\}$ which is defined in w, then w is an ancestor of v in every extension of G_0 away from A which is defined in w.*

Proof. In Lemma 6.11, take $G_1 = G_0$ and $w = y$. If G_2 is an extension of G_0 away from A, defined in w, and w is not an ancestor of v in G_2, then G is an extension of G_0 away from $A \cup \{v\}$, defined in w. ■

Proof of Lemma 6.11. Trivially one has $y \in dom\, G$ (in fact, $G(y) = G_2(y)$); moreover

$$dom\, G \subseteq (dom\, G_1 \cup dom\, G_2) \subseteq (V \setminus (A \cup \{y\})) \cup (V \setminus A) = V \setminus A.$$

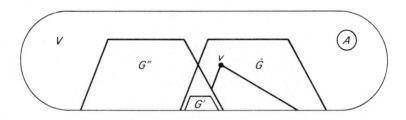

Fig. 6.3. Corollary 6.6

One also has, for all z in $dom\,G$, that either $G(z) = G_1(z)$ or $G(z) = G_2(z)$; in either case, $G(z) \in C_z$. Let us now consider an arbitrary z in $dom\,G$. Let $w \in G(z)$; we must prove that also $w \in dom\,G$. In fact: if $G_1(z)$ is defined, then $w \in G(z) = G_1(z)$ and hence $w \in dom\,G_1 \subseteq dom\,G$. If $G_1(z)$ is undefined, then there is a path $(y,\dots,z)$ in G_2 that does not touch any u in $dom\,G_1$; moreover $w \in G(z) = G_2(z)$. If $w \in dom\,G_1$, then $w \in dom\,G$; else we could obtain a path $(y,\dots,z,w)$ in G_2 by prolongating the preceding path $(y,\dots,z)$ with the edge (z,w). This path does not touch any u in $dom\,G_1$, so that $w \in dom\,G$.

Let us now consider a path $(w_0, w_1,\dots,w_n)$ in G; we must prove that no w_i appears in this path more than once. Let i_0 be the least i for which $G_1(w_i)$ is defined (if no such i exists, then put $i_0 = n + 1$). Thus

$$G(w_0) = G_2(w_0),\dots,G(w_{i_0-1}) = G_2(w_{i_0-1}),$$

and hence

$$w_1 \in G_2(w_0),\ w_2 \in G_2(w_1),\dots,\ w_{i_0-1} \in G_2(w_{i_0-2}),$$

so that $(w_0,\dots,w_{i_0-1})$ is a path in G_2 and hence it cannot contain any repeated nodes. Inductively:

> since $G_1(w_i)$ is defined, one also has $G(w_i) = G_1(w_i)$,
> and hence $w_{i+1} \in G_1(w_i)$,

for $i = i_0,\dots,n-1$. Therefore, $(w_{i_0},\dots,w_n)$ is a path in G_1, without repeated nodes. Finally note that $\{w_0,\dots,w_{i_0-1}\}$ and $\{w_{i_0},\dots,w_n\}$ are

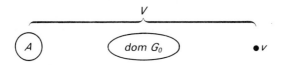

Fig. 6.4. Claim 6.1, when $h = 1$

disjoint sets. ∎

Proof of Claim 6.1. Let us proceed by induction on the cardinality $h = |(V \setminus (A \cup dom\, G_0))|$. For $h = 1$, v is the sole member of $V \setminus (A \cup dom\, G_0)$—see Figure 6.4. The innermost loop of the routine *expand* (which is the while-loop in lines 3–8) is never executed, as it follows from the two facts

$$K \cap (A \cup \{v\}) = \emptyset, \quad K \subseteq V$$

that $K \subseteq dom\, G_0$. Therefore *expand*(v, A) behaves like the instructions

> *if* $\exists K \in C_v \mid K \cap (A \cup \{v\}) = \emptyset$
> *then*
> > $G \leftarrow G \cup \{(v, K)\}$;
> > *return*;
> *end if*;

and the thesis holds trivially.

For $h > 1$: let us first prove that whenever a call *expand*(w, $A \cup \{v\}$) is made during the execution of *expand*(v, A), G is an extension of G_0 away from $A \cup \{v, w\}$, and $v \neq w \notin A$. This is clear for the first of these calls. If w_i and G_i are the values of w and G at the time when the i-th call *expand*(w, $A \cup \{v\}$) takes place, then it follows—since $v \notin dom\, G_0$ and $dom\, G_i \supseteq dom\, G_0$—that $V \setminus (A \cup \{v\} \cup dom\, G_i)$ has fewer members than $V \setminus (A \cup dom\, G_0)$, hence has less than h members. The main inductive hypothesis can be applied, because $w_i \notin (A \cup dom\, G_i)$; therefore G_{i+1} will be an extension of G_i—and hence also of G_0—away from $A \cup \{v\}$. If there is a last w_i (or, in other words, if the execution of *expand*(v, A) eventually terminates—we will prove later that this is indeed the case), this same argument reveals that G is an extension of G_0 away from A at the end.

In the second place we recognize, making use of Corollary 6.7, that every enlargement of G achieved by means of a call $expand(w, A \cup \{v\})$ is 'safe' in the sense that if G' and G'' are the respective values of G just before and immediately after this call, and if G' admits an extension away from A, defined in v, then also G'' admits an extension away from A, defined in v.

In the third place, we want to prove that v cannot belong to the domain of any extension G_∞ of G_0 if the return from the routine $expand(v, A)$ takes place at line 12. For this purpose, we prove that if $G_{i_1}, G_{i_2}, \ldots$ (respectively, $F_1, F_2, \ldots$) are the values of G (respectively, of F) the first, the second, $\ldots$ time the iteration condition in line 2 is about to be evaluated, then no extension G'_{i_j} of G_{i_j} away from A which is defined in v can have $G'_{i_j}(v)$ intersecting A, or $\{v\}$, or F_j. The fact that $G'_{i_j}(v)$ cannot intersect A or $\{v\}$ is obvious. Let us make the absurd hypothesis that it can intersect F_j. Then, since

$$F_{k+1} = \{w_{i_1}, \ldots, w_{i_k}\}$$

for $k = 0, \ldots, j-1$, there must exist a least k_0 in $\{1, \ldots, j-1\}$ for which $w_{i_{k_0}} \in G'_{i_j}(v)$. However, by the induction hypothesis, $G_{i_{k_0}}$ cannot have extensions away from $A \cup \{v\}$ defined in $w_{i_{k_0}}$ and hence (by Corollary 6.7) $w_{i_{k_0}}$ should be an ancestor of v in any extension of $G_{i_{k_0}}$ away from A which is defined for $w_{i_{k_0}}$; in particular, $w_{i_{k_0}}$ would be an ancestor of v in $\cdot G'_{i_j}$, contradicting the fact that G'_{i_j} has no cycles.

Finally, it is clear that the execution of the procedure $expand(v, A)$ will eventually terminate, because C_v is a finite set and, for every K in C_v that is tried, the execution of the innermost cycle (lines 3–8) either breaks in the middle, in which case the algorithm will never try again to make K the progeny of v; or reaches completion, in which case the algorithm will encounter the *return* instruction in line 10.   ∎

6.7   How to embed two-level syllogistic into MLS

Let us informally describe a suitable language for two-level syllogistic (slightly richer than the one treated in [FO78]). The primitive symbols of this language include, in addition to $\emptyset$, a new constant named $\mathit{1}$ and two infinite sequences of variables. The constant $\mathit{1}$ denotes an unspecified non-empty set $\mathcal{U}$ of 'individuals' (cf. [Jec78]). The variables of one sort, called *individual variables*, designate members of $\mathcal{U}$; variables of the other sort, called *set variables*, designate subsets of $\mathcal{U}$. Compound terms, which are assembled using $\cap$, $\backslash$, and $\cup$ from $\emptyset$, $\mathit{1}$ and terms of the form $\{x_1, \ldots, x_n\}$, where the x_is are individual variables, also designate subsets of $\mathcal{U}$. Having this semantics in mind, one will regard as *ill-formed* any expression of the form *Finite x*, $x \subseteq Y$, $Y \subseteq x$, or $X \in x$ where x is an individual variable, or X is a set variable, or X is a compound term, or X is in $\{\emptyset, \mathit{1}\}$. Any expression $x_0 = x_1$

where x_b is an individual variable whereas x_{1-b} is not ($b = 0$ or $b = 1$), is also regarded as an ill-formed formula.

Rather than characterizing in a precise fashion the language of two-level syllogistic, we give the following definitions, which refer to the multilevel language we have been considering so far and to the sequence of variables ξ_i that have been introduced in Definition 6.1:

Definition 6.6 *A* two-level term *is a term* t *of MLS of one of the forms:*

- $\xi_{2 \cdot j}$, *with* j *in* ω;

- $\{ \xi_{2 \cdot j_1 + 1}, \ldots, \xi_{2 \cdot j_n + 1} \}$, *with* n *and* $j_1, \ldots, j_n$ *all in* ω *(for* $n = 0$, *this term 'degenerates' to the constant* $\emptyset$*);*

- $(t_0 \cap t_1)$, $(t_0 \setminus t_1)$, $(t_0 \cup t_1)$, *where* t_0 *and* t_1 *are two-level terms.*

A two-level formula *is a formula of MLS in which only five kinds of atomic formulae occur, namely* $(\xi_{2 \cdot i+1} = \xi_{2 \cdot j+1})$, *Finite* X, $(X \subseteq Y)$, $(X = Y)$, *and* $(\xi_{2 \cdot i+1} \in X)$, *where* X *and* Y *are two-level terms.*

A two-level interpretation *is an interpretation (in the original sense of the word defined in Section 6.3) which sends:*

- ξ_0 *to a non-empty set* $\mathcal{U}$ *of individuals,*

- $\xi_{2 \cdot n}$ *to a subset of* $\mathcal{U}$, *for every* $n > 0$;

- $\xi_{2 \cdot n+1}$ *to a member of* $\mathcal{U}$, *for every* $n \geq 0$.   □

Clearly, two-level formulae can be identified with the well-formed formulae of the language for two-level syllogistic that we have described above. In this identification, ξ_0 acts as the constant $\mathbf{1}$, while the variables $\xi_{2 \cdot i+1}$ (respectively, $\xi_{2 \cdot i}$, with $i > 0$) play the role of individual (respectively, set) variables. Notice that p^φ is defined whenever φ is a two-level interpretation and p is a two-level formula. Let p be a two-level formula in which no variables ξ_j with $j > 2 \cdot n$ occur. We indicate by p_* the two-level formula:

$$\xi_0 \neq \emptyset \,\&\, \xi_1 \in \xi_0 \,\&\, \xi_2 \subseteq \xi_0 \,\&\, \cdots \,\&\, \xi_{2 \cdot n-1} \in \xi_0 \,\&\, \xi_{2 \cdot n} \subseteq \xi_0 \to p.$$

Lemma 6.12 p *is true in every two-level interpretation if and only if* p_* *is valid.*

Proof. Plainly, if p_* is valid, then $p^\psi = TRUE$ in every two-level interpretation ψ. In fact $p_*^\psi = TRUE$, and

$$(\xi_0 \neq \emptyset \,\&\, \xi_1 \in \xi_0 \,\&\, \xi_2 \subseteq \xi_0 \,\&\, \cdots \,\&\, \xi_{2 \cdot n-1} \in \xi_0 \,\&\, \xi_{2 \cdot n} \subseteq \xi_0 \,)^\psi = TRUE,$$

because ψ is two-level.

Conversely, if $p^\psi = TRUE$ in every two-level interpretation ψ, then p_* is valid. In fact, if there is an interpretation φ such that $p_*^\varphi = FALSE$, then one can establish a one-to-one correspondence, M, between ξ_0^φ and a non-empty set $\mathcal{U}$ of individuals. If, for every i such that $0 \le i \le n$, we put:

$$\xi_{2\cdot i}^\psi = \{M\, x \,:\, x \in \xi_{2\cdot i}^\varphi\}, \qquad \xi_{2\cdot i+1}^\psi = M\, \xi_{2\cdot i+1}^\varphi,$$

then ψ is a two-level interpretation such that $t^\psi = \{\, M\, x \,:\, x \in t^\varphi \,\}$ (respectively, $q_*^\psi = q^\varphi$) for every two-level term t (respectively, two-level formula q) that involves at most the variables $\xi_0,\, \xi_1,\ldots,\, \xi_{2\cdot n}$. In particular, $p^\psi = p_*^\varphi = FALSE$. ∎

6.8   Venn diagrams and their affinity with syllogistic schemes

Definition 6.7 *A* Venn diagram *is a function $\mu : Pow(\alpha) \to \{0,1\}$, with $\mu(\emptyset) = 1$ and $\alpha \in Ord$. Its* places *are the sets $A \subseteq \alpha$ with $\mu(A) = 1$.* □

Usually α is a natural number, in which case the Venn diagram is said to be *finite*.

A Venn diagram can be induced by sets, in the following sense:

Definition 6.8 *Suppose α is an ordinal and $\beta \overset{M}{\mapsto} X_\beta$ is defined for all β in α. Put:*

$$
\begin{aligned}
\mathcal{A} &= \{A \subseteq \alpha : A \ne \emptyset\}; \\
\varrho_A &= \left(\bigcap_{\beta \in A} X_\beta\right) \setminus \left(\bigcup_{\gamma \in \alpha \setminus A} X_\gamma\right), \text{ for all } A \text{ in } \mathcal{A}; \\
\mathcal{W} &= \bigcup_{A \in \mathcal{A}} \varrho_A; \\
R &= \text{first limit ordinal that exceeds } rk\,\mathcal{W}; \\
\varrho_0 &= V_R \setminus \mathcal{W}; \\
\mu(A) &= \begin{cases} 0 \text{ if } \varrho_A = \emptyset \\ 1 \text{ if } \varrho_A \ne \emptyset \end{cases}, \text{ for all } A \text{ in } Pow(\alpha).
\end{aligned}
$$

Thus μ turns out to be a Venn diagram (note in fact that $\mathcal{W} \cup \{\mathcal{W}\} \in \varrho_0$), which is said to be induced *by the function M. The sets ϱ_A are called* Venn regions *of M (ϱ_0 is sometimes included among these regions, but more often it is not). The set $\mathcal{U} = \mathcal{W} \cup \varrho_0$ is called* universal space *of M.* □

Remark 6.2 It must be admitted that the above choice of $\mathcal{W} \cup \varrho_0$ as the universal space $\mathcal{U}$ of M is somewhat arbitrary. (As an evidence of this, a simpler characterization of $\mathcal{U}$ will suffice for the purposes of Chapter 8—cf. the definition of $\mathcal{U}$ in Section 8.3). At any rate, it is generally useful to impose that $\mathcal{U}$ is a transitive set with $\mathcal{U} \setminus \bigcup_{\beta \in \alpha} X_\beta \ne \emptyset$. □

Quite often one speaks of the Venn diagram induced by a family $\mathcal{X}$, even though $\mathcal{X}$ is not a function defined on an ordinal. It is very easy, indeed, to tidy up things. For this purpose, one will define:

$$
\begin{aligned}
\alpha &= |\mathcal{X}| \text{ and} \\
X_\beta &= \text{the } \beta\text{-th element of } \mathcal{X} \text{ with respect to } <,
\end{aligned}
$$

for all β in α. It is the diagram induced by $\beta \mapsto X_\beta$ that one considers to be induced by $\mathcal{X}$. The Venn regions of $\mathcal{X}$ and its universal space are defined accordingly.

It is also rather easy to identify the finite Venn diagrams with a restricted family of syllogistic schemes. To see this, consider a Venn diagram μ whose domain has 'logarithm' α belonging to ω; let, as before, $\mathcal{A} = \{A \subseteq \alpha : A \neq \emptyset\}$. Since α belongs to ω, one can associate with every β in α the set variable ξ_β (see Definition 3.1). Next take $X = \{\xi_\beta : \beta \in \alpha\}$, and assume that Y and $\overset{\circ}{} $ are as in Section 6.2, so that the association $y \mapsto \{\beta : \xi_\beta \in \overset{\circ}{y}\}$ is a bijection between Y and $\mathcal{A}$, admitting an inverse $A \mapsto y_A$. Then put

$$F = Z = \{y_A \in Y : \mu(A) = 1\},$$

and

$$\xi_\beta \sim \xi_\gamma \text{ if and only if } \{y \in Z : \xi_\beta \in \overset{\circ}{y}\} = \{y \in Z : \xi_\gamma \in \overset{\circ}{y}\},$$

for all β and γ in α. Finally, let S be the quotient of X modulo $\sim$ as usual, and let $G = (S, \,\dot{}\,)$ be the DAG devoid of edges.

We leave it as an exercise for the reader to verify that this quadruple $(\sim, G, F, Z)$ indeed identifies a syllogistic scheme σ_μ, that the function $\mu \mapsto \sigma_\mu$ is one-to-one, and that for any model $v \mapsto v^\varphi$ of σ_μ, the function $\beta \mapsto \xi_\beta^\varphi$ defined on α induces the diagram μ.

In view of this 'parenthood' between Venn diagrams and syllogistic schemes, it is also easy to prove the following lemma, which states about Venn diagrams facts analogous to those that have already been proved about syllogistic schemes:

Lemma 6.13 *Assume that the sets α, $\mathcal{A}$, and the functions $\beta \mapsto X_\beta$, $A \mapsto \varrho_A$, and μ be as in Definition 6.8. Then, putting*

$$\Xi_\beta = \{A \in \mathcal{A} : \beta \in A \text{ and } \mu(A) = 1\}$$

for all β in α, every X_β becomes decomposable as a disjoint union

$$X_\beta = \bigcup_{A \in \Xi_\beta} \varrho_A$$

of non-empty sets. As a consequence,

$$
\begin{array}{lll}
X_\beta \subseteq X_\gamma & \text{if and only if} & \Xi_\beta \subseteq \Xi_\gamma, \\
X_\beta = X_\gamma \otimes X_\vartheta & \text{if and only if} & \Xi_\beta = \Xi_\gamma \otimes \Xi_\vartheta,
\end{array}
$$

for all β, γ, ϑ in α, where $\otimes$ ranges over the operators $\cap$, $\backslash$, $\cup$. $\square$

This lemma suggests the following way of determining sets that induce *any given* Venn diagram μ. Associate with each A such that $\mu(A) = 1$ a non-empty set ϱ_A, in such a manner that the ϱ_As are pairwise disjoint (it will suffice, for instance, to take $\varrho_A = \{A\}$). Then define the Ξ_β as before, and put $X_\beta = \bigcup_{A \in \Xi_\beta} \varrho_A$ for all β, *by definition*. The proof that these sets induce μ is straightforward, and we omit it. Conclusion:

Lemma 6.14 *Every Venn diagram μ is induced by suitable sets X_β.* $\square$

Structures much more general than Venn diagrams are introduced in Chapter 11, where a theory of *syllogistic diagrams* is developed. The latter theory (anticipations of which have been drafted in Section 1.4) subsumes, at least for finite Venn diagrams, both lemmas of this section.

7. RESTRICTED QUANTIFIERS, ORDINALS, AND ω

7.1 Introduction

This book deals almost exclusively with *unquantified* subtheories of set theory. The present chapter, which is largely independent of the rest of the book, is the only major exception: satisfiability decision algorithms are described for some classes of *quantified* formulae in the language $\mathcal{L}$ that includes $\emptyset$, $=$, $\in$ together with the constant ω, which denotes the set of all natural numbers, and with a unary predicate, $ORD(x)$, having the meaning 'x is an ordinal'. Quantifiers are all restricted (i.e. of the form $\forall x \in y$ or $\exists x \in y$), with no alternations of the form $\forall \exists$ allowed. The subject matter of this chapter was first studied in [BFOS81].

More precisely, the class of formulae we will deal with is the propositional closure of basic formulae of type $(\forall x_1 \in y_1)(\forall x_2 \in y_2) \cdots (\forall x_n \in y_n)\, p$, where p is a quantifier-free formula in the language $\mathcal{L}$. We will show that if in each prefix $(\forall x_1 \in y_1) \cdots (\forall x_n \in y_n)$ the y_js are all free, i.e. no x_i is a y_j, then the corresponding quantified class is decidable. Cases in which this hypothesis can be relaxed are discussed in [PP88b]. The variables y_j will be called *conditional variables*.

In [CCF88] it is shown that finite satisfiability can be decided also in presence of rank comparison. Moreover in [CC88] cases in which the construct $Finite(\bullet)$ can be added are presented. Specifically, it is shown that the above quantified theory remains decidable when the operator $Finite(\bullet)$ is applied only to free variables and no equality relation between bound variables occurs.

In order to keep our initial presentation of the satisfiability algorithm simple, we assume in the next two sections that neither ω nor ORD appears in the formulae. Under this restriction, the following *reflection result* will be proved: if a formula p containing n free variables and m conditional variables has a model, then it has a hereditarily finite model M with

$$rk(Mx) < 2 \cdot \left(n + \binom{m+1}{2} \right)$$

for every variable x in the formula p.

The decision algorithm will consist in eliminating the universal quantifiers by interpreting $(\forall x \in y)$ as: 'for all sets x directly designated by free variables and constants in the formula...'. This step is of course always sound but, in general, not complete. Nonetheless, one can achieve completeness of the elimination procedure in the case under consideration, by suitably forcing the set of conditional variables to become extensional (cf. Definition 2.6).

The treatment of *ORD* would be out of reach of the decision algorithm just outlined if one 'eliminated' this construct in terms of restricted quantifiers. A definition of *ORD* in such terms, although feasible, requires in fact relaxing the hypothesis of freeness of the conditional variables. As we will see in Section 4, with so much expressive freedom available one can describe infinite sets; which proves that the above-mentioned reflection result does not hold any more.

In Sections 5 and 6 an application of Lemma 2.11 shows how to cope with the general case in which ω and *ORD* may be present.

The treatment of operations whose definition requires quantifier alternation, e.g. *Un*, is beyond today's possibilities of the approach discussed in this chapter.

7.2 The elementary syllogistic including only $=$, $\in$, $\emptyset$

Let us consider the language $\mathbf{F}'_0$ consisting of all formulae of the theory $\mathbf{F}'_2$ treated in Chapter 5 that do not contain η. This $\mathbf{F}'_0$ is just the propositional closure of atoms of type $x = y$, $x \in y$ where each x, y is either a variable or the constant $\emptyset$.

From Chapter 5 we know that this theory is decidable. Nonetheless we want to describe a satisfiability algorithm for $\mathbf{F}'_0$ which, if used in connection with a suitable quantifier elimination procedure, will induce a decision algorithm for a significantly wide class of quantified formulae in the language of $\mathbf{F}'_0$. To this end, thanks to disjunctive normal forms (see Theorem 3.1), it will suffice to produce a satisfiability algorithm for formulae of $\mathbf{F}'_0$ which are conjunctions of literals of the forms

$$
\begin{array}{ll}
(=) & x = y, \\
(\neq) & x \neq y, \\
(\in) & x \in y, \\
(\notin) & x \notin y.
\end{array}
$$

where each x, y is either a variable or the constant $\emptyset$.

Let q be a conjunction of literals of the above types $(=)$, $(\neq)$, $(\in)$, $(\notin)$, and let

- $V = \{\text{variables in } q\} \cup \{\emptyset\}$;

- $\sim$ be the smallest equivalence relation on V such that $x = y$ in q implies $x \sim y$ (cf. Definition 5.1).

The following algorithm decides the satisfiability of any such conjunction:

Algorithm 7.1 Step 1. *Choose a representative $\tilde{x}$ in each class $\{y : y \sim x\}$ in such a way that $\tilde{\emptyset} = \emptyset$. Let $\tilde{q}$ be the simplified form of q obtained by* *replacing each variable x in q by $\tilde{x}$, dropping the conjuncts of the form $\tilde{x} = \tilde{x}$, and let $\tilde{V} = \{variables\ in\ \tilde{q}\} \cup \{\emptyset\}$.*

Step 2. *Check that q includes no explicit contradiction of the forms:*

(i) $x \neq x$,

(ii) $x \in y$ *and* $x \notin y$.

Step 3. *Check that there exists an ordering $\lhd$ of $\tilde{V}$ such that*

(a) $\emptyset$ *is minimal in the ordering;*

(b) $x \lhd y$ *whenever* $x \in y$ *is in* $\tilde{q}$.

Step 4. *If both* **Step 2** *and* **Step 3** *have been successful (in the sense that no explicit contradiction was found and an ordering $\lhd$ exists) declare that q is satisfiable; otherwise declare that q is unsatisfiable.* □

Lemma 7.1 (Soundness) *Algorithm 7.1 is sound.*

Proof. The soundness of **Step 1** is plain, since every model M of q is also a model of $\tilde{q}$ and $x \sim y$ implies $Mx = My$. Moreover $\tilde{q}$ cannot contain an explicit contradiction of type (i) or (ii) because M is a model of $\tilde{q}$. This yields the soundness of **Step 2**. Finally, by putting $x \lhd y$ if $rk(Mx) < rk(My)$, we get an ordering $\lhd$ of $\tilde{V}$ satisfying conditions (a) and (b) of **Step 3**. ■

The converse is established by the following:

Lemma 7.2 (Completeness) *Algorithm 7.1 is complete. More precisely, if in* **Step 4** *satisfiability of q is declared, then a model M for q can be obtained in the following way. Let*

- $G = (\tilde{V}, \dot{})$ *be the WFG where $\dot{x} = \{y \in \tilde{V} : x \in y$ appears in $\tilde{q}\}$;*

- S *be any extensional set of nodes of G satisfying condition (i) of Definition 2.6;*

- $(U, \ddot{})$ *be a grafting of G possessing properties (0), (1), (2), (4), and (3') of Lemma 2.11;*

- $x \mapsto x'$ *be the function induced by $(U, \ddot{})$ as in Example 2.6.*

Then, by putting $Mx = (\tilde{x})'$ for every variable x in V, one defines a model of q such that:

$$(\star) \qquad Mx = \{Mz : \tilde{z} \in \tilde{x}\ appears\ in\ q\},$$

for every x in V such that $\tilde{x} \in S$.

Proof. Since satisfiability is detected in **Step 4**, an ordering $\lhd$ of $\tilde{V}$ is found in **Step 3** such that conditions (a) and (b) are satisfied. This implies that $G = (\tilde{V}, \,^{\cdot})$ is a WFG. By recourse to Lemma 2.11, we obtain a grafting $(U, \,^{\cdot\cdot})$ of G meeting the properties (0), (1), (2), (4), and (3') of Lemma 2.11. So let $x \mapsto x'$ be the induced function, i.e.

$$x' = \{z' : x \in \dot{z}\} \cup \bigcup_{x \in \ddot{u}} u = \{z' : z \in x \text{ appears in } \tilde{q}\} \cup \bigcup_{x \in \ddot{u}} u \,.$$

We want to show that the map M defined by

$$Mx = (\tilde{x})' = \{Mz : \tilde{z} \in \tilde{x} \text{ appears in } \tilde{q}\} \cup \bigcup_{\tilde{x} \in \ddot{u}} u$$

is a model of q.

Indeed, if $x = y$ appears in q then $\tilde{x}$ is identical to $\tilde{y}$, yielding $Mx = (\tilde{x})' = (\tilde{y})' = My$. Therefore M is a model of all literals of type $(=)$ in q. Moreover, if $x \neq y$ appears in q then $\tilde{x}$ is different from $\tilde{y}$ by (i) of **Step 2**. It follows by (3') of Lemma 2.11 that $Mx = (\tilde{x})' \neq (\tilde{y})' = My$, showing that M models correctly all literals of type $(\neq)$ in q.

When $x \in y$ is in q, then $\tilde{x} \in \tilde{y}$ appears in $\tilde{q}$, and therefore $Mx = (\tilde{x})' \in (\tilde{y})' = My$, because $x \mapsto x'$ is the induced function. To finish the proof of M being a model of q, it remains to show that all literals of type $(\notin)$ are correctly modelled. Indeed, assuming that $x \notin y$ is in q, it turns out that $\tilde{x} \notin \tilde{y}$ is in $\tilde{q}$. Hence, by (ii) of **Step (2)**, $\tilde{x} \in \tilde{y}$ is not in q. Using (4) of Lemma 2.11 we get

$$Mx = (\tilde{x})' \notin (\tilde{y})' = My \,,$$

showing that M is a model of q.

Finally, property $(\star)$ is just an immediate consequence of (1b) of Lemma 2.11. This yields that $M\emptyset$ is the empty set, completing the proof. ∎

This lemma, together with Corollary 2.7, yields immediately the following.

Corollary 7.1 *A finite conjunction q of literals of types $(=)$–$(\notin)$ with n variables is satisfiable if and only if it has a model M such that $rk(Mx) \leq 2 \cdot n$ for every variable x in q.* ∎

7.3 A satisfiability algorithm for a class of quantified formulae

In this section we describe a sound and complete satisfiability test for the theory $\mathbf{G}_1$ which is the propositional closure of prenex formulae $Q_1 Q_2 \cdots Q_n p$ satisfying the following conditions:

(1) p is an unquantified formula of the theory $\mathbf{F}_0'$ discussed in the preceding section (i.e. p is an unquantified formula in the language $=, \in, \emptyset$).

(2) Every quantifier Q_i is of type $(\forall x_i \in y_i)$. The variables y_i appearing as right-hand sides of these restricted quantifiers are called *conditional variables*.

(3) Every conditional variable y_i is free. This means that in every prefix, which by (2) is of type $(\forall x_1 \in y_1) \cdots (\forall x_n \in y_n)$, no x_i can be a y_j. $\square$

To illustrate this definition we notice that the prenex formula

$$(\forall x_1 \in y_1)(\forall x_2 \in y_1)(x_1 \neq \emptyset \ \& \ x_1 \in x_2)$$

satisfies conditions (1)–(3) whereas

$$(\forall x_1 \in y_1)(\exists x_2 \in y_2)(x_2 \in x_1 \ \& \ x_2 \neq \emptyset)$$

does not satisfy (2), and

$$(\forall x_1 \in y_1)(\forall x_2 \in x_1)(x_2 \in y_1)$$

does not satisfy (3).

Many common set-theoretic constructs can be expressed by formulae of the theory $\mathbf{G}_1$. For example

$(\subseteq)$ $x \subseteq y$ iff $(\forall z \in x)z \in y$,

$(\cup)$ $x = y \cup z$ iff $y \subseteq x \ \& \ z \subseteq x \ \& \ (\forall u \in x)(u \in y \lor u \in z)$,

$(\setminus)$ $x = y \setminus z$ iff
$$(\forall u \in x)(u \in y \& u \notin z) \& (\forall v \in y)(v \notin z \to v \in x),$$

$(\{\bullet, \ldots, \bullet\})$ $x = \{y_1, y_2, \ldots, y_n\}$ iff
$$y_1 \in x \ \& \ y_2 \in x \ \& \cdots \& \ y_n \in x \ \&$$
$$(\forall z \in x)(z = y_1 \lor z = y_2 \lor \cdots \lor z = y_n),$$

$(\in Un)$ $x \in Un \, y$ iff $(\exists z \in y)(x \in z)$.

This shows that multilevel syllogistic extended with $\{\bullet, \ldots, \bullet\}$ and $\in Un$ is a subtheory of $\mathbf{G}_1$.

Next we describe a satisfiability algorithm for the theory $\mathbf{G}_1$. To this end, by considering the disjunctive normal form, it is sufficient to show how to decide finite conjunctions of 'literals' of type

$(+)$ $(\forall x_1 \in y_1)(\forall x_2 \in y_2) \cdots (\forall x_n \in y_n)p$,
$(-)$ $\neg(\forall x_1 \in y_1)(\forall x_2 \in y_2) \cdots (\forall x_n \in y_n)p$

such that p is an unquantified formula in the language $=$, $\in$, $\emptyset$, and all conditional variables y_i are free.

Formulae of type $(-)$ can be eliminated by replacing each of them by

$$\bar{x}_1 \in y_1 \ \& \ \bar{x}_2 \in y_2 \ \& \cdots \& \ \bar{x}_n \in y_n \ \& \ \neg\bar{p},$$

where $\bar{x}_1, \ldots, \bar{x}_n$ are newly introduced variables and $\bar{p}$ is the result of replacing x_i by $\bar{x}_i$ in p for $i = 1, 2, \ldots, n$. (Notice that an analogous elimination process could be carried out even if some x_i were a y_j in a conjunct of the form $(-)$—this remark plainly indicates how one could weaken condition (3) above in order to obtain a broader class $\mathbf{G}_1$ still decidable by the same method that we are about to describe.)

We are now left with the problem of deciding the satisfiability of a finite conjunction q of 'atoms' of type $(+)$, where, of course, some sequence of quantifiers can be empty.

The following algorithm solves this problem:

Algorithm 7.2 *Let*

$$C = \{\, conditional\ variables\ in\ q\,\} \cup \{\emptyset\}.$$

Step 1 (Preprocessing making C an extensional set of nodes). *For each pair of distinct elements a, b in C non-deterministically choose one of the three formulae*

$$a = b, \quad z_{ab} \in a\ \&\ z_{ab} \notin b, \quad z_{ab} \notin a\ \&\ z_{ab} \in b,$$

where z_{ab} is a newly introduced variable, and add the chosen formula to q. Let

$$\bar{q}\ be\ the\ resulting\ conjunction;$$
$$\Phi = \{\, free\ variables\ in\ \bar{q}\,\} \cup \{\emptyset\}.$$

Step 2 (Quantifier elimination). *Replace each conjunct of type*

$$(\forall x_1 \in y_1)\,(\forall x_2 \in y_2)\cdots(\forall x_n \in y_n)\,p$$

by the finite conjunction of all formulae

$$z_1 \in y_1\ \&\ z_2 \in y_2\ \&\ \cdots\ \&\ z_n \in y_n\ \rightarrow\ p^{x_1 x_2 \ldots x_n}_{z_1 z_2 \ldots z_n}$$

where $(z_1, z_2, \ldots, z_n)$ ranges over Φ^n and $p^{x_1 x_2 \ldots x_n}_{z_1 z_2 \ldots z_n}$ denotes the result of uniformly replacing $x_1, x_2, \ldots, x_n$ in p by $z_1, z_2, \ldots, z_n$ respectively.

Step 3 (Ground case). *Let*

- *Q be the resulting unquantified formula of the theory F_0' discussed in the preceding section. This formula is called a residue of q.*

- *$q_1 \vee q_2 \vee \cdots \vee q_m$ be a disjunctive normal form of Q.*

If the satisfiability algorithm described in the preceding section declares that q_i is unsatisfiable, for each $i = 1, 2, \ldots, m$, then declare that the input conjunction q is unsatisfiable.

Otherwise declare that q is satisfiable. $\square$

Lemma 7.3 (Soundness) *The above Algorithm 7.2 is sound. That is, if unsatisfiability is declared in* **Step 3** *then the input formula q is really unsatisfiable.*

Proof. First notice that soundness and completeness of **Step 1** is an immediate consequence of the axiom of extensionality.

Soundness of **Step 2** comes from the evident observation that the assertion that p holds for all possible n-tuples in the cartesian product $y_1 \times y_2 \times \cdots \times y_n$ is stronger than asserting that p holds for those n-tuples which are values of elements in Φ^n.

Soundness and completeness of **Step 3** come from the soundness and completeness of the satisfiability algorithm described in the preceding section (cf. Lemmas 7.1 and 7.2).   ∎

The converse is also true:

Theorem 7.1 (Completeness) *Algorithm 7.2 is also complete.*

Moreover a model of the input formula q can be found in the following way. Let q_i be a satisfiable conjunction in the disjunctive normal form of the unquantified residue Q in **Step 3**. *Let $\tilde{q}_i$ be the result of replacing each variable x in q_i by its representative $\tilde{x}$ in the class $\{y : y \sim x\}$, where $\sim$ is the smallest equivalence relation over Φ such that $x = y$ in q_i implies $x \sim y$ (cf. the preceding section).*

By **Step 1** *of Algorithm 7.2 the set $S = \{\tilde{x} : x \in C\}$ is an extensional set of nodes in the WFG $(\tilde{\Phi}, \dot{\ })$, with*

$$\dot{x} = \{y \in \tilde{\Phi} \ : \ x \in y \text{ belongs to } \tilde{q}_i\}.$$

Let M be the corresponding model of q_i as in Lemma 7.2.
Then M is also a model of the input formula q.

Proof. By $(\star)$ of Lemma 7.2 we have that for every x in C every member of Mx is of type Mz for some z in Φ. It follows that if M satisfies all formulae of type

$$(a_1 \in y_1 \& \cdots \& a_n \in y_n) \to p_{a_1 a_2 \ldots a_n}^{x_1 x_2 \ldots x_n}$$

for some $y_1, \ldots, y_n$ in C, where $(a_1, \ldots, a_n)$ takes all possible values in Φ^n, then M satisfies $(\forall x_1 \in y_1) \cdots (\forall x_n \in y_n) p$ too. This shows the completeness of **Step 2**, concluding the proof of the theorem.   ∎

Corollary 7.2 (Decidability) *A finite conjunction q of prenex formulae $Q_1 Q_2 \cdots Q_n p$ satisfying conditions (1), (2), and (3) above with n free variables and m conditional variables has a model if and only if it has a model M such that $rk(Mx) \leq 2 \cdot n + \left(n + \binom{m+1}{2}\right)$ for every free variable x in q.*   □

7.4   A quantified formula expressing infinite sets

In this section we will show that condition (3) of the preceding section, asserting that all conditional variables must be free, is necessary for the reflection principle expressed in Corollary 7.2 to hold. To show this, we exhibit an example of a conjunction q_∞ of prenex formulae containing non-free conditional variables, such that q_∞ is satisfiable but has no models in which every free variable is mapped into a finite set. Our example is a variant of another one described in [PP88c], which is only slightly more intricate than ours, and is usable as a statement of the existence of infinite sets even when the well-foundedness of membership is not assumed—cf. [PP89].

The formula q_∞ is the following:

$a \neq b \,\&\, a \not\subseteq b \,\&\, b \not\subseteq a \,\&\, (\forall x \in a)(\forall y \in b)(x \in y \lor y \in x) \,\&$
$((\forall x \in a)(\forall y \in x)y \in b) \,\&\, ((\forall x \in b)(\forall y \in x)y \in a)\,,$

equivalent to

$a \neq b \,\&\, a \not\subseteq b \,\&\, b \not\subseteq a \,\&\, (\forall x \in a)(\forall y \in b)(x \in y \lor y \in x) \,\&$
$Un(\,a\,) \subseteq b \,\&\, Un(\,b\,) \subseteq a\,.$

Notice that the fourth conjunct implies $a \cap b = \emptyset$. To see that q_∞ is satisfiable put

$$
\begin{aligned}
s_0 &= \emptyset, \\
t_n &= \{\,s_0, \ldots, s_n\,\}, \\
s_{n+1} &= \{\,t_0, \ldots, t_n\,\},
\end{aligned}
$$

and

$$Ma = \{\,s_n \,:\, n = 0, 1, 2, \ldots\,\}, \quad Mb = \{\,t_n \,:\, n = 0, 1, 2, \ldots\,\}\,.$$

It is straightforward to verify that M is indeed a model of q_∞. Moreover we have the following.

Lemma 7.4 *In any model M of q_∞, and for every t in $Ma \cup Mb$, there exists t' in $Ma \cup Mb$ such that $t \in t'$ and t, t' do not belong to the same set Ma or Mb (indeed, we have already noticed that $Ma \cap Mb = \emptyset$).*

Proof. Without loss of generality we can assume $t \in Ma$. This, by the fifth conjunct of q_∞, implies $t \subseteq Mb$ and $t \neq Mb$ since $Mb \not\subseteq Ma$. Hence there exists t' in Mb such that $t' \not\subseteq t$. By the fourth conjunct of q_∞ we get $t \in t'$, completing the proof of the lemma.   ∎

Corollary 7.3 *In any model M of q_∞, Ma and Mb are both infinite and have rank the same limit ordinal.*

Proof. It is sufficient to notice that $Ma \neq Mb$, $Ma \cup Mb \neq \emptyset$ and to exploit the preceding lemma.   ∎

7.5   An elementary syllogistic including integers and ordinals

The decision algorithm for $\mathbf{F}'_0$ introduced in Section 7.2 will be enhanced now and made adequate for the treatment of another theory, again unquantified, denoted $\mathbf{F}_0$ in what follows. $\mathbf{F}_0$ is obtained by just adding the constant ω and the unary predicate ORD to $\mathbf{F}'_0$.

Reasoning as in Section 7.2, we simply have to decide the satisfiability of any given conjunction q of literals of type

$$
\begin{array}{ll}
(=) & x = y, \\
(\neq) & x \neq y, \\
(\in) & x \in y, \\
(\notin) & x \notin y, \\
(ORD) & ORD(x), \\
(\neg ORD) & \neg ORD(x),
\end{array}
$$

where each x, y is either a variable or one of the two constants $\emptyset$, ω.

The following generalized variant of Algorithm 7.1 is able to decide the satisfiability of q:

Algorithm 7.3 *Let*

- $V = \{\,variables\ in\ q\,\} \cup \{\,\emptyset, \omega\,\}$;

- $x\widehat{\in}y$ *(respectively, $x\widehat{\notin}y$) denote that $x \in y$ (respectively, $x \notin y$) appears in q.*

Declare satisfiability if and only if there exist an equivalence relation $\sim$ on V and a subset $\mathcal{O}$ of V such that:

(i) $x = y$ in q implies $x \sim y$;

(ii) $x \neq y$ in q implies $x \not\sim y$;

(iii) $ORD(x)$ in q implies $x \in \mathcal{O}$;

(iv) $\neg ORD(x)$ in q implies $x \notin \mathcal{O}$;

(v) $\emptyset \in \mathcal{O}$, $\omega \in \mathcal{O}$, $\emptyset \not\sim \omega$;

(vi) $\mathcal{O}$ is a union of equivalence classes, that is:
$$x \sim y \text{ implies that } x \in \mathcal{O} \text{ if and only if } y \in \mathcal{O};$$

(vii) if $x \in y$ and $x' \notin y'$ both appear in q, then either $x \not\sim x'$ or $y \not\sim y'$;

(viii) $z\widehat{\in}x$ and $x \in \mathcal{O}$ implies $z \in \mathcal{O}$;

(ix) there is no cycle of type
$$x\widehat{\in}x_1\widehat{\in}x_2\widehat{\in}\cdots\widehat{\in}x_n\widehat{\in}x \quad (n \geq 0)$$
involving (distinct) variables x, $x_1, \ldots, x_n$ not in $\mathcal{O}$;

(x) there exists a total ordering $\lhd$ of $\mathcal{O}/\!\sim = \{\{y \in V : y \sim x\} : x \in \mathcal{O}\}$ such that:

(a) $\{x : x \sim \emptyset\}$ is the least element of such an ordering;

(b) $z \widehat{\in} z'$, with z' in $\mathcal{O}$, implies $\{x : x \sim z\} \lhd \{y : y \sim z'\}$;

(c) $z \widehat{\notin} z'$, with z and z' in $\mathcal{O}$, implies that either $z \sim z'$ or $\{y : y \sim z'\} \lhd \{x : x \sim z\}$. □

Lemma 7.5 (Soundness) *Algorithm 7.3 is sound.*

Proof. Let M be a model of q. Then put

- $x \sim y$ if and only if $Mx = My$;

- $x \in \mathcal{O}$ if and only if Mx is an ordinal;

- $x \lhd y$ if and only if $Mx \in My$ and My is an ordinal.

It is plain to verify that conditions (i)–(x) are all satisfied. ∎

The converse is established by the following:

Lemma 7.6 (Completeness) *Algorithm 7.3 is complete. More precisely, if satisfiability of q is declared then a model M of q can be obtained by the following construction.*
Assume that $\sim$, $\mathcal{O}$, $\lhd$ satisfy conditions (i)–(x); furthermore let

- $[x]$ *denote* $\{z : z \sim x\}$;

- Ω *denote* $\{[x] : x \in \mathcal{O}\} = \mathcal{O}/\sim$;

- $[N]$ *be the immediate predecessor of* $[\omega]$ *in the ordering* $\lhd$ *of* Ω;

- n *be the number of elements preceding* $[\omega]$ *in this ordering;*

- $\tilde{V} = (V/\sim) \cup \{x_k : k \text{ integer}, k > n\}$, *where the* x_k*s are newly introduced variables;*

- $a \widehat{\in} b$, *with a, b in $\tilde{V}$ be true if and only if one of the following holds:*

    (1) $x \in y$ *is in q for some $x \in a$, $y \in b$,*

    (2) $a, b \in \Omega$ & $a \lhd b$,

    (3) $a = x_k$ & $k > n$ & $[N] \widehat{\in} b$,

    (4) $b = x_k$ & $k > n$ & $(a = [N] \lor a \widehat{\in} [N])$,

    (5) $a = x_k$ & $b = x_j$ & $n < k < j$;

- $G = (\tilde{V}, \,\dot{\ })$ *be the WFG defined by* $\dot{a} = \{b : b \widehat{\in} a\}$;

- S *be any extensional set of nodes of G such that:*

    (a) S *satisfies condition (i) of Definition 2.6,*

    (b) $S \supseteq \Omega \cup \{x_k : k \text{ integer}, k > n\}$,

(c) *for every a in $S \setminus (\Omega \cup \{ x_k : k \text{ integer}, k > n \})$ there exists $b \,\widehat{\in}\, a$ such that either $b \notin \Omega \cup \{ x_k : k \text{ integer}, k > n \}$ or, for some c, $c \,\widehat{\in}\, b \;\&\; \neg c \,\widehat{\in}\, a$,*

(the existence of such an extensional set of nodes is an immediate consequence of $\widehat{\in}$ being a well-ordering of $\Omega \cup \{ x_k : k \text{ integer}, k > n \}$ in view of the properties (2)–(5) of $\widehat{\in}$);

- $(U, \ ^{\cdot \cdot})$ *be a grafting of G possessing properties (0), (1), (2), (4), and (3') of Lemma 2.11;*

- $x \mapsto x'$ *be the function induced by this grafting as described in Example 2.6.*

Then, by putting $Mx = [x]'$ for every x in V, one defines a model of q such that

$$(\star) \qquad\qquad Mx = \{a' : a \in \tilde{V} \;\&\; a \,\widehat{\in}\, [x]\} \text{ for every } x \text{ in } S.$$

Moreover, if no equality literal $N = x$ involving N and some other variable x distinct from N appears in q, and if M is extended by putting $Mx_k = x'_k = k$ for every $k > n$, then M is a model of $q \cup \bigcup_{k>n}(q)^N_{x_k}$, where $(q)^N_{x_k}$ is the result of replacing every occurrence of N in q by x_k.

Proof. To show that $G = (\tilde{V}, \ ^{\cdot})$ is a WFG, we reason as follows. By condition (ix) of Algorithm 7.3 there cannot be an infinite descending chain of relations $\widehat{\in}$ involving only nodes of $\tilde{V} \setminus (\Omega \cup \{ x_k : k \text{ integer}, k > n \})$.

Similarly there cannot be an infinite descending chain of $\widehat{\in}$ involving only variables of $\Omega \cup \{ x_k : k \text{ integer}, k > n \}$, since $\Omega \cup \{ x_k : k \text{ integer}, k > n \}$ is well-ordered by $\widehat{\in}$ as a consequence of the properties (2)–(5) of $\widehat{\in}$.

Finally, if an infinite descending chain of $\widehat{\in}$ contains an element λ of $\Omega \cup \{ x_k : k \text{ integer}, k > n \}$ then, by (viii) of Algorithm 7.3, every element following λ in the chain is also in $\Omega \cup \{ x_k : k \text{ integer}, k > n \}$, which is impossible by the preceding argument. This shows that $G = (\tilde{V}, \ ^{\cdot})$ is a WFG.

The proof of M being a model of all literals of type $(=)$, $(\neq)$, $(\in)$, $(\notin)$ is completely analogous to that of Lemma 7.2.

Moreover, since $\Omega \cup \{ x_k : k \text{ integer}, k > n \}$ is well-ordered by $\widehat{\in}$, it can be easily proved by transfinite induction that for every b in $\Omega \cup \{ x_k : k > n \}$, $b' = \{ a' : a \in \Omega \cup \{ x_k : k \text{ integer}, k > n \} \;\&\; a \,\widehat{\in}\, b \}$ is an ordinal.

Therefore $[x]$ belongs to Ω, and $Mx = [x]'$ is an ordinal, whenever $ORD(x)$ appears in q.

Notice also that $M\omega = \omega$.

Conversely let us show that if $a \notin \Omega \cup \{ x_k : k \text{ integer}, k > n \}$ then a' is not an ordinal.

Case $a \notin S$.

Then by property (0) of Lemma 2.11, a' contains a singleton set $\{d\}$ with $rk\{d\} > 1$. Plainly $\{d\}$ cannot be an ordinal and a' is not an ordinal.

Case $a \in S$.

We proceed by induction on the height of a in $G = (\tilde{V}, \cdot)$. By the property (c) of S we have that for some $b \hat{\in} a$ either $b \notin \Omega \cup \{x_k : k \text{ integer}, k > n\}$ or, for some c, $c \hat{\in} b$ & $\neg c \hat{\in} a$.

In the latter case $c' \in b' \in a'$ whereas $c' \notin a'$, so that a' cannot be an ordinal.

On the other hand, if $b \notin \Omega \cup \{x_k : k \text{ integer}, k > n\}$ then by the induction hypothesis b' is not an ordinal. Since $b' \in a'$, it follows that a' cannot be an ordinal.

We have thus shown that if $a \notin \Omega \cup \{x_k : k \text{ integer}, k > n\}$, then a' is not an ordinal.

Thus let $\neg ORD(x)$ appear in q. Then $[x] \notin \Omega \cup \{x_k : k \text{ integer}, k > n\}$ and $Mx = [x]'$ is not an ordinal. This concludes the proof of M being a model of q showing the completeness of Algorithm 7.3.

To finish the proof of the present lemma, assume that no equality literal $N = x$ appears in q and that $Mx_k = x_k' = k$ for every $k > n$.

Let $(q)_{x_k}^N$ be the result of replacing N by x_k in q. If $x \neq x_k$ appears in $(q)_{x_k}^N$ but not in q, since $[x]$ is distinct from x_k, then by (3') of Lemma 2.11 $Mx = [x]' \neq x_k' = Mx_k$.

If $x_k \in x$ appears in $(q)_{x_k}^N$, then $N \in x$ appears in q. It follows by the properties (1) and (3) of $\hat{\in}$ that $[N]\hat{\in}[x]$ and $x_k \hat{\in} [x]$. This yields $Mx_k = x_k' \in [x]' = Mx$.

Similarly if $x \in x_k$ appears in $(q)_{x_k}^N$, then by the properties (1) and (4) of $\hat{\in}$ it follows that $Mx \in Mx_k$.

Furthermore if $x_k \notin x$ appears in q_k^N, then $N \notin x$ appears in q. By (vii) of Algorithm 7.3 we obtain $\neg N \hat{\in} x$ and by the property (3) of $\hat{\in}$ we reach $\neg x_k \hat{\in} [x]$. This, again by (3') of Lemma 2.11, gives $Mx_k = x_k' \notin [x]' = Mx$.

Analogously if $x \notin x_k$ appears in $(q)_{x_k}^N$, then by using the property (4) of $\hat{\in}$ we can obtain $Mx \notin Mx_k$. Finally, $\neg ORD(N)$ cannot appear in q whereas if $ORD(N)$ appears in q then $ORD(x_k)$ appears in $(q)_{x_k}^N$. But $Mx_k = k$ is an ordinal, completing the proof of M being a model of $q \cup \bigcup_{k>n}(q)_{x_k}^N$. This concludes the verification of the present lemma. ∎

7.6   A quantified syllogistic including integers and ordinals

In this final section we strengthen the decidability result of Section 7.3 by replacing the ground theory $\mathbf{F}_0'$ described in Section 7.2 by the theory $\mathbf{F}_0$ discussed in the preceding section.

We consider the theory $\mathbf{G}_2$ which is the propositional closure of prenex formulae $Q_1 Q_2 \cdots Q_n p$ satisfying properties (2) and (3) of Section 7.3 together with

(1') p is an unquantified formula of the theory $\mathbf{F}_0$ in the language $=$, $\in$, $\emptyset$, ω, $ORD(\bullet)$.

Some basic set-theoretic concepts can be expressed by formulae in this class. For example

(ORD)	x is an ordinal	iff $ORD(x)$;
(INT)	x is an integer	iff $x \in \omega$;
(SET_OF_ORD)	x is a set of ordinals	iff $(\forall z \in x)ORD(z)$;
(SET_OF_INT)	x is a set of integers	iff $(\forall z \in x)z \in \omega$.

Interesting elementary theorems on ordinals can be stated by formulae of the same class. For example (cf. [Pas78]), to show that 'the successor of any ordinal is an ordinal' one can simply prove the unsatisfability of

$$ORD(x) \,\&\, y = \{x\} \,\&\, z = x \cup y \,\&\, \neg ORD(z),$$

which can be expressed by a formula of $\mathbf{G}_2$.

In much the same way can be treated the following theorems.

- Two ordinals with the same successor are equal;

- there are no ordinals between an ordinal and its successor;

- ordinals are closed with respect to binary intersection and union.

Moreover, in proving automatically more complicated theorems in the style highlighted in Section 2.1, one is often led to generating as subgoals formulae of the present class. Thus, for example (cf. [Pas78]), in proving that

- if x is a limit ordinal then $x = Un(x)$,

one might try to prove that there cannot be an element z belonging to only one of the two sets x, $Un(x)$. This could be done by showing that the two formulae

- $z \notin Un(x) \,\&\, z \in z' \,\&\, z' \in x$,

- $z \in Un(x) \,\&\, z \notin x \,\&\, ORD(x)$,

are both unsatisfiable. The first of these formulae belongs to the theory $\mathbf{G}_1$, the second to the present theory $\mathbf{G}_2$.

This illustrates a possible use of decision algorithms for the class $\mathbf{G}_2$. We go on to show that such an algorithm exists by slightly modifying Algorithm 7.2, described in Section 7.3. Let q be a conjunction of atoms of type $(\forall x_1 \in y_1) \cdots (\forall x_n \in y_n) p$ in $\mathbf{G}_2$.

Algorithm 7.4 *Let*

- $C = \{\, conditional\ variables\ in\ q \,\} \cup \{\emptyset\}$,

- N be a newly introduced variable to be interpreted intuitively as the successor of the largest integer which is the value of a free variable or constant in q in the sought model of q.

Step 1. *Add to q the formula*

- $\emptyset \in N \& N \in \omega$.

Moreover for every free variable x in q do the following:

- *Add $N \neq x$ to q;*
- *non-deterministically choose one of the two formulae*

  (a) $x \in \omega \& x \in N$,

  (b) $x \notin \omega$,

  and add it to q.

Step 2. *Is the same of* **Step 1** *of Algorithm 7.2. Non-deterministically add to q one of the formulae*

$$a = b, \quad z_{ab} \in a \& z_{ab} \notin b, \quad z_{ab} \notin a \& z_{ab} \in b,$$

for all a, b in C.

Step 3. *For each variable x in C, non-deterministically choose one of the three formulae*

$$ORD(x),$$
$$\neg ORD(x) \& z_x \in x \& \neg ORD(z_x),$$
$$\neg ORD(x) \& z'_x \in z_x \& z_x \in x \& z'_x \notin x,$$

where z_x, z'_x are newly introduced variables, and add the chosen formula to q. Let

- $\overline{q}$ *be the resulting conjunction,*
- $\Phi = \{$ *free variables in $\overline{q}$* $\} \cup \{\emptyset, \omega\}$.

Step 4. *Is completely analogous to* **Step 2** *of Algorithm 7.2.*

Step 5. *Is completely analogous to* **Step 3** *of Algorithm 7.2, but with the extended syllogistic $\mathbf{F}_0$, including ω and ORD, in place of the theory $\mathbf{F}'_0$.* $\square$

Lemma 7.7 (Soundness) *Algorithm 7.4 is sound.*

Proof. The proof is analogous to that of Lemma 7.3. The soundness of **Step 1** is immediate by interpreting N as the successor of the largest integer which is the value of a free variable or constant in some model of q. Soundness of **Step 2** is completely analogous to that of **Step 1** in Lemma 7.3.

To show that **Step 3** is also sound notice that a set x is not an ordinal if it is either not transitive or it contains a non-ordinal element.

Soundness of **Step 4** is proved as soundness of **Step 2** in Lemma 7.3.

Finally, soundness and completeness of **Step 5** is a consequence of Lemmas 7.5 and 7.6. ∎

We conclude by showing that the converse also holds.

Theorem 7.2 (Completeness) *Algorithm 7.4 is complete.*

Furthermore a model of the input formula q can be generated in the following way.

*Let q_i be a satisfiable conjunction in the disjunctive normal form of the unquantified residue Q in **Step 5**.*

Let $\sim$ and $\mathcal{O}$ be, respectively, the equivalence relation and the set of variables and constants whose existence is ensured by Lemma 7.6 as a consequence of the satisfiability of q_i.

Put $\Omega = \mathcal{O}/\sim$ and $S = \Omega \cup \{x_k : k \text{ integer}, k > n\} \cup \{[x] : x \in C\}$.

*Then by **Steps 2** and **3**, S is an extensional set of nodes of $G = (V, \cdot)$ satisfying properties (a), (b) and (c) of Lemma 7.6.*

*Moreover by **Step 1** no equality literal $N = x$ appears in q_i.*

Let M be the model of $q_i \cup \bigcup_{k>n}(q_i)^N_{x_k}$ constructed in Lemma 7.6.

Then M is also a model of the input formula q.

Proof. By $(\star)$ of Lemma 7.6 we have that for every conditional variable x in C, every element of Mx is of type Mz for some

$$z \text{ in } \Phi \cup \{x_k : k \text{ integer}, k > n\}.$$

It follows that since M satisfies all formulae of type

$$a_1 \in y_1 \& \cdots \& a_n \in y_n \to p^{x_1 x_2 \ldots a_n}_{a_1 a_2 \ldots a_n}$$

for some $y_1, \ldots, y_n$ in C, where $(a_1, \ldots, a_n)$ ranges over $(\Phi \cup \{x_k : k \text{ integer}, k > n\})^n$, then M also satisfies $(\forall x_1 \in y_1) \cdots (\forall x_n \in y_n)p$. This shows the completeness of the quantifier elimination step (namely **Step 4**), concluding the proof of the theorem and establishing the decidability of the quantified theory $\mathbf{G}_2$. ∎

Part II

EXTENDED MULTILEVEL SYLLOGISTICS

8. THE POWERSET OPERATOR

8.1 Introduction

In this chapter we show that the class of unquantified formulae of set theory involving boolean operators, the powerset and singleton operators, and the equality and membership predicates has a solvable satisfiability problem. In particular we will show that whenever a formula P in the above class is satisfiable, it has a hereditarily finite model of finite and bounded rank (cf. [Can87] and [Can88]).

In Chapter 6, we have already proved that the theory MLS (multilevel syllogistic) consisting of formulae built using the boolean connectives (conjunction, disjunction, implication and negation) from set-theoretic atoms of the types

$$x = y \cup z, x = y \setminus z, x \in y$$

has a solvable satisfiability problem. The extension MLSS of MLS obtained by allowing the singleton operator to appear has been proved decidable in the same chapter (see also [FOS80a]). In his doctoral dissertation [Fer81], Ferro showed that the class MLS of formulae remains still decidable even if at most *two* occurrences of the powerset operator are allowed; see also the earlier result of [Bre82]. This result was later extended in [CFS85] where it was shown that the class MLSP of formulae obtained from MLS by allowing an unrestricted number of occurrences of the powerset has a solvable satisfiability problem.

In this chapter we describe a decision procedure for a class of set theoretic formulae involving singleton and powerset operators. Specifically, it is shown that the family of unquantified formulae of set theory built up using binary union and intersection, set difference, powerset, singleton, the binary predicates set-membership and equality, and the propositional boolean connectives has a decidable satisfiability problem.

The technique used in our solution of the aforementioned satisfiability problem involves an interplay of syntactic and model-theoretic arguments. Firstly, each variable occurring in a normalized conjunction P is—so to say— split into disjoint parts (the *places* of P), which are essentially syntactic counterparts of the Venn regions associated to a model of P. Then it is proved that under the assumption that P is satisfiable, P must admit a canonical model of rank bounded solely by a function of the size of the conjunction P whose satisfiability is to be tested. This is achieved by exhibiting a non-deterministic standardization algorithm consisting of an *initialization phase* and a *stabilization loop*. During the initialization phase, places are provisionally assigned empty models. The subsequent stabilization phase enlarges the sets corresponding to places in such a way as to ensure that all clauses in P are correctly modelled. To prove that such completion can be carried out in an *a priori* bounded number of steps whenever P is satisfiable, an assumed model M of P is used as an oracle within a nondeterministic standardization algorithm which associates final models with all places of P.

As a corollary of this construction, it follows that a conjunction P is satisfiable if and only if it has a hereditarily finite model of rank bounded by a doubly exponential expression in the number of variables occurring in P.

8.2   Preliminaries

We denote by MLSSP the propositional closure of atoms of the types

$$x = y \cup z, \quad x = y \setminus z, \quad x \in y, \quad x = \{y\}, \quad x = Pow(y), \tag{8.1}$$

where we recall that $\{y\}$ stands for the set having y has its unique element, and where $Pow(y) = \{z : z \subseteq y\}$.

By using a normalization process of the kind described in Section 3.8, the satisfiability problem for MLSSP can be reduced to the satisfiability problem for the subtheory MLSSP$'$ consisting of the *conjunctions* of literals of type (8.1).

In addition, in view of Lemma 3.3 and by making use of the notion of *injective* satisfiability (cf. Definition 3.9), we have the following reduction corollary.

Corollary 8.1 *The satisfiability problem and the injective satisfiability problem for* MLSSP$'$ *are equivalent.*   □

It is also helpful to consider the subtheory MLSSP$''$ of MLSSP$'$ consisting of those conjunctions P which contain clauses $q_0 = q_0 \setminus q_0$, $p_0 = Pow(q_0)$, $p_v = Pow(q_v)$, $q_v \setminus p_v = q_0$, and $x \in q_v$ for all variables x in P distinct from q_v and p_v, where q_0, p_0, q_v, and p_v are four distinct variables, occurring in P only within the above clauses.

Then we have:

Lemma 8.1 *The injective satisfiability problem for* MLSSP$'$ *is reducible to the injective satisfiability problem for* MLSSP$''$.

Proof. It is enough to show that every MLSSP$'$ conjunction can be effectively transformed into an equisatisfiable MLSSP$''$ conjunction. So, let P be an MLSSP$'$ conjunction. We let $\tau(P)$ denote the conjunction

$$P \,\&\, q_0 = q_0 \setminus q_0 \,\&\, p_0 = Pow(q_0) \,\&\, p_v = Pow(q_v)$$
$$\&\, q_v \setminus p_v = q_0 \,\&\, \left(\underset{\substack{x \ occurs \\ in \ P}}{\&} x \in q_v \right),$$

where q_0, p_0, q_v and p_v are pairwise distinct variables not occurring in P. Clearly, if $\tau(P)$ is satisfiable, so is P. Conversely, assume P is satisfiable and let M be one of its models. Put $Mq_0 = \emptyset$, $Mp_0 = \{\emptyset\}$,

$$Mq_v = transitive\_closure(\{Mx : x \text{ occurs in } P\}), \quad Mp_v = Pow(Mq_v),$$

where

$$transitive\_closure(s) = \bigcap_{\substack{s \subseteq t \\ \& \ t \ is \ transitive}} t$$

and where a set t is said to be transitive if $u \subseteq t$ for all $u \in t$.

It is an easy matter to show that the assignment so extended models correctly $\tau(P)$, thus proving that P and $\tau(P)$ are equisatisfiable. ∎

Next we will introduce some concepts of fundamental relevance in the following sections.

Suppose that a conjunction P of MLSSP′ is given.

Definition 8.1 *A place π of P is a 0/1-valued function defined on the set of all variables in P such that $\pi(x) = \pi(y) \vee \pi(z)$ (respectively, $\pi(x) = \pi(y) \ \& \ \neg\pi(z)$) if $x = y \cup z$ (respectively, $x = y \setminus z$) appears in P.* □

Notice that we are identifying here 0 and 1 with the truth values *FALSE* and *TRUE*, respectively.

Definition 8.2 *Given a variable x, the place π is said to be a place at x of P if $\pi(y) = 1$ whenever $x \in y$ appears in P.* □

In the next section we will see that any model of P defines naturally a set of places for P and places to variables which go a long way toward describing the structure of the model itself.

The following definition introduces a concept of central importance to our purposes.

Definition 8.3 *Let s be a set. Then we put*

$$Pow^*(s) = \{t : t \subseteq Un(s) \ \& \ (\forall s' \in s)(t \cap s' \neq \emptyset)\}$$

(where we recall that Un denotes the unary union defined by $Un(s) = \{u : u \in r, \text{ for some } r \in s\}$ —cf. [Jec78]). □

Some obvious properties of the operator Pow^* are listed below.

Lemma 8.2 *(a) $Pow^*(\emptyset) = \{\emptyset\}$.*

(b) Let s_1 and s_2 be two sets such that $s_1 \cup s_2$ is a collection of pairwise disjoint non-empty sets. If $Pow^*(s_1) = Pow^*(s_2)$, then $s_1 = s_2$.

(c) Let $\{s_i : i \in I\}$ and $\{t_i : i \in I\}$ be collections of sets such that $s_i \subseteq t_i$ for all $i \in I$. Then $Pow^*(\{s_i : i \in I\}) \subseteq Pow^*(\{t_i : i \in I\})$.

(d) For each set s,

$$Pow(Un(s)) = Un\{Pow^*(t) : t \in Pow(s)\}.$$

Proof. (a) is an immediate consequence of the definition of Pow^*.

Concerning (b), it suffices to observe that if $Pow^*(s_1) = Pow^*(s_2)$ then $Un(s_1) = Un(s_2)$. The latter equality combined with the disjointness hypothesis implies $s_1 = s_2$.

Concerning (c), let $\{s_i : i \in I\}$ and $\{t_i : i \in I\}$ be such that $s_i \subseteq t_i$ for all $i \in I$. Plainly, $Un(\{s_i : i \in I\}) \subseteq Un(\{t_i : i \in I\})$. Moreover, if $u \cap s_i \neq \emptyset$ for all $i \in I$, then, a fortiori, $u \cap t_i \neq \emptyset$ for all $i \in I$. Hence $Pow^*(\{s_i : i \in I\}) \subseteq Pow^*(\{t_i : i \in I\})$.

Finally, as regards (d), let $u \in Pow(Un(s))$, and let $t_u = \{t \in s : t \cap u \neq \emptyset\}$. Plainly $u \in Pow^*(t_u)$ and $t_u \in Pow(s)$, which prove $Pow(Un(s)) \subseteq Un\{Pow^*(t) : t \in Pow(s)\}$. Also, as $Pow^*(t) \subseteq Pow(Un(t)) \subseteq Pow(Un(s))$ for all $t \in Pow(s)$, the converse inclusion follows at once, thereby proving (d) and in turn concluding the proof of the lemma. ■

Remark 8.1 Throughout this chapter, for any given mapping $f : X \to Y$, we denote by $\overline{f}$ the mapping from $Pow(X)$ into $Pow(Y)$ defined by

$$\overline{f}(A) = f[A], \text{ for all } A \text{ in } Pow(X).$$

□

The following properties hold.

Lemma 8.3 Let f be a one-to-one function. Then,

(a) $\overline{f}$ is one-to-one.

(b) If $s \subseteq Pow(Dom(f))$, then

    (b.1) $Pow^*(s) \subseteq Dom(\overline{f})$, and

    (b.2) $\overline{f}[Pow^*(s)] = Pow^*(\overline{f}[s])$.

Proof. Concerning (a), let s_1 and s_2 be two distinct subsets of $Dom(f)$, and let $u \in (s_1 \setminus s_2) \cup (s_2 \setminus s_1)$. Then $f(u) \in f[(s_1 \setminus s_2) \cup (s_2 \setminus s_1)] = (f[s_1] \setminus f[s_2]) \cup (f[s_2] \setminus f[s_1])$, which implies $f[s_1] \neq f[s_2]$, i.e. $\overline{f}$ is injective.

As regards (b), let $s \subseteq Pow(Dom(f))$. Then for each $t \in Pow^*(s)$, $t \subseteq Un(s) \subseteq Dom(f)$, i.e. $t \in Dom(\overline{f})$. Therefore $Pow^*(s) \subseteq Dom(\overline{f})$, which proves (b.1). To prove (b.2), let $t \in Pow^*(s)$. As $t \subseteq Un(s)$, we have $\overline{f}(t) = f[t] \subseteq f[Un(s)] = Un(\overline{f}[s])$. Moreover, $t \cap s' \neq \emptyset$ for each $s' \in s$. Hence $\overline{f}(t) \cap u \neq \emptyset$, for all $u \in \overline{f}[s]$. Thus $\overline{f}(t) \in Pow^*(\overline{f}[s])$, which proves $\overline{f}[Pow^*(s)] \subseteq$

$Pow^*(\overline{f}[s])$. Conversely, let $w \in Pow^*(\overline{f}[s])$. Then $w \subseteq Un(\overline{f}[s]) \subseteq f[Un(s)]$. Put $u = f^{-1}[w]$. Plainly $u \subseteq Un(s)$. Moreover, for all $s' \in s$, $w \cap f[s'] = w \cap \overline{f}(s') \neq \emptyset$, so that $u \cap s' = f^{-1}[w] \cap f^{-1}[f[s']] = f^{-1}[w \cap f[s']] \neq \emptyset$. Hence $u \in Pow^*(s)$, and since $w = f[f^{-1}[w]] = f[u] = \overline{f}(u)$, we obtain $w \in \overline{f}[Pow^*(s)]$ which in turn implies $Pow^*(\overline{f}[s]) \subseteq \overline{f}[Pow^*(s)]$. This concludes the proof of (b.2), so that the lemma is completely proved.   ∎

8.3   The main result

In the preceding section we have shown that the *satisfiability* problem for the class MLSSP of formulae can be reduced to the *injective satisfiability* problem for the narrower class MLSSP″. This section solves this latter problem.

The following theorem gives decidable conditions for a conjunction P in MLSSP″ to be injectively satisfiable.

Theorem 8.1 (Decidability) *Let P be a conjunction of literals each of which has one of the following types,*

$$
\begin{array}{ll}
(=) & x = y \cup z, x = y \setminus z \\
(\in) & x \in y \\
(\{\cdot\}) & x = \{y\} \\
(Pow) & x = Pow(y).
\end{array}
$$

Assume also that P contains the clauses

$$q_0 = q_0 \setminus q_0, \quad p_0 = Pow(q_0), \quad p_v = Pow(q_v), \quad q_v \setminus p_v = q_0, \quad x \in q_v, \quad (8.2)$$

for all variables x in P distinct from q_v and p_v, where q_0, p_0, q_v, and p_v occur in P only within the clauses (8.2). Also, let $p_0 = Pow(q_0)$, $p_1 = Pow(q_1), \ldots, p_{v-1} = Pow(q_{v-1})$, $p_v = Pow(q_v)$ be all powerset clauses in P, and let V be the collection of distinct variables occurring in P. Then P is injectively satisfiable, i.e. satisfiable by a model which maps distinct variables into distinct sets, if and only if there exist

(i) a set $\Pi = \{\pi_1, \ldots, \pi_n\}$ of places of P;

(ii) a mapping $x \mapsto \pi^x$ from $V \setminus \{p_v\}$ into Π;

(iii) a mapping $\pi \mapsto \overline{\pi}$ from Π to sets;

such that

(C1) no two distinct variables in P are Π-equivalent, i.e. for every pair of distinct variables x, y in P, there is a place $\pi \in \Pi$ such that $\pi(x) \neq \pi(y)$;

(C2) for each variable x in $V \setminus \{p_v\}$ the place π^x is at the variable x;

(C3.a) $\overline{\pi} \neq \emptyset$, *for all places* $\pi \in \Pi$;

(C3.b) $\overline{\alpha} \cap \overline{\beta} = \emptyset$, *for all places* $\alpha, \beta \in \Pi$, *with* $\alpha \neq \beta$;

(C3.c) $\bigcup_{\pi(x)=1} \overline{\pi} \in \overline{\pi^x}$, *for all variables* x *in* $V \setminus \{p_v\}$;

(C3.d) $\bigcup_{\alpha(p)=1} \overline{\alpha} = Pow(\bigcup_{\beta(q)=1} \overline{\beta})$, *for all powerset clauses* $p = Pow(q)$ *in* P;

(C3.e) $\bigcup_{\alpha(x)=1} \overline{\alpha} = \{\bigcup_{\beta(y)=1} \overline{\beta}\}$, *for all singleton clauses* $x = \{y\}$ *in* P;

(C4) it must be possible to produce the mapping $\pi \mapsto \overline{\pi}$ *in (iii) by an execution of the following nondeterministic association algorithm, in which Step 2 is executed at most* $(\varrho - 2) \cdot (n-1) \cdot 2^{n-2} + 3 \cdot 2^{n-1} - 2$ *times, where* $n = |\Pi|$, *and* ϱ *is any natural number greater than 1 and such that* $2^{\varrho - 1} > \varrho \cdot (n-1) + 1$.

Algorithm 8.1 (Association Algorithm)

Step 1. *Put*

$$\overline{\pi} \leftarrow \emptyset,$$

for all places $\pi \in \Pi$.

Step 2. *Pick a set* $\{\alpha_1, \ldots, \alpha_\ell\} \subseteq \Pi$ *and choose sets* Δ_π, *with* $\pi \in \Pi$, *such that*

$$\bigcup_{\pi \in \Pi} \Delta_\pi \subseteq Pow^*(\{\overline{\alpha_1}, \ldots, \overline{\alpha_\ell}\}).$$

Enlarge each set $\overline{\pi}$ *by putting*

$$\overline{\pi} \leftarrow \overline{\pi} \cup \Delta_\pi$$

Step 3. *Stop or to go Step 2.*

$\square$

Proof. Sufficiency. Assume, first, that there exist $\Pi, x \mapsto \pi^x, \pi \mapsto \overline{\pi}$ as in (i), (ii), and (iii) and such that conditions (C1)–(C4) are all satisfied. Then we will prove that the assignment M^* defined on the variables of P by

$$M^*x = \bigcup_{\pi(x)=1} \overline{\pi} \tag{8.3}$$

is injective and satisfies all the conjuncts in P.

Let x, y be variables occurring in P, and assume that $M^*x = M^*y$. We will show that x coincides with y. Indeed, since $\bigcup_{\pi(x)=1} \overline{\pi} = \bigcup_{\pi(y)=1} \overline{\pi}$, conditions (C3.a) and (C3.b) imply that x and y are Π-equivalent, which by (C1) gives $x = y$.

Next we prove that M^* satisfies each conjunct in P. Let $x = y \cup z$ (respectively, $x = y \setminus z$) occur in P. In view of (8.3) and conditions (C3.a) and (C3.b), in order to prove that $M^*x = M^*y \cup M^*z$ (respectively, $M^*x = M^*y \setminus M^*z$), it suffices to show that $\pi(x) = \pi(y) \vee \pi(z)$ (respectively, $\pi(x) = \pi(y) \& \neg\pi(z)$), for all $\pi \in \Pi$. But this follows immediately from Definition 8.1.

Having proved that M^* models correctly all literals in P of type (=), we show that membership relations are also satisfied. Let the clause $x \in y$ belong to P. By (C2) and Definition 8.2, $\pi^x(y) = 1$. Hence (8.3) and (C3.c) imply $M^*x = \bigcup_{\pi(x)=1} \overline{\pi} \in \overline{\pi^x} \subseteq \bigcup_{\pi(y)=1} \overline{\pi} = M^*y$, i.e. $M^*x \in M^*y$.

Finally, in view of (8.3), conditions (C3.d) and (C3.e), respectively, imply that powerset and singleton clauses in P are satisfied by M^*, thereby completing the proof that M^* injectively satisfies P.

Necessity. Next assume that there exists an injective model M of P. The following lemma lists some properties of the model M which are immediate consequences of clauses (8.2).

Lemma 8.4 *(1)* $Mq_0 = \emptyset, Mp_0 = \{\emptyset\};$

(2) Mq_v *and* Mp_v *are transitive sets;*

(3) $Mx \subseteq Mp_v,$ *for all* $x \in V;$

(4) $My \in Mp_v$ *and* $My \subseteq Mq_v,$ *for all* $y \in V \setminus \{p_v\};$

(5) $Mz \in Mq_v,$ *for all* $z \in V \setminus \{q_v, p_v\}.$ □

We proceed to the construction of Π, $x \mapsto \pi^x$, and $\pi \mapsto \overline{\pi}$ as from (i)–(iii).

Let $\sigma_1, \sigma_2, \ldots, \sigma_n$ be the (non-empty, disjoint) regions of the Venn diagram of the collection of sets $\{My : y \in V \setminus \{p_v\}\}$ in the universal space

$$\mathcal{U} = \left(\bigcup_{y \in V} My \right) \cup \bigcup_{y \in V \setminus \{p_v\}} \{My\} = Mp_v.$$

Then each set σ_i is either wholly contained in Mx or wholly disjoint from it, for each x in V, and we can define

$$\pi_i(y) = \begin{cases} 0 & \text{if } \sigma_i \cap My = \emptyset \\ 1 & \text{if } \sigma_i \subseteq My, \end{cases}$$

where y ranges over V. We put $\Pi = \{\pi_1, \ldots, \pi_n\}$. Furthermore, given a variable y in $V \setminus \{p_v\}$, we put $\pi^y = \pi_{i_y}$, where $My \in \sigma_{i_y}$, for some $i_y \in \{1, 2, \ldots, n\}$.

Remark 8.2 For each $\pi \in \Pi$, we designate by σ^π the region of the Venn diagram relative to M and P which induces the place π. □

It is an easy matter to verify that the πs are places of P and that for each y in $V \setminus \{p_v\}$, the place π^y is at the variable y. Thus (i), (ii), and (C2) hold.

Next we prove that (C1) holds too. Let x, y be two distinct variables occurring in P. Since $Mx \neq My$, there exists $s \in (Mx \setminus My) \cup (My \setminus Mx)$, so that $s \in \mathcal{U}$ (where $\mathcal{U}$ is the universe). Let σ_{i_0} be the region of the Venn diagram which contains the set s, whence $\pi_{i_0}(x) \neq \pi_{i_0}(y)$, showing that x and y are not Π-equivalent.

To complete the proof of the necessity of conditions (C1)–(C4) we need to exhibit an instantiation of the Association Algorithm which produces the sets $\overline{\pi}$ in at most

$$(\varrho - 2) \cdot (n - 1) \cdot 2^{n-2} + 3 \cdot 2^{n-1} - 2$$

executions of Step 2 and such that (C3.a)–(C3.e) are satisfied. This is accomplished by using the given model M as an oracle which permits us to 'extract' a transformed canonical model of P. Such canonical model will have rank bounded by a doubly exponential expression in the size of V.

The following lemma expresses some useful facts about the regions of the Venn diagram relative to M and P and their corresponding places.

Lemma 8.5 (1) $\sigma^{\pi^{q_0}} = \{\emptyset\}$ and π^{q_0} is the unique place $\pi \in \Pi$ such that $\pi(p) = 1$ for all powerset clauses $p = Pow(q)$ in P. Furthermore, π^{q_0} is the unique place $\pi \in \Pi$ such that $\pi(p_0) = 1$.

(2) $\sigma^{\pi^{q_v}} = Mp_v \setminus Mq_v$. Hence $\pi^{q_v}(p_v) = 1$ and $\pi^{q_v}(q_v) = 0$.

(3) $\pi(q_v) = 1$, for each $\pi \in \Pi \setminus \{\pi^{q_v}\}$.

(4) $n = |\Pi| \leq 2^{|V|-4} + 2$.

(5) If $x = \{y\}$ occurs in P, then π^y is the unique place $\pi \in \Pi$ such that $\pi(x) = 1$.

Proof. As $\emptyset = Mq_0 \in \sigma^{\pi^{q_0}} \cap Mp_0$, then $\sigma^{\pi^{q_0}} = Mp_0 = \{\emptyset\}$, which implies $\sigma^{\pi^{q_0}} = \{\emptyset\}$. Therefore $\sigma^{\pi^{q_0}} \subseteq Pow(s)$ for all sets s and in particular $\sigma^{\pi^{q_0}} \subseteq Mp$, i.e. $\pi^{q_0}(p) = 1$, for all powerset clauses $p = Pow(q)$ in P. The uniqueness of π^{q_0} follows plainly since $q_0 = q_0 \setminus q_0 \ \& \ p_0 = Pow(q_0)$ occurs in P. Finally, since $Mp_0 = \{\emptyset\}$, π^{q_0} is the unique place $\pi \in \Pi$ such that $\pi(p_0) = 1$. Thus (1) is proved.

Concerning (2), since $Mq_v \in \sigma^{\pi^{q_v}}$, we have $\sigma^{\pi^{q_v}} \subseteq Mp_v \setminus Mq_v$. If $\sigma^\pi \subseteq Mp_v \setminus Mq_v$ for some $\pi \neq \pi^{q_v}$, then there would exist a variable y in P such that

$$\sigma^{\pi_0} \cap My = \emptyset \text{ and } \sigma^{\pi_1} \subseteq My,$$

where $\{\pi_0, \pi_1\} = \{\pi, \pi^{q_v}\}$. Hence from Lemma 8.4(3) it would follow $y \neq p_v$ which by (4) of the same lemma would imply $My \subseteq Mq_v$. This inclusion in turn would give $\sigma^{\pi_1} \subseteq Mq_v$, a contradiction. Therefore $\sigma^{\pi^{q_v}} = Mp_v \setminus Mq_v$, and (2) is proved.

Next, to prove (3) we consider $\pi \in \Pi \setminus \{\pi^{q_v}\}$. Lemma 8.4(3) implies $\sigma^\pi \subseteq Mp_v \setminus \sigma^{\pi^{q_v}} = Mq_v$, which yields $\pi(q_v) = 1$.

Concerning (4) we note that $\pi(q_0) = 0$ for all $\pi \in \Pi$, since $q_0 = q_0 \setminus q_0$ is in P, and $\pi(p_v) = 1$ for all $\pi \in \Pi$, since by (4) of Lemma 8.4 $Mx \subseteq Mp_v$ for all $x \in V$. Thus in view of (1) and (3) above it follows that $|\Pi \setminus \{\pi^{q_0}, \pi^{q_v}\}| \leq 2^{|V|-4}$, which yields (4).

Finally, to prove (5) we observe that $My \in \sigma^{\pi^y} \subseteq Mx$, so that $\sigma^{\pi^y} = \{My\} = Mx$, showing that π^y is the unique place $\pi \in \Pi$ such that $\pi(x) = 1$. Thus the proof of the lemma is completed.   ■

Next we introduce the concepts of P-nodes and their P-targets, which will play a central role in the proof of the necessity of conditions (C4) and (C3).

Let $Pow^*(\{\sigma^{\alpha_1}, \ldots, \sigma^{\alpha_\ell}\}) \cap \sigma^\beta \neq \emptyset$, for some regions $\sigma^{\alpha_1}, \ldots, \sigma^{\alpha_\ell}, \sigma^\beta$ of the Venn diagram relative to M and P. Then there exists

$$s \in Pow^*(\{\sigma^{\alpha_1}, \ldots, \sigma^{\alpha_\ell}\}) \cap \sigma^\beta,$$

and we can write $s = \bigcup_{i=1}^\ell (s \cap \sigma^{\alpha_i})$, where every one of the sets $s \cap \sigma^{\alpha_i}$ is non-empty. For every powerset clause $p = Pow(q)$ in P such that $\beta(p) = 1$, we have $\sigma^\beta \subseteq Mp$ and consequently $s \in Pow(Mq)$. Thus $s \cap \sigma^{\alpha_i} \subseteq Mq$ and hence $\sigma^{\alpha_i} \subseteq Mq \subseteq Mq_v$ for all $i = 1, \ldots, \ell$, implying that $\alpha_i(q) = 1$ (and $\alpha_i(q_v) = 1$) for all $i = 1, \ldots, \ell$. In the same way it can be shown that for a given powerset clause $p = Pow(q)$ in P, if $\alpha_i(q) = 1$, $i = 1, \ldots, \ell$, then $\beta(p) = 1$. These semantic considerations suggest the following definition, which is purely syntactic.

Definition 8.4   (a) Any subset of $\Pi \setminus \{\pi^{q_v}\}$ is called a P-node.

(b) Let A be a P-node. A place β is called a P-target (or simply a target) of A if for every powerset clause $p = Pow(q)$ in P we have

$$\beta(p) = 1 \text{ if and only if } \alpha(q) = 1, \text{ for all } \alpha \in A.$$

□

Remark 8.3 In the following, for each P-node A we will write $T(A)$ to denote the set of P-targets of A, so that T maps the set $Pow(\Pi \setminus \{\pi^{q_v}\})$ into $Pow(\Pi)$.   □

The discussion preceding Definition 8.4 is now restated in terms of the new concepts just introduced.

Lemma 8.6 Let $Pow^*(\{\sigma^{\alpha_1}, \ldots, \sigma^{\alpha_\ell}\}) \cap \sigma^\beta \neq \emptyset$, for some places $\alpha_1, \ldots, \alpha_\ell, \beta$ of P. Then $\{\alpha_1, \ldots, \alpha_\ell\}$ is a P-node and β is a target of $\{\alpha_1, \ldots, \alpha_\ell\}$, i.e. $\beta \in T(\{\alpha_1, \ldots, \alpha_\ell\})$.   □

The following lemma states other useful facts about P-nodes and their targets.

Lemma 8.7 *(a)* $T(\emptyset) = \{\pi^{q_0}\}$;

(b) $\displaystyle\bigcup_{A \subseteq \Pi \setminus \{\pi^{q_v}\}} T(A) = \Pi$;

(c) $\displaystyle\sigma^\beta \subseteq \bigcup_{\substack{A \subseteq \Pi \setminus \{\pi^{q_v}\} \\ \&\ \beta \in T(A)}} Pow^*(\{\sigma^\alpha : \alpha \in A\})$, *for all places $\beta \in \Pi$*;

(d) $\displaystyle Pow^*(\{\sigma^\alpha : \alpha \in A\}) \subseteq \bigcup_{\beta \in T(A)} \sigma^\beta$, *for all P-nodes A*;

(e) $|T(A)| \leq n - 1$, *for all P-nodes A.*

Proof. From Definition 8.4(b), $\pi \in T(\emptyset)$ if and only if $\pi(p) = 1$ for all powerset clauses $p = Pow(q)$ in P. Therefore (a) follows at once from Lemma 8.5(1).

Concerning (b), it is enough to show that every place $\beta \in \Pi$ is target of some P-node A. Let $s \in \sigma^\beta$ and consider $A_s = \{\pi \in \Pi : \sigma^\pi \cap s \neq \emptyset\}$. As $s \in Mp_v$, then $s \subseteq Mq_v$, implying that A_s is a P-node and that $s \in Pow^*(\{\sigma^\pi : \pi \in A_s\}) \cap \sigma^\beta$. By Lemma 8.6, the latter membership implies $\beta \in T(A_s)$, proving (b). Moreover, it follows that

$$\sigma^\beta \subseteq \bigcup_{s \in \sigma^\beta} Pow^*(\{\sigma^\pi : \pi \in A_s\}) \subseteq \bigcup_{\substack{A \subseteq \Pi \setminus \{\pi^{q_v}\} \\ \&\ \beta \in T(A)}} Pow^*(\{\sigma^\alpha : \alpha \in A\}),$$

thus establishing (c).

Concerning (d), let A be a P-node. Since by definition $A \subseteq \Pi \setminus \{\pi^{q_v}\}$, use of Lemma 8.5(3) gives $\alpha(q_v) = 1$, for all $\alpha \in A$, i.e. $\bigcup_{\alpha \in A} \sigma^\alpha \subseteq Mq_v$. Hence $Pow^*(\{\sigma^\alpha : \alpha \in A\}) \subseteq Pow\left(\bigcup_{\alpha \in A} \sigma^\alpha\right) \subseteq Pow(Mq_v) = Mp_v = \bigcup_{i=1}^{n} \sigma_i$. Let $\sigma^{\beta_1}, \ldots, \sigma^{\beta_k}$ be the regions of the Venn diagram that have non-empty intersection with $Pow^*(\{\sigma^\alpha : \alpha \in A\})$. Hence $Pow^*(\{\sigma^\alpha : \alpha \in A\}) \subseteq \bigcup_{j=1}^{k} \sigma^{\beta_j}$. Moreover, the preceding lemma implies that $\{\beta_1, \ldots, \beta_k\} \subseteq T(A)$, proving (d).

Finally, from (a) $|T(\emptyset)| = |\{\pi^{q_0}\}| = 1 < n - 1$, because $n > 3$. On the other hand, if $A \neq \emptyset$, then $\alpha(q_0) = 0$ for all $\alpha \in A$, so that $\beta(p_0) = 0$ for all targets of A. Since $\pi^{q_0}(p_0) = 1$, it follows $T(A) \subseteq \Pi \setminus \{\pi^{q_0}\}$, thus showing $|T(A)| \leq n - 1$ in all cases and concluding the proof of the lemma. ∎

To prepare for the required construction of the map $\pi \mapsto \bar{\pi}$ in (iii), we need a bit of additional terminology.

As above, let ϱ be an integer such that

$$\varrho \geq 2 \quad \text{and} \quad 2^{\varrho-1} > \varrho \cdot (n-1) + 1,$$

where $n = |\Pi|$.

Definition 8.5 *A place $\pi \in \Pi$ is called M-ϱ-trapped if $|\sigma^\pi| < \varrho$. A P-node $\{\alpha_1, \ldots, \alpha_\ell\}$ is called M-ϱ-trapped if for $i = 1, \ldots, \ell$ each α_i is M-ϱ-trapped.*   □

Remark 8.4 In what follows, M-ϱ-trapped places and M-ϱ-trapped P-nodes will be referred to simply as trapped places and trapped P-nodes, since the model M and the constant ϱ will not change in the course of our proof.   □

Definition 8.6 *For places $\alpha, \beta \in \Pi$ we write $\alpha \lhd \beta$ if $rk\, \sigma^\alpha < rk\, \sigma^\beta$.*   □

Remark 8.5 Since the relation $\lhd$ over Π defined above is clearly acyclic, it can be extended to a linear ordering in Π, which we will designate by the same symbol $\lhd$.   □

We have the following lemma.

Lemma 8.8   *(a) $\pi^{q_0} = \min \Pi$ and $\pi^{q_v} = \max \Pi$.*

(b) Let $\{\alpha_1, \ldots, \alpha_\ell\}$ be a non-trapped P-node. Then $\{\alpha_1, \ldots, \alpha_\ell\}$ has at least one non-trapped target β such that $\max\limits_{i=1,\ldots,\ell} \alpha_i \lhd \beta$.

Proof. Concerning (a), it is enough to observe that by Lemma 8.5 $rk\, \sigma^{\pi^{q_0}} < rk\, \sigma^\pi$, for all $\pi \in \Pi \setminus \{\pi^{q_0}\}$, and $rk\, \sigma^{\pi^{q_v}} > rk\, \sigma^\pi$, for all $\pi \in \Pi \setminus \{\pi^{q_v}\}$.

To prove (b), let $A = \{\alpha_1, \ldots, \alpha_\ell\}$ be a non-trapped P-node. This means that $|\sigma^{\alpha_{j_0}}| \geq \varrho$ for some $j_0 \in \{1, \ldots, \ell\}$. Hence it is easy to see that

$$|\{t \in Pow^*(\{\sigma^{\alpha_1}, \ldots, \sigma^{\alpha_\ell}\}) : rk\, t = rk\, (\sigma^{\alpha_1} \cup \ldots \cup \sigma^{\alpha_\ell})\}| \geq 2^{\varrho - 1}.$$

From this inequality, Lemma 8.7(d) and the pigeon-hole principle we deduce that there must exist a place $\beta \in T(A)$ such that

$$|\{t \in Pow^*(\{\sigma^{\alpha_1}, \ldots, \sigma^{\alpha_\ell}\}) : rk\, t = rk\, (\sigma^{\alpha_1} \cup \ldots \cup \sigma^{\alpha_\ell})\} \cap \sigma^\beta| \geq \varrho,$$

since by Lemma 8.7(e) $2^{\varrho - 1} > \varrho \cdot (n - 1) + 1 > \varrho \cdot |T(A)|$.
Hence $|\sigma^\beta| \geq \varrho$, i.e. β is non-trapped, and

$$rk\, \sigma^{\alpha_i} \leq rk\, (\sigma^{\alpha_1} \cup \ldots \cup \sigma^{\alpha_\ell}) < rk\, \sigma^\beta, \quad \text{for all } i = 1, \ldots, \ell,$$

i.e. $\alpha_i \lhd \beta$, for all $i = 1, \ldots, \ell$. This completes the proof of the lemma.   ■

Let $A = \{\alpha_1, \ldots, \alpha_\ell\}$ be a P-node. Then since $Un(\{\sigma^\alpha : \alpha \in A\}) \in Pow^*(\{\sigma^\alpha : \alpha \in A\})$, by Lemma 8.7(d) we have $Un(\{\sigma^\alpha : \alpha \in A\}) \in \sigma^\tau$, for some place $\tau \in T(A)$. The preceding discussion justifies the following definition.

Definition 8.7 *Let A be a P-node. The place $\tau \in T(A)$ such that $Un(\{\sigma^\alpha : \alpha \in A\}) \in \sigma^\tau$ is called the principal target of A and is denoted by π^A.*   □

The following two lemmas state useful properties of principal targets.

Lemma 8.9 *Let A be a P-node and let x be a variable in P it such that $A = \{\pi \in \Pi : \pi(x) = 1\}$. Then the principal target of A is the place π^x at the variable x.*

Proof. It is enough to observe that under the hypotheses of the lemma we have $Un(\{\sigma^\alpha : \alpha \in A\}) = Mx \in \sigma^{\pi^x}$. ∎

Lemma 8.10 *Let A be a P-node and let π^A be its principal target. Then $\alpha \lhd \pi^A$ for all $\alpha \in A$, where $\lhd$ is the ordering relation introduced in Definition 8.6.*

Proof. By definition we have $Un(\{\sigma^\alpha : \alpha \in A\}) \in \sigma^{\pi^A}$, so that $rk\, \sigma^\alpha < rk\, \sigma^{\pi^A}$, i.e. $\alpha \lhd \pi^A$, for all $\alpha \in A$. ∎

We will prove Theorem 8.1 by showing how to use the model M to find an Instantiated Variant of the Association Algorithm (IVAA). This instantiation, shown below, will consist of:

(a) an *Initialization Phase*, followed by

(b) a *Stabilization Loop*, which is in turn divided into

    (b1) a *Blocking Phase*, and

    (b2) a *Propagation Phase*;

(c) a procedure called *Distribute*.

The Instantiated Algorithm which we present uses an arbitrary, perhaps infinite, model of P to build a canonical model of P. This is done in the following way:

(I) For each place $\pi \in \Pi$, the set $\overline{\pi}$ is initialized to the nullset.

(II) The sets $\overline{\pi}$ are then enlarged progressively (always by calls to the Distribute subprocedure). As enlargement proceeds, an auxiliary one-to-one map f is maintained. The domain of f is always a subset of $\bigcup_{\pi \in \Pi} \overline{\pi}$, and f always maps $\overline{\pi}$ into σ^π; moreover f is defined on all of $\overline{\pi}$ as long as $|\overline{\pi}|$ remains less than the critical size ϱ. We always have $|\overline{\pi}| < \varrho$ if π is trapped, so that when π is trapped f will result to be an 'isomorphism' from $\overline{\pi}$ into σ^π. Moreover, the sets $\overline{\pi}$ remain mutually disjoint as they are enlarged.

(III) As the computation proceeds, places π successively become 'blocked'; once a place π becomes blocked, the set $\overline{\pi}$ ceases to expand. Places become blocked in increasing sequence, according to the ordering relationship $\alpha \lhd \beta$, i.e. if $\alpha \lhd \beta$ then α must already have become blocked when β becomes blocked.

(IV) The condition that $Pow^*(\{\overline{\alpha}_1,\ldots,\overline{\alpha}_\ell\}) \cap \overline{\beta} \neq \emptyset$ only when β is a target of $\{\alpha_1,\ldots,\alpha_\ell\}$ (cf. Lemma 8.6) is maintained, and moreover the Instantiated Algorithm operates in such a way that elements $t \in Pow^*(\{\overline{\alpha}_1,\ldots,\overline{\alpha}_\ell\}) \cap \overline{\beta}$ are only introduced by calls $Distribute(\alpha_1,\ldots,\alpha_\ell)$, with $\alpha_1,\ldots,\alpha_\ell$ as argument list.

The Main Inductive Lemma (Lemma 8.12) to be proved in the next section will show that the Instantiated Algorithm maintains these invariants, and will also establish various other properties of the Instantiated Algorithm required for the necessary inductive proof. In Section 8.5 it will be proved that the IVAA terminates and a bound on its execution length will be established. After the rather lengthy detailed proofs of the Main Inductive Lemma and the Termination Lemma, it will become relatively easy to show that the sets $\overline{\pi}$ generated by the Instantiated Algorithm define a model for P the rank of each of whose sets satisfies an *a priori* fixed bound. This additional conclusion, which clearly establishes satisfiability of the decision problem for MLSSP will be proved at the end of this section.

In full detail, the Instantiated Variant of the Association Algorithm is as follows.

Algorithm 8.2 (Instantiated Variant of the Association Algorithm)

[INITIALIZATION PHASE]	Put $$\overline{\pi} \leftarrow \emptyset$$ for all places $\pi \in \Pi$. Put $$f \leftarrow \emptyset .$$ Mark all places as *'unblocked'*. Mark all P-nodes as *'unblocked'*. Mark all P-nodes as *'unvisited'*. *[END INITIALIZATION PHASE]*
[STABILIZATION LOOP]	**WHILE** there exist *unblocked* places **DO**
[BLOCKING PHASE]	
try_block_next_item:	Let ϑ_0 be the minimum *unblocked* place in the ordering $\lhd$ of places. *[Comment: ϑ_0 is the next candidate to be blocked.]* **IF** (ϑ_0 is trapped **and** $|\overline{\vartheta_0}| = |\sigma^{\vartheta_0}|$)

or (ϑ_0 is non-trapped **and** $|\overline{\vartheta_0}| \geq \varrho$

and $|\overline{\vartheta_0} \cap Pow^*(\{\overline{\alpha_1}, \ldots, \overline{\alpha_\ell}\})| = |\sigma^{\vartheta_0} \cap Pow^*(\{\sigma^{\alpha_1}, \ldots, \sigma^{\alpha_\ell}\})|$ for every P-node $\{\alpha_1, \ldots, \alpha_\ell\}$ which is such that $|\overline{\alpha_j}| < \varrho$, for all $j = 1, \ldots, \ell$) **THEN**

mark ϑ_0 as 'blocked';

FOR ALL P-nodes $\{\vartheta_0, \vartheta_1, \ldots, \vartheta_k\}$ such that all of $\vartheta_1, \ldots, \vartheta_k$ are blocked **DO**

mark $\{\vartheta_0, \vartheta_1, \ldots, \vartheta_k\}$ as 'blocked';
Distribute $(\vartheta_0, \vartheta_1, \ldots, \vartheta_k)$;
[Note: The code for the procedure involved here is shown below.]

END FOR ALL;

GOTO *try_block_next_item*;

END IF;

[END BLOCKING PHASE]

[Comment: When the preceding IF test fails, we enter the Propagation Phase.]

[PROPAGATION PHASE]   Pick a P-node $A = \{\alpha_1, \ldots, \alpha_\ell\}$ marked *unblocked* and such that either $(0 < |\overline{\alpha_1}|, \ldots, |\overline{\alpha_\ell}| \leq \varrho$ **and** $Pow^*(\{\overline{\alpha_1}, \ldots, \overline{\alpha_\ell}\}) \setminus \bigcup_{\beta \in T(A)} \overline{\beta} \neq \emptyset)$ or $(\overline{\alpha_i} \neq \emptyset$, for $i = 1, \ldots, \ell$, **and** $|\overline{\alpha_j}| \geq \varrho$ for some $j \in \{1, \ldots, \ell\})$, if
any such *unblocked* P-nodes exist, and call
Distribute$(\alpha_1, \ldots, \alpha_\ell)$;
[END PROPAGATION PHASE]
END WHILE;
[END STABILIZATION LOOP]

[END IVAA]

Next we show the details of the 'Distribute' subprocedure used in the preceding code.

PROCEDURE *Distribute*$(\alpha_1, \ldots, \alpha_\ell)$;

ASSERT   Assertion A: $\{\alpha_1, \ldots, \alpha_\ell\}$ is a P-node which is either marked *unblocked* or *blocked*, but not *visited*.
END ASSERT *[Assertion A]*;

IF $|\overline{\alpha_i}| < \varrho$ for all $i = 1, \ldots, \ell$ **THEN**
    GOTO *update_for_small_alphas*;
ELSE   [Comment: $|\overline{\alpha_j}| \geq \varrho$ for some $j \in \{1, \ldots, \ell\}$]
    GOTO *update_for_some_large_alpha*;
END IF;

update_for_small_alphas:

For all $\pi \in \Pi$ let

$$\Delta_\pi = \overline{f}^{-1}[Pow^*(\{f[\overline{\alpha_1}], \ldots, f[\overline{\alpha_\ell}]\}) \cap \sigma^\pi] \setminus \bigcup_{\beta \in T(A)} \overline{\beta}$$

(where we recall that $\overline{f}$ denotes the function defined on the parts of the domain of f by $\overline{f}(B) = f[B]$, for $B \subseteq Dom(f)$).

ASSERT   *Assertion B*:

(B.1) $\Delta_\pi = \emptyset$, for all $\pi \in \Pi \setminus T(A)$.

(B.2) $\{\Delta_\pi : \pi \in \Pi\}$ is a partition of $Pow^*(\{\overline{\alpha_1}, \ldots \overline{\alpha_\ell}\}) \setminus \bigcup_{\beta \in T(A)} \overline{\beta}$.

(B.3) If $\Delta_\beta \neq \emptyset$ then the place β is an *unblocked* target of $\{\alpha_1, \ldots, \alpha_\ell\}$, for all $\beta \in \Pi$.

END ASSERT [*Assertion B*];

Put

$$f \leftarrow \overline{f} \cup f|_{Pow^*(\{\overline{\alpha_1}, \ldots, \overline{\alpha_\ell}\})}.$$

Also put

$$\overline{\pi} \leftarrow \overline{\pi} \cup \Delta_\pi,$$

for each place $\pi \in \Pi$.

GOTO *exit*;

update_for_some_large_alpha:

Let $\beta_1, \ldots, \beta_g, \tau_1, \ldots, \tau_k, \nu_1, \ldots, \nu_h$ be all the targets of $\{\alpha_1, \ldots, \alpha_\ell\}$, where the βs are the *blocked* targets, the τs are the *unblocked* trapped targets and the νs are the *unblocked* non-trapped targets of $\{\alpha_1, \ldots, \alpha_\ell\}$, respectively. Let

$$\Delta' = Pow^*(\{\overline{\alpha_1}, \ldots, \overline{\alpha_\ell}\}) \setminus \bigcup_{\beta \in T(A)} \overline{\beta} \setminus \{\overline{\alpha_1} \cup \ldots \cup \overline{\alpha_\ell}\},$$

$$\Delta'' = \begin{cases} \emptyset & \text{if A is not } blocked \\ \{\overline{\alpha_1} \cup \ldots \cup \overline{\alpha_\ell}\} & \text{if A is } blocked \end{cases}$$

$$\Delta = \Delta' \cup \Delta''.$$

Also let

$$\sigma^A = \begin{cases} \emptyset & \text{if A is } blocked \\ \{\sigma^{\alpha_1} \cup \ldots \cup \sigma^{\alpha_\ell}\} & \text{if A is not } blocked. \end{cases}$$

In addition, for every *unblocked* trapped target τ_j of A, let

$$n_j = |\sigma^{\tau_j} \cap Pow^*(\{\sigma^{\alpha_1}, \ldots, \sigma^{\alpha_\ell}\}) \setminus \sigma^A| - |\overline{\tau_j} \cap Pow^*(\{\overline{\alpha_1}, \ldots, \overline{\alpha_\ell}\})|.$$

ASSERT   *Assertion C*:

(C.1) $\overline{\alpha_1} \cup \ldots \cup \overline{\alpha_\ell} \notin \bigcup_{\beta \in T(A)} \overline{\beta}.$

(C.2) $\sigma^{\alpha_1} \cup \ldots \cup \sigma^{\alpha_\ell} \in \sigma^{\pi^A} \cap Pow^*(\{\sigma^{\alpha_1}, \ldots, \sigma^{\alpha_\ell}\}) \setminus range(f)$, where π^A is the principal target of the P-node A.

(C.3) There exists a partition $\Delta_{\tau_1}, \ldots, \Delta_{\tau_k}, \Delta_{\nu_1}, \ldots, \Delta_{\nu_h}$ of the set Δ such that

(C.3.a) $|\Delta_{\tau_j}| = n_j$ for each $j = 1, \ldots, k$;

(C.3.b) if $\Delta' \neq \emptyset$ then $|\Delta_{\nu_i}| \geq \varrho$, for each $i = 1, \ldots, h$; and

(C.3.c) $\Delta'' \subseteq \Delta_{\pi^A}.$

(C.4) For each $j = 1, \ldots, k$,

$$n_j = |\sigma^{\tau_j} \cap Pow^*(\{\sigma^{\alpha_1}, \ldots, \sigma^{\alpha_\ell}\}) \setminus \sigma^A \setminus range(f)|$$

(so that the sets Δ_{τ_j} and $\sigma^{\tau_j} \cap Pow^*(\{\sigma^{\alpha_1}, \ldots, \sigma^{\alpha_\ell}\}) \setminus \sigma^A \setminus range(f)$ can be put in one-to-one correspondence).

END ASSERT [*Assertion C*];

Let $\{\Delta_\pi : \pi \in \Pi\}$ be a partition of the set Δ such that

- $|\Delta_{\tau_j}| = n_j$, for each $j = 1, \ldots, k$;
- if $\Delta' \neq \emptyset$ then $|\Delta_{\nu_i}| \geq \varrho$, for each $i = 1, \ldots, h$;
- $\Delta'' \subseteq \Delta_{\pi^A}$;
- $\Delta_\pi = \emptyset$, if $\pi \in \Pi \setminus \{\tau_1, \ldots, \tau_k, \nu_1, \ldots, \nu_h\}.$

Also, for each $j = 1, \ldots, k$, let f_{τ_j} denote a one-to-one correspondence between Δ_{τ_j} and the set $\sigma^{\tau_j} \cap Pow^*(\{\sigma^{\alpha_1}, \ldots, \sigma^{\alpha_\ell}\}) \setminus \sigma^A \setminus range(f)$, such that if A is *blocked* and the principal target π^A of A is trapped, then $f_{\pi^A}(\overline{\alpha_1} \cup \ldots \cup \overline{\alpha_\ell}) = \sigma^{\alpha_1} \cup \ldots \cup \sigma^{\alpha_\ell}$.

[Comment: From the preceding assertion, it follows immediately that such a partition $\{\Delta_\pi : \pi \in \Pi\}$ and functions $f_{\tau_1}, \ldots, f_{\tau_k}$ exist.]

Put

$$\overline{\pi} \leftarrow \overline{\pi} \cup \Delta_\pi$$

for all $\pi \in \Pi$.

Also put

$$f \leftarrow f \cup f_{\tau_1} \cup \ldots \cup f_{\tau_k}.$$

IF $\{\alpha_1, \ldots, \alpha_\ell\}$ is marked *unblocked* **THEN**
    mark $\{\alpha_1, \ldots, \alpha_\ell\}$ as *'visited'*
END IF;

ASSERT   *Assertion D*: For every place π, if π is trapped, then

$$|\sigma^\pi \cap Pow^*(\{\sigma^{\alpha_1}, \ldots, \sigma^{\alpha_\ell}\}) \setminus \sigma^A| = |\overline{\pi} \cap Pow^*(\{\overline{\alpha_1}, \ldots, \overline{\alpha_\ell}\})|;$$

END ASSERT *[Assertion D]*;

exit:

ASSERT   *Assertion E*:

    (E.1) f is a one-to-one function.

    (E.2) $Dom(f) \subseteq \bigcup_{\pi \in \Pi} \overline{\pi}$.

    (E.3) $f[\overline{\pi}] \subseteq \sigma^\pi$, for all $\pi \in \Pi$.

    (E.4) If $|\overline{\pi}| < \varrho$ then $\overline{\pi} \subseteq Dom(f)$.

    (E.5) If π is trapped, then $|\overline{\pi}| \leq |\sigma^\pi|$.

    (E.6) $\overline{\alpha} \cap \overline{\beta} = \emptyset$, for all $\alpha, \beta \in \Pi$ with $\alpha \neq \beta$.

    (E.7) If $u \in Pow^*(\{\overline{\gamma} : \gamma \in \Gamma\}) \cap (\bigcup_{\pi \in \Pi} \overline{\pi})$, with Γ any P-node such that $|\overline{\gamma}| < \varrho$ for each $\gamma \in \Gamma$, then $u \in Dom(f)$, $u \subseteq Dom(f)$ and $f(u) = f[u]$.

    (E.8) If $u \in Pow^*(\{\overline{\gamma} : \gamma \in \Gamma\}) \cap Dom(f)$, with Γ any P-node, then $f(u) \in Pow^*(\{\sigma^\gamma : \gamma \in \Gamma\})$.

    (E.9) If $t \in Pow^*(\{\overline{\gamma_1}, \ldots, \overline{\gamma_h}\}) \cap \overline{\delta}$, for some places $\gamma_1, \ldots, \gamma_h, \delta \in \Pi$, then

(E.9.a) $\{\gamma_1, \ldots, \gamma_h\}$ is a P-node;

(E.9.b) δ is a target of $\{\gamma_1, \ldots, \gamma_h\}$;

(E.9.c) t must have been introduced into $\bar{\delta}$ during the execution of an earlier call to the procedure *Distribute*, and the argument to this prior call must have been the same P-node $\{\gamma_1, \ldots, \gamma_h\}$.

END ASSERT [*Assertion E*];

END PROCEDURE *Distribute*.   □

Remark 8.6 As just presented, the IVAA is still nondeterministic. Indeed, the Propagation Phase code does not specify how the unblocked P-node is to be chosen among all nodes that satisfy the condition appearing in the code. Moreover, the body of the procedure *Distribute* does not specify what partition $\{\Delta_\pi : \pi \in \Pi\}$ of the set Δ is to be used when a P-node $\{\alpha_1, \ldots, \alpha_\ell\}$, with $\overline{\alpha_i} \neq \emptyset$ for all $i = 1, \ldots, \ell$ and $|\overline{\alpha_j}| \geq \varrho$ for some $j \in \{1, \ldots, \ell\}$, is processed. Nevertheless it is unnecessary to specify these fine details since we will show that *every* possible instantiation of the IVAA represents an acceptable instance of the Association Algorithm.   □

The following lemma lists some immediate properties of the IVAA which can be proved just by inspection of its code.

Lemma 8.11 *For all possible computations of the IVAA we have:*

(a) *After the Initialization Phase the sets $\overline{\pi}$ are modified only by calls to the procedure* Distribute, *and each such modification of a set $\overline{\pi}$ enlarges it.*

(b) *Once $|\overline{\alpha_j}| \geq \varrho$ for some $j \in \{1, \ldots, \ell\}$, a P-node $\{\alpha_1, \ldots, \alpha_\ell\}$ can be processed at most once by a call to the procedure* Distribute *made from the Propagation Phase.*

(c) *Every P-node can be processed at most once by a call to the procedure* Distribute *made from the Blocking Phase.*

(d) *When a P-node $\{\alpha_1, \ldots, \alpha_\ell\}$ is processed by the procedure* Distribute, *all places β such that $\alpha_i \lhd \beta$ for all $i = 1, \ldots, \ell$ are unblocked.*

(e) *Once a place, or a P-node, becomes blocked, it cannot subsequently become unblocked again.*

(f) *At each call of the procedure* Distribute, *if all ASSERT-statements are executed successfully then Assertion E is executed.*

(g) *When a call* Distribute$(\alpha_1, \ldots, \alpha_\ell)$ *takes place, $\overline{\alpha_j} \neq \emptyset$ for $j = 1, \ldots, \ell$.*
                                      □

Let K be an execution of the IVAA, and let $C_1, C_2, C_3, \ldots$ be the sequence of calls to the procedure *Distribute* arranged in the order in which they occur during the computation K.

For each place $\pi \in \Pi$, let $\overline{\pi}^{(r)}$ (respectively, $\Delta_\pi^{(r)}$), $r \geq 1$, designate the value of $\overline{\pi}$ (respectively, the value of Δ_π) just after (respectively, during) the execution of the r-th call C_r. Analogously, we will denote by $f^{(r)}$ the value of f after completion of C_r. Moreover, for a given ASSERT instruction labelled 'Assertion X' and executed during the processing of the call C_r, we denote by 'Assertion $X^{(r)}$' the result of substituting in it each program variable by its corresponding value at the time the ASSERT statement is executed. Finally, we put $\overline{\pi}^{(0)} = f^{(0)} = \emptyset$, for all $\pi \in \Pi$.

The following basic lemmas, which will be proved in the next two sections, express the correctness of the IVAA.

Lemma 8.12 (Main Inductive Lemma: Partial correctness) *All AS-SERT statements encountered during the computation K are executed successfully.*   □

Lemma 8.13 (Termination) *The number of calls $C_1, C_2, C_3, \ldots$ in the computation K is bounded by $(\varrho - 2) \cdot (n-1) \cdot 2^{n-2} + 3 \cdot 2^{n-1} - 2$. Moreover, when the last call to the procedure* Distribute *is made, all places $\pi \in \Pi$ are already blocked.*   □

Temporarily assuming that Lemmas 8.12 and 8.13 hold, we will prove below that the sets $\overline{\pi}$ produced by the computation K satisfy both conditions (C3) and (C4).

Let C_ξ be the last call to Distribute in the computation K. For simplicity, we will often write $\overline{\pi}$ in place of $\overline{\pi}^{(\xi)}$, for all $\pi \in \Pi$, and f in place of $f^{(\xi)}$.

In order to prove that condition (C3.a) is fulfilled, it is enough to observe that since all places $\pi \in \Pi$ are blocked at the end of the computation K, then either $|\overline{\pi}| = |\sigma^\pi|$, if π is trapped, or $|\overline{\pi}| \geq \varrho$, if π is non-trapped, and in any case $\overline{\pi} \neq \emptyset$.

Condition (C3.b) follows immediately from Assertion $(E.6)^{(\xi)}$.

To prove that condition (C3.c) is satisfied, we will need the lemma below.

Lemma 8.14 *(1) Let $|\overline{\alpha_i}^{(r-1)}| < \varrho$, for $i = 1, \ldots, \ell$ and $1 \leq r \leq \xi$. Then*

$$\overline{f^{(r-1)}[Pow^*(\{\overline{\alpha_1}^{(r-1)}, \ldots, \overline{\alpha_\ell}^{(r-1)}\})]}$$
$$= Pow^*(\{f^{(r-1)}[\overline{\alpha_1}^{(r-1)}], \ldots, f^{(r-1)}[\overline{\alpha_\ell}^{(r-1)}]\})$$

(2) Let C_r be the call Distribute$(\alpha_1, \ldots, \alpha_\ell)$. Then

$$\bigcup_{\pi \in \Pi} (\overline{\pi}^{(r)} \setminus \overline{\pi}^{(r-1)}) \subseteq Pow^*(\{\overline{\alpha_1}^{(r-1)}, \ldots, \overline{\alpha_\ell}^{(r-1)}\}).$$

(3) $Pow^*(\{\overline{\alpha_1}, \ldots, \overline{\alpha_\ell}\}) \subseteq \bigcup_{\beta \in T(A)} \overline{\beta}$, for all P-nodes $A = \{\alpha_1, \ldots, \alpha_\ell\}$.

Proof. Concerning (1), if $|\overline{\alpha_i}^{(r-1)}| < \varrho$, for $i = 1, \ldots, \ell$, then from Assertion $E^{(r-1)}$, $\overline{\alpha_i}^{(r-1)} \subseteq Dom(f^{(r-1)})$, $i = 1, \ldots, \ell$, and $f^{(r-1)}$ is one-to-one. Therefore Lemma 8.3(b.2) implies (1).

Assertions $(B.2)^{(r)}$ and $(C.3)^{(r)}$ imply (2).

Finally to prove (3) let C_{r_0} be the last call to Distribute with argument the P-node $A = \{\alpha_1, \ldots, \alpha_\ell\}$. Then (3) is an immediate consequence of Assertion $(B.2)^{(r_0)}$ and $(C.3)^{(r_0)}$.   ■

We are now ready to prove (C3.c). To this end, let x be a variable in $V \setminus \{p_v\}$, and let $\{\alpha_1, \ldots, \alpha_\ell\} = \{\pi \in \Pi : \pi(x) = 1\}$. Notice that by Lemma 8.4 the set $A = \{\alpha_1, \ldots, \alpha_\ell\}$ is a P-node. Therefore clause (3) of the preceding lemma implies that $\bigcup_{\pi(x)=1} \overline{\pi} = \bigcup_{i=1}^{\ell} \overline{\alpha_i} \in \overline{\beta}$, for some place $\beta \in T(A)$. Let C_{r_0} be the call $Distribute(\alpha_1, \ldots, \alpha_\ell)$ during whose execution the element $\bigcup_{\pi(x)=1} \overline{\pi}$ is introduced into $\overline{\beta}$. If $|\overline{\alpha_i}| < \varrho$ for all $i = 1, \ldots, \ell$, then all places α_i are trapped. Therefore by combining Assertions $(E.4)^{(\xi)}$, $(E.3)^{(\xi)}$, and the fact that all the α_is are blocked, we obtain

$$\overline{f^{(r_0-1)}}(\overline{\alpha_1}^{(r_0-1)} \cup \ldots \cup \overline{\alpha_\ell}^{(r_0-1)})$$
$$= \quad f[\overline{\alpha_1} \cup \ldots \cup \overline{\alpha_\ell}]$$
$$= \quad f[\overline{\alpha_1}] \cup \ldots \cup f[\overline{\alpha_\ell}]$$
$$= \quad \sigma^{\alpha_1} \cup \ldots \cup \sigma^{\alpha_\ell}$$
$$\in \quad Pow^*(\{f[\overline{\alpha_1}], \ldots, f[\overline{\alpha_\ell}]\}) \cap \sigma^{\pi^x}$$
$$= \quad Pow^*(\{f^{(r_0-1)}[\overline{\alpha_1}^{(r_0-1)}], \ldots, f^{(r_0-1)}[\overline{\alpha_\ell}^{(r_0-1)}]\}) \cap \sigma^{\pi^x},$$

which shows that $\overline{\alpha_1}^{(r_0-1)} \cup \ldots \cup \overline{\alpha_\ell}^{(r_0-1)} \in \Delta_{\pi^x}^{(r_0)} \subseteq \overline{\pi^x}$, i.e. $\beta \equiv \pi^x$. On the other hand, if $|\overline{\alpha_j}| \geq \varrho$ for some $j \in \{1, \ldots, \ell\}$, our conclusion follows from Assertion $(C3.c)^{(r_0)}$, thus fully establishing condition (C3.c).

Next we prove that condition (C3.d) is also satisfied, i.e. that if $p = Pow(q)$ is a powerset clause in P, then $\bigcup_{\alpha(p)=1} \overline{\alpha} = Pow(\bigcup_{\beta(q)=1} \overline{\beta})$. Let $\alpha \in \Pi$ such that $\alpha(p) = 1$. Lemma 8.11(a) implies that for each $t \in \overline{\alpha}$ there exists a P-node B_t such that $\alpha \in T(B_t)$ and $t \in Pow^*(\{\overline{\beta} : \beta \in B_t\})$. Notice also that $\beta(q) = 1$ for each $\beta \in B_t$, so that $\bigcup_{\alpha(p)=1} \overline{\alpha} \subseteq Pow(\bigcup_{\beta(q)=1} \overline{\beta})$. To show the converse inclusion, let $t \in Pow(\bigcup_{\beta(q)=1} \overline{\beta})$. Then $t \in Pow^*(\{\overline{\beta_1}, \ldots, \overline{\beta_k}\})$ for some places $\beta_1, \ldots, \beta_k$, with $\beta_j(q) = 1$ for all $j = 1, \ldots, k$. In view of Lemma 8.14(3) we have $Pow^*(\{\overline{\beta_1}, \ldots, \overline{\beta_k}\}) \subseteq \bigcup_{\alpha \in T(B)} \overline{\alpha} \subseteq \bigcup_{\alpha(p)=1} \overline{\alpha}$, where $B = \{\beta_1, \ldots, \beta_k\}$. Therefore $Pow(\bigcup_{\beta(q)=1} \overline{\beta}) \subseteq \bigcup_{\alpha(p)=1} \overline{\alpha}$, thus proving (C3.d).

Finally, to complete the proof of condition (C3) we only need to verify (C3.e). To this end, let $x = \{y\}$ be a singleton clause in P. From (C3.c) and Lemma 8.5(5), we have $\bigcup_{\beta(y)=1} \overline{\beta} \in \overline{\pi^y} = \bigcup_{\alpha(x)=1} \overline{\alpha}$. But $|\sigma^{\pi^y}| = 1$, i.e.

π^y is trapped. Therefore Assertion (E.5)$^{(\xi)}$ implies $|\overline{\pi^y}| = 1$, which in turn yields $\bigcup_{\alpha(x)=1} \overline{\alpha} = \{\bigcup_{\beta(y)=1} \overline{\beta}\}$. This completes the proof of the necessity of condition (C3).

In view of Lemma 8.13, to prove (C4) it is enough to show that each call to the procedure *Distribute* is an instantiation of Step 2 of the Association Algorithm. But this follows plainly from Lemma 8.14(2). Hence, up to the proofs of Lemmas 8.12 and 8.13 (which will be provided in the next two sections), the proof of the necessity of conditions (C1)–(C4) of Theorem 8.1 is completed.   ∎

Notice that for each conjunction P in the class MLSSP″ there are only finitely many and *a priori* determinable sets of places of P and mappings $x \mapsto \pi^x$ and $\pi \mapsto \overline{\pi}$ as in (i)–(iii) of Theorem 8.1 and satisfying conditions (C1)–(C4). In other words, Theorem 8.1 contains a decidability test for the MLSSP″ injective satisfiability problem. Therefore, Corollary 8.1, Lemma 8.1, and the discussion at the beginning of Section 8.2 yield the following corollary.

Corollary 8.2 *The class* MLSSP *of formulae has a solvable satisfiability problem.*   □

Actually, Theorem 8.1 implies a slightly stronger result.

Corollary 8.3 (Decidability) *Let P be a conjunction in* MLSSP″ *in which only m distinct variables occur. Then P is satisfiable if and only if it has a model of rank less than*

$$2^{2^{m-4}+2\cdot m} .$$

Proof. The above condition for the satisfiability of P is trivially sufficient.

Conversely, if P is (injectively) satisfiable, by Theorem 8.1 there exist $\Pi, x \mapsto \pi^x$ and $\pi \mapsto \overline{\pi}$ as from (i)–(iii) and satisfying (C1)–(C4). As shown in the sufficiency proof of Theorem 8.1, the assignment $M^*x = \bigcup_{\pi(x)=1} \overline{\pi}$, for all x occurring in P, is a model of P.

We will prove the corollary by showing that for every x in P,

$$rk\, M^*x \leq 2^{2^{m-4}+2\cdot m} .$$

For this, observe that by Lemma 8.5(4), $n = |\Pi| \leq 2^{m-4} + 2$. Moreover, if we put $\varrho = m + \lceil log_2 m \rceil$, since $m \geq 4$ then $\varrho \geq 8$, so that $\varrho > 1$ and $2^{\varrho-1} > \varrho \cdot (n-1) + 1$ are both satisfied. Therefore we may assume that the ϱ appearing in (C4) has been chosen in this way.

During each execution of Step 2 of the Association Algorithm, the rank of any set $\overline{\pi}$ of maximal rank is augmented by at most 1. Since by (C4) $\pi \mapsto \overline{\pi}$ is produced by a computation of the Association Algorithm in which Step 2 is executed at most $(\varrho - 2) \cdot (n-1) \cdot 2^{n-2} + 3 \cdot 2^{n-1} - 2$ times, the corollary follows.   ∎

An immediate consequence is the following.

Corollary 8.4 *Given a conjunction P in* MLSSP$''$, *P is satisfiable if and only if it is hereditarily finitely satisfiable.* $\Box$

As a consequence of Lemma 3.4, the decidability result contained in Corollary 8.2 can be lifted to the quantified case as stated in the following

Corollary 8.5 *The family of left formulae (cf Definition 3.10) over the language $\emptyset$, $\cap$, $\backslash$, $\cup$, $\{\bullet\}$, Pow, $=$, $\in$ has a solvable satisfiability problem.* $\Box$

8.4   Proof of the Main Inductive Lemma

In this section we will prove the partial correctness of the IVAA.

As in the previous section, let K be an execution of the IVAA, and let $C_1, C_2, C_3, \ldots$ be the sequence of calls to the procedure *Distribute*, arranged in the order in which they occur during the computation K. We will also make use of the notation $\overline{\pi}^{(r)}$, $\Delta_\pi^{(r)}$, $f^{(r)}$, Assertion X$^{(r)}$ as explained in the previous section. For the sake of completeness, we restate the Main Inductive Lemma.

Lemma 8.12 (Main Inductive Lemma: Partial Correctness) *All AS-SERT statements encountered during the computation* K *are executed successfully.*

Proof. We will proceed by induction on $r \geq 1$, by proving that all AS-SERT statements executed during the processing of the call C_r are completed successfully, i.e., they hold at the time they are encountered.

Base case $r = 1$. Notice that $f^{(0)} = \emptyset$ and $\overline{\pi}^{(0)} = \emptyset$ for all places $\pi \in \Pi$. Moreover, when the call C_1 is made, all places and P-nodes are marked unblocked. Indeed all places and P-nodes are marked unblocked during the Initialization Phase, so that it is enough to show that no place and no P-node can become blocked during the subsequent Blocking Phase. Observe that by Lemma 8.8(a) π^{q_0} is the minimum unblocked place after the execution of the Initialization Phase. But $\overline{\pi^{q_0}}^{(0)} = \emptyset$, so that the IF-test of the Blocking Phase fails. As $\overline{\pi}^{(0)} = \emptyset$, for all $\pi \in \Pi$, the only P-node satisfying the conditions in the Propagation Phase is the empty P-node. Indeed, it is immediately seen (cf. Lemma 8.2(a)) that $Pow^*(\emptyset) \setminus \bigcup_{\beta \in T(\emptyset)} \overline{\beta}^{(0)} = \{\emptyset\}$. Therefore C_1 is the call *Distribute*$(\emptyset)$. Assertion $A^{(0)}$ is plain. Notice that $|\overline{\alpha}^{(0)}| < \varrho$, for all α in the empty P-node, is vacuously true, so that the empty P-node is processed at *update_for_small_alphas*. It is immediate to see that

$$\Delta_\pi^{(1)} = \begin{cases} \emptyset & \text{if } \pi \neq \pi^{q_0} \\ \{\emptyset\} & \text{if } \pi = \pi^{q_0} \end{cases}$$

and obviously Assertion $B^{(1)}$ is true (cf. Lemma 8.7(a)). After the execution of the assignment statements, we have

$$\overline{\pi}^{(1)} = \begin{cases} \emptyset & \text{if } \pi \neq \pi^{q_0} \\ \{\emptyset\} & \text{if } \pi = \pi^{q_0} \end{cases} \quad \text{and } f^{(1)} = \{(\emptyset, \emptyset)\},$$

where by (x, y) we denote any suitable representation of the ordered pair having x and y as first and second element, respectively.

To complete the proof in the base case, we only have to prove that Assertion $E^{(1)}$ holds. Assertions $(E.1)^{(1)}$, $(E.2)^{(1)}$, $(E.4)^{(1)}$-$(E.9)^{(1)}$ are immediate. Concerning Assertion $(E.3)^{(1)}$ it is enough to observe that by Lemma 8.5(1), $\sigma^{\pi^{q_0}} = \{\emptyset\} = range(f^{(1)})$. This completes our first induction step.

Inductive step. Next we assume that all ASSERT statements encountered before the execution of the call C_{r_0}, with $r_0 > 1$, are completed successfully, and we prove that all ASSERT statements met during the processing of the call C_{r_0} are valid as well.

Let C_{r_0} be the call $Distribute(\alpha_1, \ldots, \alpha_\ell)$. Clearly the P-node $A = \{\alpha_1, \ldots \alpha_\ell\}$ cannot be the empty P-node. Notice also that according to whether the call C_{r_0} is made from the Blocking Phase or from the Propagation Phase, we have respectively that either A is blocked or A is unblocked. In any case A cannot be marked visited, showing that Assertion $A^{(r_0)}$ holds.

To prove the validity of the remaining assertions, we will distinguish two cases according to whether $|\overline{\alpha_i}^{(r_0-1)}| < \varrho$ for all $i = 1, \ldots, \ell$, or $|\overline{\alpha_{i_0}}^{(r_0-1)}| \geq \varrho$ for some $j_0 \in \{1, \ldots, \ell\}$.

Case: $|\overline{\alpha_i}^{(r_0-1)}| < \varrho$ *for all* $i = 1, \ldots, \ell$. In this case we have

$$\Delta_\pi^{(r_0)} = (\overline{f^{(r_0-1)}})^{-1}[Pow^*(\{f^{(r_0-1)}[\overline{\alpha_1}^{(r_0-1)}], \ldots \\ \ldots, f^{(r_0-1)}[\overline{\alpha_\ell}^{(r_0-1)}]\}) \cap \sigma^\pi] \setminus \bigcup_{\beta \in T(A)} \overline{\beta}^{(r_0-1)}.$$

for all $\pi \in \Pi$.

The following lemma states some useful properties.

Lemma 8.15 *Let* $\gamma_1, \ldots, \gamma_k$ *be places of P such that* $|\overline{\gamma_j}^{(r_0-1)}| < \varrho$ *for $j = 1, \ldots, k$. Then*

(i) $\overline{\gamma_j}^{(r_0-1)} \subseteq Dom(f^{(r_0-1)})$, *for each $j = 1, \ldots, k$;*

(ii) $f^{(r_0-1)}[\overline{\gamma_j}^{(r_0-1)}] \subseteq \sigma^{\gamma_j}$, *for each $j = 1, \ldots, k$;*

(iii) $Pow^*(\{f^{(r_0-1)}[\overline{\gamma_1}^{(r_0-1)}], \ldots, f^{(r_0-1)}[\overline{\gamma_k}^{(r_0-1)}]\})$
$\subseteq Pow^*(\{\sigma^{\gamma_1}, \ldots, \sigma^{\gamma_k}\}).$

Proof. By inductive hypothesis Assertion $E^{(r_0-1)}$ we have plainly (i) and (ii). By Lemma 8.2(c) and (ii) we have also (iii). ∎

Let $\pi \in \Pi \setminus T(A)$. Then, by Lemma 8.7(d), $Pow^*(\{\sigma^\alpha : \alpha \in A\}) \cap \sigma^\pi = \emptyset$, which by (iii) of the preceding lemma yields $\Delta_\pi^{(r_0)} = \emptyset$, thus proving Assertion $(B.1)^{(r_0)}$.

Next we prove that $(B.2)^{(r_0)}$ is also satisfied. First of all we notice that as $f^{(r_0-1)}$ is one-to-one, the sets $\Delta_\pi^{(r_0)}$ are pairwise disjoint, since so are the sets σ^π. Moreover from Lemma 8.3, Lemma 8.15, and the injectivity of $f^{(r_0-1)}$, we have

$$
\bigcup_{\pi \in \Pi} \Delta_\pi^{(r_0)} = \bigcup_{\pi \in \Pi} (\overline{f^{(r_0-1)}})^{-1}[Pow^*(\{f^{(r_0-1)}[\overline{\alpha_1}^{(r_0-1)}], \dots
$$

$$
\dots, f^{(r_0-1)}[\overline{\alpha_\ell}^{(r_0-1)}]\}) \cap \sigma^\pi] \setminus \bigcup_{\beta \in T(A)} \overline{\beta}^{(r_0-1)}
$$

$$
= (\overline{f^{(r_0-1)}})^{-1}[Pow^*(\{f^{(r_0-1)}[\overline{\alpha_1}^{(r_0-1)}], \dots, f^{(r_0-1)}[\overline{\alpha_\ell}^{(r_0-1)}]\})
$$

$$
\cap \bigcup_{\pi \in \Pi} \sigma^\pi] \setminus \bigcup_{\beta \in T(A)} \overline{\beta}^{(r_0-1)}
$$

$$
= (\overline{f^{(r_0-1)}})^{-1}[Pow^*(\{f^{(r_0-1)}[\overline{\alpha_1}^{(r_0-1)}], \dots, f^{(r_0-1)}[\overline{\alpha_\ell}^{(r_0-1)}]\})
$$

$$
\setminus \bigcup_{\beta \in T(A)} \overline{\beta}^{(r_0-1)}
$$

$$
= Pow^*(\{\overline{\alpha_1}^{(r_0-1)}, \dots, \overline{\alpha_\ell}^{(r_0-1)}\}) \setminus \bigcup_{\beta \in T(A)} \overline{\beta}^{(r_0-1)}.
$$

This completes the proof of Assertion $(B.2)^{(r_0)}$.

Finally, to prove $(B.3)^{(r_0)}$, let β_0 be a place of P such that $\Delta_{\beta_0}^{(r_0)} \neq \emptyset$. From $(B.1)^{(r_0)}$ we have $\beta_0 \in T(A)$, so that we only need to show that the place β_0 is unblocked at the time the call C_{r_0} has been made. We distinguish two subcases according to whether β_0 is trapped or not.

Subcase: β_0 *is trapped.* If β_0 were blocked when the call C_{r_0} is made, then $|\overline{\beta_0}^{(r_0-1)}| = |\sigma^{\beta_0}| < \varrho$, which by Lemma 8.15 gives $f^{(r_0-1)}[\overline{\beta_0}^{(r_0-1)}] = \sigma^{\beta_0}$. Therefore, again from Lemma 8.15 and the inductive hypothesis $(E.8)^{(r_0-1)}$, we would have

$$
\Delta_{\beta_0}^{(r_0)} \subseteq (\overline{f^{(r_0-1)}})^{-1}\left[\overline{f^{(r_0-1)}}[Pow^*(\{\overline{\alpha_1}^{(r_0-1)}, \dots, \overline{\alpha_\ell}^{(r_0-1)}\})]\right.
$$

$$
\left. \cap f^{(r_0-1)}[\overline{\beta_0}^{(r_0-1)}]\right]
$$

$$
\subseteq (f^{(r_0-1)})^{-1} f^{(r_0-1)}[\overline{\beta_0}^{(r_0-1)}]
$$

$$
= \overline{\beta_0}^{(r_0-1)}
$$

$$
\subseteq \bigcup_{\beta \in T(A)} \overline{\beta}^{(r_0-1)}.
$$

The above inclusion chain implies that $\Delta_{\beta_0}^{(r_0)} = \emptyset$, contradicting our assumption $\Delta_{\beta_0}^{(r_0)} \neq \emptyset$, and consequently showing that the place β_0 must be unblocked, at least in the case in which β_0 is trapped.

Subcase: β_0 *is non-trapped.* Suppose that β_0 has become blocked just before the call $C_{r'}$, with $1 \leq r' \leq r_0$. Therefore

$$|Pow^*(\{\overline{\alpha_1}^{(r'-1)}, \ldots, \overline{\alpha_\ell}^{(r'-1)}\}) \cap \overline{\beta_0}^{(r'-1)}| = |Pow^*(\{\sigma^{\alpha_1}, \ldots, \sigma^{\alpha_\ell}\}) \cap \sigma^{\beta_0}|,$$

since $\overline{\alpha_i}^{(r'-1)} \subseteq \overline{\alpha_i}^{(r_0-1)}$ for all $i = 1, \ldots, \ell$.

Notice also that

$$Pow^*(\{\overline{\alpha_1}^{(r'-1)}, \ldots, \overline{\alpha_\ell}^{(r'-1)}\}) \cap \overline{\beta_0}^{(r'-1)}$$
$$\subseteq Pow^*(\{\overline{\alpha_1}^{(r_0-1)}, \ldots, \overline{\alpha_\ell}^{(r_0-1)}\}) \cap \overline{\beta_0}^{(r_0-1)}.$$

Moreover, let $u \in Pow^*(\{\overline{\alpha_1}^{(r_0-1)}, \ldots, \overline{\alpha_\ell}^{(r_0-1)}\}) \cap \overline{\beta_0}^{(r_0-1)}$, and let $C_{r''}$ with $1 \leq r'' \leq r_0$, be the call $Distribute(\alpha_1, \ldots, \alpha_\ell)$ during whose execution the element u is introduced into $\overline{\beta_0}$ (cf. inductive hypothesis $(E.9)^{(r_0-1)}$). Since

$$f^{(r'')} = f^{(r''-1)} \cup \overline{f^{(r''-1)}}|_{Pow^*(\{\overline{\alpha_1}^{(r''-1)}, \ldots, \overline{\alpha_\ell}^{(r''-1)}\})},$$

we have

$$f^{(r_0-1)}(u) = f^{(r'')}(u) = \overline{f^{(r''-1)}}(u) = \overline{f^{(r_0-1)}}(u).$$

Therefore $\overline{f^{(r_0-1)}}(u) = f^{(r_0-1)}(u) \in f^{(r_0-1)}[\overline{\beta_0}^{(r_0-1)}] \subseteq \sigma^{\beta_0}$, so that

$$\overline{f^{(r_0-1)}}(u) \in \overline{f^{(r_0-1)}}[Pow^*(\{\overline{\alpha_1}^{(r_0-1)}, \ldots, \overline{\alpha_\ell}^{(r_0-1)}\})] \cap \sigma^{\beta_0},$$

thus proving

$$\overline{f^{(r_0-1)}}[Pow^*(\{\overline{\alpha_1}^{(r_0-1)}, \ldots, \overline{\alpha_\ell}^{(r_0-1)}\}) \cap \overline{\beta_0}^{(r_0-1)}]$$
$$\subseteq \overline{f^{(r_0-1)}}[Pow^*(\{\overline{\alpha_1}^{(r_0-1)}, \ldots, \overline{\alpha_\ell}^{(r_0-1)}\})] \cap \sigma^{\beta_0}.$$

From Lemmas 8.3 and 8.15 we have

$$\overline{f^{(r_0-1)}}[Pow^*(\{\overline{\alpha_1}^{(r_0-1)}, \ldots, \overline{\alpha_\ell}^{(r_0-1)}\})]$$
$$\subseteq Pow^*(\{f^{(r_0-1)}[\overline{\alpha_1}^{(r_0-1)}], \ldots, f^{(r_0-1)}[\overline{\alpha_\ell}^{(r_0-1)}]\})$$
$$\subseteq Pow^*(\{\sigma^{\alpha_1}, \ldots, \sigma^{\alpha_\ell}\}).$$

Hence

$$\overline{f^{(r_0-1)}}[Pow^*(\{\overline{\alpha_1}^{(r'-1)}, \ldots, \overline{\alpha_\ell}^{(r'-1)}\}) \cap \overline{\beta_0}^{(r'-1)}]$$
$$\subseteq \overline{f^{(r_0-1)}}[Pow^*(\{\overline{\alpha_1}^{(r_0-1)}, \ldots, \overline{\alpha_\ell}^{(r_0-1)}\}) \cap \overline{\beta_0}^{(r_0-1)}]$$
$$\subseteq Pow^*(\{f^{(r_0-1)}[\overline{\alpha_1}^{(r_0-1)}], \ldots, f^{(r_0-1)}[\overline{\alpha_\ell}^{(r_0-1)}]\}) \cap \sigma^{\beta_0}$$
$$\subseteq Pow^*(\{\sigma^{\alpha_1}, \ldots, \sigma^{\alpha_\ell}\}) \cap \sigma^{\beta_0}.$$

As $\overline{f^{(r_0-1)}}$ is injective and $Pow^*(\{\overline{\alpha_1}^{(r'-1)}, \ldots, \overline{\alpha_\ell}^{(r'-1)}\})$ has finite cardinality, the latter chain of inclusions combined with above cardinality equality yields

$$\overline{f^{(r_0-1)}}[Pow^*(\{\overline{\alpha_1}^{(r_0-1)}, \ldots, \overline{\alpha_\ell}^{(r_0-1)}\}) \cap \overline{\beta_0}^{(r_0-1)}]$$
$$= Pow^*(\{f^{(r_0-1)}[\overline{\alpha_1}^{(r_0-1)}], \ldots, f^{(r_0-1)}[\overline{\alpha_\ell}^{(r_0-1)}]\}) \cap \sigma^{\beta_0}.$$

Therefore

$$\Delta_{\beta_0}^{(r_0)} \subseteq (\overline{f^{(r_0-1)}})^{-1}[Pow^*(\{f^{(r_0-1)}[\overline{\alpha_1}^{(r_0-1)}], \ldots, f^{(r_0-1)}[\overline{\alpha_\ell}^{(r_0-1)}]\}) \cap \sigma^{\beta_0}]$$
$$\subseteq Pow^*(\{\overline{\alpha_1}^{(r_0-1)}, \ldots, \overline{\alpha_\ell}^{(r_0-1)}\}) \cap \overline{\beta_0}^{(r_0-1)}$$
$$\subseteq \overline{\beta_0}^{(r_0-1)}$$
$$\subseteq \bigcup_{\beta \in T(A)} \overline{\beta}^{(r_0-1)}.$$

The latter inclusion contradicts our earlier assumption $\Delta_{\beta_0}^{(r_0)} \neq \emptyset$, thereby showing that the place β_0 must be unblocked even in the present case. Thus Assertion (B.3)$^{(r_0)}$ is fully proved.

After the assignment statements have been executed, we have

$$\overline{\pi}^{(r_0)} = \overline{\pi}^{(r_0-1)} \cup \Delta_{\pi}^{(r_0)}, \text{ for each } \pi \in \Pi,$$

and

$$f^{(r_0)} = f^{(r_0-1)} \cup \overline{f^{(r_0-1)}}|_{Pow^*(\{\overline{\alpha_1}^{(r_0-1)}, \ldots, \overline{\alpha_\ell}^{(r_0-1)}\})} \quad .$$

To complete our analysis of the inductive step in the case in which $|\overline{\alpha_i}^{(r_0-1)}| < \varrho$ for all $i = 1, \ldots \ell$, we only need to verify Assertion E$^{(r_0)}$. This is done as follows.

Concerning (E.1)$^{(r_0)}$, we begin by showing that $f^{(r_0)}$ is indeed a function, i.e. that for each u in $Dom(f^{(r_0)})$ there is exactly one pair $(u, t) \in f^{(r_0)}$ having u as first element. As by induction $f^{(r_0-1)}$ and $\overline{f^{(r_0-1)}}$ are functions, it is enough to prove that for each $u \in Dom(f^{(r_0-1)}) \cap Pow^*(\{\overline{\alpha_1}^{(r_0-1)}, \ldots, \overline{\alpha_\ell}^{(r_0-1)}\})$, $f^{(r_0-1)}(u) = f^{(r_0-1)}[u]$. But this follows plainly from Assertion (E.5)$^{(r_0-1)}$. Next we prove that $f^{(r_0)}$ is injective. As by induction and by Lemma 8.3 both $f^{(r_0-1)}$ and $\overline{f^{(r_0-1)}}$ are injective, it is enough to prove the following lemma.

Lemma 8.16 *Let* $u_1 \in Dom(f^{(r_0-1)})$ *and* $u_2 \in Pow^*(\{\overline{\gamma}^{(r_0-1)} : \gamma \in \Gamma\})$ *such that* $f^{(r_0-1)}(u_1) = \overline{f^{(r_0-1)}}(u_2)$, *where* Γ *is a P-node with* $|\overline{\gamma}^{(r_0-1)}| < \varrho$ *for all* $\gamma \in \Gamma$. *Then* $u_1 = u_2$.

Proof. Let u_1, u_2 and Γ be as in the above hypotheses. Then from Lemma 8.3 and Lemma 8.15(iii) it follows that $f^{(r_0-1)}(u_1) \in Pow^*(\{\sigma^\gamma : \gamma \in \Gamma\})$, which by (E.8)$^{(r_0-1)}$, by the disjointness of the sets σs and Lemma 8.2(c) implies $u_1 \in Pow^*(\{\overline{\gamma}^{(r_0-1)} : \gamma \in \Gamma\})$. Therefore (E.2)$^{(r_0-1)}$ and (E.7)$^{(r_0-1)}$ yield $f^{(r_0-1)}(u_1) = \overline{f^{(r_0-1)}}(u_1) = \overline{f^{(r_0-1)}}(u_2)$, which by the injectivity of $\overline{f^{(r_0-1)}}$ implies $u_1 = u_2$, thus proving the lemma. ∎

Having proved the preceding lemma, and thus established (E.1)$^{(r_0)}$, we turn to (E.2)$^{(r_0)}$. To prove that (E.2)$^{(r_0)}$ is satisfied it is enough to observe that, by (B.2)$^{(r_0-1)}$,

$$Pow^*(\{\overline{\alpha_1}^{(r_0-1)}, \ldots, \overline{\alpha_\ell}^{(r_0-1)}\}) \subseteq \bigcup_{\beta \in T(A)} \overline{\beta}^{(r_0-1)} \cup \bigcup_{\beta \in T(A)} \Delta_\beta^{(r_0)} = \bigcup_{\beta \in T(A)} \overline{\beta}^{(r_0)}.$$

Next we prove that (E.3)$^{(r_0)}$ holds. Let $\pi \in \Pi$. Since $f^{(r_0)}[\overline{\pi}^{(r_0)}] = f^{(r_0-1)}[\overline{\pi}^{(r_0-1)}] \cup \overline{f^{(r_0-1)}}[\Delta_\pi^{(r_0)}]$, by inductive hypothesis we can limit ourselves to proving $\overline{f^{(r_0-1)}}[\Delta_\pi^{(r_0)}] \subseteq \sigma^\pi$. But this follows immediately from the very definition of $\Delta_\pi^{(r_0)}$. Since $\Delta_\pi^{(r_0)} \subseteq Dom(f^{(r_0)})$, for all $\pi \in \Pi$, by induction we obtain (E.4)$^{(r_0)}$.

Next we verify Assertion (E.5)$^{(r_0)}$. Let π be a trapped place. From (B.2)$^{(r_0)}$ and the inductive hypothesis (E.5)$^{(r_0-1)}$, we have $\overline{\pi}^{(r_0)} \subseteq Dom(f^{(r_0)})$. Moreover (E.3)$^{(r_0)}$ implies $f^{(r_0)}[\overline{\pi}^{(r_0)}] \subseteq \sigma^\pi$. Therefore, as $f^{(r_0)}$ is injective we have $|\overline{\pi}^{(r_0)}| \leq |\sigma^\pi|$, which proves (E.5)$^{(r_0)}$.

Concerning (E.6)$^{(r_0)}$, we need to prove that if α and β are two distinct places of P, then $\overline{\alpha}^{(r_0)} \cap \overline{\beta}^{(r_0)} = \emptyset$. But $\overline{\alpha}^{(r_0)} \cap \overline{\beta}^{(r_0)} = (\overline{\alpha}^{(r_0-1)} \cap \Delta_\beta^{(r_0)}) \cup (\overline{\beta}^{(r_0-1)} \cap \Delta_\alpha^{(r_0)})$ therefore it is enough to show that $\overline{\gamma}^{(r_0-1)} \cap \Delta_\delta^{(r_0)} = \emptyset$, for any two distinct places γ and δ. Suppose by contradiction that there exists $u \in \overline{\gamma}^{(r_0-1)} \cap \Delta_\delta^{(r_0)}$. Then, by (B.2)$^{(r_0)}$, $u \in Pow^*(\{\overline{\alpha_1}^{(r_0-1)}, \ldots, \overline{\alpha_\ell}^{(r_0-1)}\})$. In view of (E.9.b)$^{(r_0-1)}$, this implies that $\Delta_\delta^{(r_0)} \cap \overline{\gamma}^{(r_0-1)} = \emptyset$, thus completing the proof that Assertion (E.6)$^{(r_0)}$ is valid.

Concerning (E.7)$^{(r_0)}$, let $u \in Pow^*(\{\overline{\gamma}^{(r_0)} : \gamma \in \Gamma\}) \cap \bigcup_{\pi \in \Pi} \overline{\pi}^{(r_0)}$, with Γ a P-node such that $|\overline{\gamma}^{(r_0)}| < \varrho$ for each $\gamma \in \Gamma$. Let $C_{r'}$, with $r' \leq r_0$, be the call to Distribute during whose execution the element u is introduced into $\bigcup_{\pi \in \Pi} \overline{\pi}$. Notice that by the disjointness of the sets $\overline{\pi}^{(r'-1)}$, $C_{r'}$ must be the call $Distribute(\Gamma)$. As $|\overline{\gamma}^{(r'-1)}| < \varrho$ for each $\gamma \in \Gamma$, it follows that $u \in Pow^*(\{\overline{\gamma}^{(r'-1)} : \gamma \in \Gamma\}) \subseteq Dom(f^{(r'-1)}) \cap Dom(f^{(r')}) \subseteq Dom(\overline{f^{(r_0)}}) \cap Dom(f^{(r_0)})$, and also $f^{(r_0)}(u) = f^{(r')}(u) = \overline{f^{(r'-1)}}[u] = f^{(r'-1)}[u] = f^{(r_0)}[u]$, which proves Assertion (E.7)$^{(r_0)}$.

Concerning (E.8)$^{(r_0)}$, let $u \in Pow^*(\{\overline{\gamma}^{(r_0)} : \gamma \in \Gamma\}) \cap Dom(f^{(r_0)})$, where Γ is any P-node. From (E.2)$^{(r_0)}$, we have $u \in \overline{\pi_0}^{(r_0)}$, for some $\pi \in \Pi$. Let $C_{r'}$, with $r' \leq r_0$, be the call $Distribute(\Gamma)$ during whose execution the element u is put into $\bigcup_{\pi \in \Pi} \overline{\pi}$. If $|\overline{\gamma}^{(r'-1)}| \geq \varrho$ for some $\gamma \in \Gamma$, then $f^{(r')}(u) \in \sigma^{\pi_0} \cap Pow^*(\{\sigma^\gamma : \gamma \in \Gamma\}) \setminus (\sigma^\Gamma)^{(r')} \setminus range(f^{(r'-1)}))$. On the other hand, if $|\overline{\gamma}^{(r'-1)}| < \varrho$ for all $\gamma \in \Gamma$, then by Lemma 8.15(iii) $f^{(r')}(u) = \overline{f^{(r'-1)}}(u) \in$

$Pow^*(\{\sigma^\gamma : \gamma \in \Gamma\})$. In any case we have $f^{(r_0)}(u) = f^{(r')}(u) \in Pow^*(\{\sigma^\gamma : \gamma \in \Gamma\})$, which proves (E.8)$^{(r_0)}$.

Finally, in order to complete our analysis of the inductive step, at least in the case in which $|\overline{\alpha_i}^{(r_0-1)}| < \varrho$ for all $i = 1, \ldots, \ell$, we only need to prove that Assertion (E.9)$^{(r_0)}$ is satisfied too. Let $t \in Pow^*(\{\overline{\gamma_1}^{(r_0)}, \ldots, \overline{\gamma_h}^{(r_0)}\}) \cap \overline{\delta}^{(r_0)}$, for some places $\gamma_1, \ldots, \gamma_h, \delta$. Let $C_{r'}$ be the call to the procedure *Distribute* during whose execution the element t is introduced into the set $\overline{\delta}$. Suppose that $C_{r'}$ is the call Distribute$(\beta_1, \ldots, \beta_k)$, with $\{\beta_1, \ldots, \beta_k\}$ a P-node. Therefore by inductive hypothesis (B.2)$^{(r')}$ and (C.3)$^{(r')}$ $t \in Pow^*(\{\overline{\beta_1}^{(r'-1)}, \ldots, \overline{\beta_k}^{(r'-1)}\}) \subseteq Pow^*(\{\overline{\beta_1}^{(r_0)}, \ldots, \overline{\beta_k}^{(r_0)}\})$. The latter relationships imply $\{\beta_1, \ldots, \beta_k\} = \{\gamma_1, \ldots, \gamma_h\}$, since the sets $\overline{\beta_1}^{(r_0)}, \ldots, \overline{\beta_k}^{(r_0)}$, $\overline{\gamma_1}^{(r_0)}, \ldots, \overline{\gamma_h}^{(r_0)}$ are pairwise disjoint, if not identical (cf. Lemma 8.2(b)). Therefore by (B.3)$^{(r')}$ and (C.3)$^{(r')}$ we have that $\{\gamma_1, \ldots, \gamma_h\}$ is a P-node having δ among its targets, thus proving (E.9.a)$^{(r_0)}$ and (E.9.b)$^{(r_0)}$. Furthermore (E.9.c)$^{(r_0)}$ follows immediately from the above discussion, so that Assertion (E.9)$^{(r_0)}$ holds. This completes the analysis of our first case. The second, and last, case to be considered is the following.

Case: $|\overline{\alpha_{j_0}}^{(r_0-1)}| \geq \varrho$ *for some* $j_0 \in \{1, \ldots, \ell\}$. Let $\beta_1, \ldots, \beta_g, \tau_1, \ldots, \tau_k, \nu_1, \ldots, \nu_h$ be all the targets of the P-node A, where at the time the call C_{r_0} is made the βs are the blocked targets, the τs are the unblocked trapped targets, and the νs are the unblocked untrapped targets of $\{\alpha_1, \ldots, \alpha_\ell\}$, respectively.

We have

$$\Delta'^{(r_0)} = Pow^*(\{\overline{\alpha_1}^{(r_0-1)}, \ldots, \overline{\alpha_\ell}^{(r_0-1)}\}) \setminus \bigcup_{\beta \in T(A)} \overline{\beta}^{(r_0-1)}$$
$$\setminus \{\overline{\alpha_1}^{(r_0-1)} \cup \ldots \cup \overline{\alpha_\ell}^{(r_0-1)}\},$$

$$\Delta'^{(r_0)} = \begin{cases} \emptyset & \text{if } A \text{ is not blocked just} \\ & \text{before the call } C_{r_0} \text{ is made} \\ \{\overline{\alpha_1}^{(r_0-1)} \cup \ldots \cup \overline{\alpha_\ell}^{(r_0-1)}\} & \text{otherwise.} \end{cases}$$

$$\Delta^{(r_0)} = \Delta'^{(r_0)} \cup \Delta''^{(r_0)},$$

$$(\sigma^A)^{(r_0)} = \begin{cases} \emptyset & \text{if } A \text{ is blocked just before the call} \\ & C_{r_0} \text{ is made} \\ \{\sigma^{\alpha_1} \cup \ldots \cup \sigma^{\alpha_\ell}\} & \text{otherwise.} \end{cases}$$

In addition, for all unblocked targets $\tau_1, \ldots, \tau_k$ of A, we have

$$n_j^{(r_0)} = |\sigma^{\tau_j} \cap Pow^*(\{\sigma^{\alpha_1}, \ldots, \sigma^{\alpha_\ell}\}) \setminus (\sigma^A)^{(r_0)}|$$
$$- |\overline{\tau_j}^{(r_0-1)} \cap Pow^*(\{\overline{\alpha_1}^{(r_0-1)}, \ldots, \overline{\alpha_\ell}^{(r_0-1)}\})|.$$

We begin by proving that Assertion C$^{(r_0)}$ is satisfied.

Concerning (C.1)$^{(r_0)}$, assume that $\overline{\alpha_1}^{(r_0-1)} \cup \ldots \cup \overline{\alpha_\ell}^{(r_0-1)} \in \overline{\beta}^{(r_0-1)}$, for some place $\beta_0 \in T(A)$. Then the element $\overline{\alpha_1}^{(r_0)} \cup \ldots \cup \overline{\alpha_\ell}^{(r_0)}$ has been introduced into the set $\overline{\beta}_0$ during the execution of a prior call $C_{r'}$, Distribute$(\alpha_1, \ldots,$

α_ℓ), and furthermore $\overline{\alpha_1}^{(r'-1)} \cup \ldots \cup \overline{\alpha_\ell}^{(r'-1)} = \overline{\alpha_1}^{(r_0-1)} \cup \ldots \cup \overline{\alpha_\ell}^{(r_0-1)}$. As $\overline{\alpha_1}^{(r'-1)} \cup \ldots \cup \overline{\alpha_\ell}^{(r'-1)} \in \Delta^{(r')}$, the P-node A must have been blocked when the call $C_{r'}$ was made, preventing it from being processed again by procedure *Distribute*. But this is a contradiction, since the call C_{r_0} has as argument the P-node A. Therefore

$$\overline{\alpha_1}^{(r_0-1)} \cup \ldots \cup \overline{\alpha_\ell}^{(r_0-1)} \notin \bigcup_{\beta \in T(A)} \overline{\beta}^{(r_0-1)},$$

proving $(C.1)^{(r_0)}$.

Concerning $(C.2)^{(r_0)}$, suppose by contradiction that $\sigma^{\alpha_1} \cup \ldots \cup \sigma^{\alpha_\ell} \in range(f^{(r_0-1)})$, and let $u \in Dom(f^{(r_0-1)})$ such that $f^{(r_0-1)}(u) = \sigma^{\alpha_1} \cup \ldots \cup \sigma^{\alpha_\ell}$. Notice that the inductive hypothesis $(E.8)^{(r_0-1)}$ implies $u \in Pow^*(\{\overline{\alpha_1}^{(r_0-1)}, \ldots, \overline{\alpha_\ell}^{(r_0-1)}\})$, so that the element u must have been introduced during a prior call $C_{r'}$, $Distribute(\alpha_1, \ldots, \alpha_\ell)$. If $|\overline{\alpha_i}^{(r'-1)}| < \varrho$ for all $i = 1, \ldots \ell$, then $\sigma^{\alpha_1} \cup \ldots \cup \sigma^{\alpha_\ell} = f^{(r')}(u) = f^{(r'-1)}[u] \subseteq f^{(r'-1)}[\overline{\alpha_1}^{(r'-1)}] \cup \ldots \cup f^{(r'-1)}[\overline{\alpha_\ell}^{(r'-1)}]$, and since by Lemma 8.15(ii) $f^{(r'-1)}[\overline{\alpha_i}^{(r'-1)}] \subseteq \sigma^{\alpha_i}$, for $i = 1, \ldots, \ell$, then $f^{(r'-1)}[\overline{\alpha_i}^{(r'-1)}] = \sigma^{\alpha_i}$, $i = 1, \ldots, \ell$. In view of the injectivity of $f^{(r'-1)}$, the latter equalities imply $|\sigma^{\alpha_i}| = |\overline{\alpha_i}^{(r'-1)}|$, i.e. the places α_i are trapped, for $i = 1, \ldots, \ell$. But then by inductive hypothesis $(E.5)^{(r_0-1)}$, we should have $|\overline{\alpha_i}^{(r'-1)}| < \varrho$, for all $i = 1, \ldots, \ell$, contradicting our assumption $|\overline{\alpha_{j_0}}^{(r_0-1)}| \geq \varrho$ for some $j_0 \in \{1, \ldots, \ell\}$. This contradiction shows that $|\overline{\alpha_j}^{(r'-1)}| \geq \varrho$ for some $j \in \{1, \ldots, \ell\}$. But then, an argument similar to the one given for the proof of Assertion $(C.1)^{(r_0)}$ allows us to conclude that $(C.2)^{(r_0)}$ is also satisfied.

Next we show that Assertions $(C.3)^{(r_0)}$ and $(C.4)^{(r_0)}$ hold too. If the P-node $\{\alpha_1, \ldots, \alpha_\ell\}$ has never been processed by *Distribute* prior to the call C_{r_0}, we put $r' = 1$, otherwise we let $C_{r'}$ denote the latest call *Distribute* $(\alpha_1, \ldots, \alpha_\ell)$, with $r' < r_0$. We will distinguish two subcases, according to whether the set $\bigcup_{\alpha \in A}(\overline{\alpha}^{(r_0-1)} \setminus \overline{\alpha}^{(r'-1)})$ is empty or not.

Subcase: $\bigcup_{\alpha \in A}(\overline{\alpha}^{(r_0-1)} \setminus \overline{\alpha}^{(r'-1)}) = \emptyset$. Notice that in this case $r' > 1$, and $\overline{\alpha}^{(r_0-1)} = \overline{\alpha}^{(r'-1)}$, for all $\alpha \in A$. Therefore after the execution of $C_{r'}$, the P-node A is marked as 'visited', preventing it from being further processed from the Propagation Phase. Thus the call C_{r_0} must have been made from the Blocking Phase, and in particular the P-node A must be marked 'blocked' when the call C_{r_0} is made. Therefore, $\Delta''^{(r_0)} = \{\overline{\alpha_1}^{(r_0-1)} \cup \ldots \cup \overline{\alpha_\ell}^{(r_0-1)}\}$, $(\sigma^A)^{(r_0)} = \emptyset$, and, by the inductive hypothesis $C^{(r')}$, $\Delta'^{(r_0)} = \emptyset$ and $\Delta^{(r_0)} = \Delta''^{(r_0)}$. Assertion $(C.3.b)^{(r_0)}$ is vacuously true. As regards $(C.3.c)^{(r_0)}$, it is sufficient to notice that as $\pi^A > \alpha$, for all $\alpha \in A$ (cf. Lemma 8.10), then Lemma 8.11(d) assures us that the place π^A is unblocked during the execution fo the call C_{r_0} (we recall that π^A denotes the principal target of the P-node A). In order to prove $(C.3.a)^{(r_0)}$, all we have to show is that $n_j^{(r_0)} \leq 1$, $j = 1, \ldots, k$, where the equality holds if and only if τ_j coincides with the place π^A. We have

$$
\begin{aligned}
n_j^{(r')} &= |\sigma^{\tau_j} \cap Pow^*(\{\sigma^{\alpha_1}, \ldots, \sigma^{\alpha_\ell}\}) \setminus \{\sigma^{\alpha_1} \cup \ldots \cup \sigma^{\alpha_\ell}\}| \\
&\quad - |\overline{\tau_j}^{(r'-1)} \cap Pow^*(\{\overline{\alpha_1}^{(r'-1)}, \ldots, \overline{\alpha_\ell}^{(r'-1)}\})|
\end{aligned}
$$

and $\overline{\tau_j}^{(r')} = \overline{\tau_j}^{(r'-1)} \cup \Delta_{\tau_j}^{(r')}$. Therefore, by inductive hypothesis,

$$
\begin{aligned}
&\quad |\sigma^{\tau_j} \cap Pow^*(\{\sigma^{\alpha_1}, \ldots, \sigma^{\alpha_\ell}\})| - n_j^{(r_0)} \\
&= |\overline{\tau_j}^{(r_0-1)} \cap Pow^*(\{\overline{\alpha_1}^{(r_0-1)}, \ldots, \overline{\alpha_\ell}^{(r_0-1)}\})| \\
&= |\overline{\tau_j}^{(r')} \cap Pow^*(\{\overline{\alpha_1}^{(r'-1)}, \ldots, \overline{\alpha_\ell}^{(r'-1)}\})| \\
&= |(\overline{\tau_j}^{(r'-1)} \cup \Delta_{\tau_j}^{(r')}) \cap Pow^*(\{\overline{\alpha_1}^{(r'-1)}, \ldots, \overline{\alpha_\ell}^{(r'-1)}\})| \\
&= |\overline{\tau_j}^{(r'-1)} \cap Pow^*(\{\overline{\alpha_1}^{(r'-1)}, \ldots, \overline{\alpha_\ell}^{(r'-1)}\})| + |\Delta_{\tau_j}^{(r')}| \\
&= |\overline{\tau_j}^{(r'-1)} \cap Pow^*(\{\overline{\alpha_1}^{(r'-1)}, \ldots, \overline{\alpha_\ell}^{(r'-1)}\})| + n_j^{(r')} \\
&= |\sigma^{\tau_j} \cap Pow^*(\{\sigma^{\alpha_1}, \ldots, \sigma^{\alpha_\ell}\}) \setminus \{\sigma^{\alpha_1} \cup \ldots \cup \sigma^{\alpha_\ell}\}|,
\end{aligned}
$$

so that

$$
n_j^{(r_0)} = \begin{cases} 1 & \text{if } \tau_j \equiv \pi^A \\ 0 & \text{otherwise.} \end{cases}
$$

concluding the proof of $(C.3.a)^{(r_0)}$ and in turn of Assertion $(C.3)^{(r_0)}$, at least in the present case in which $\bigcup_{\alpha \in A}(\overline{\alpha}^{(r_0-1)} \setminus \overline{\alpha}^{(r'-1)}) = \emptyset$.

Subcase: $\bigcup_{\alpha \in A}(\overline{\alpha}^{(r_0-1)} \setminus \overline{\alpha}^{(r'-1)}) = \emptyset$. Notice that $n_j^{(r_0)} \le |\sigma^{\tau_j}| < \varrho$, for each trapped target τ_j of A. Hence, by Lemma 8.8(b) and since by Lemma 8.7(e) each P-node can have at most $n - 1$ targets, to prove $(C.3)^{(r_0)}$ it is enough to show that $|\Delta^{(r_0)}| \ge \varrho \cdot (n - 1)$.

Claim 8.1 *For each $t_0 \in \bigcup_{\alpha \in A}(\overline{\alpha}^{(r_0-1)} \setminus \overline{\alpha}^{(r'-1)})$ we have:*

(I) $Pow^*(\{\{t_0\}, \overline{\alpha_1}^{(r_0-1)}, \ldots, \overline{\alpha_\ell}^{(r_0-1)}\}) \subseteq \Delta^{(r_0)} \cup \{\overline{\alpha_1}^{(r_0-1)} \cup \ldots \cup \overline{\alpha_\ell}^{(r_0-1)}\}$

(II) $|Pow^*(\{\{t_0\}, \overline{\alpha_1}^{(r_0-1)}, \ldots, \overline{\alpha_\ell}^{(r_0-1)}\})| > \varrho \cdot (n - 1)$.

Proof. As $\{t_0\} \subseteq \bigcup_{\alpha \in A} \overline{\alpha}^{(r_0-1)}$, we have clearly

$$
Pow^*(\{\{t_0\}, \overline{\alpha_1}^{(r_0-1)}, \ldots, \overline{\alpha_\ell}^{(r_0-1)}\}) \subseteq Pow^*(\{\overline{\alpha_1}^{(r_0-1)}, \ldots, \overline{\alpha_\ell}^{(r_0-1)}\}).
$$

Therefore, to prove (I) it suffices to show that

$$
Pow^*(\{\{t_0\}, \overline{\alpha_1}^{(r_0-1)}, \ldots, \overline{\alpha_\ell}^{(r_0-1)}\}) \cap \bigcup_{\beta \in T(A)} \overline{\beta}^{(r_0-1)} = \emptyset.
$$

We do this as follows. Let $u_0 \in Pow^*(\{\{t_0\}, \overline{\alpha_1}^{(r_0-1)}, \ldots, \overline{\alpha_\ell}^{(r_0-1)}\}) \cap \overline{\beta}^{(r_0-1)}$, for some $\beta_0 \in T(A)$. By inductive hypothesis $(E.9)^{(r_0-1)}$, the element u_0 has been introduced $\overline{\beta}$ during the execution of a call $C_{r''}$, $Distribute(\alpha_1, \ldots, \alpha_\ell)$, with $r'' \leq r'$. In particular, $t_0 \in \bigcup_{\alpha \in A} \overline{\alpha}^{(r'-1)} \subseteq \bigcup_{\alpha \in A} \overline{\alpha}^{(r'-1)}$, contradicting the fact that $t_0 \notin \bigcup_{\alpha \in A} \overline{\alpha}^{(r'-1)}$, and thereby completing the proof of (I).

To prove (II) we just observe that as $|\overline{\alpha_{j_0}}^{(r_0-1)}| \geq \varrho$ for some $j_0 \in \{1, \ldots \ell\}$, then obviously

$$\left|Pow^*\left(\left\{\{t_0\}, \overline{\alpha_1}^{(r_0-1)}, \ldots, \overline{\alpha_\ell}^{(r_0-1)}\right\}\right)\right| \geq \min(2^\varrho - 1, 2^{\varrho-1} - 1) = 2^{\varrho-1} - 1.$$

Then (II) follows plainly from the assumption $2^{\varrho-1} - 1 > \varrho \cdot (n-1)$.

Thus the claim is fully established. $\blacksquare$

The above claim obviously implies that $|\Delta^{(r_0)}| \geq \varrho \cdot (n-1)$, which concludes the proof of Assertion $(C.3)^{(r_0)}$ even in the case in which $\bigcup_{\alpha \in A}(\overline{\alpha}^{(r_0-1)} \setminus \overline{\alpha}^{(r'-1)}) \neq \emptyset$.

Finally we show that Assertion $(C.4)^{(r_0)}$ is satisfied too. It is enough to prove

$$|\sigma^{\tau_j} \cap Pow^*(\{\sigma^{\alpha_1}, \ldots, \sigma^{\alpha_\ell}\}) \cap range(f^{(r_0-1)}) \setminus (\sigma^A)^{(r_0)}|$$
$$= |\overline{\tau_j}^{(r_0-1)} \cap Pow^*(\{\overline{\alpha_1}^{(r_0-1)}, \ldots, \overline{\alpha_\ell}^{(r_0-1)}\})|,$$

for each $j = 1, \ldots, k$. As τ_j is trapped, by inductive hypothesis $(E.4)^{(r_0-1)}$ and $(E.5)^{(r_0-1)}$, $\overline{\tau_j}^{(r_0-1)} \subseteq Dom(f^{(r_0-1)})$. Also from $(C.2)^{(r_0)}$, $(\sigma^A)^{(r_0)} \cap range(f^{(r_0-1)}) = \emptyset$. Therefore we can limit ourselves to proving that

$$f^{(r_0-1)}[\overline{\tau_j}^{(r_0-10} \cap Pow^*(\{\overline{\alpha_1}^{(r_0-1)}, \ldots, \overline{\alpha_\ell}^{(r_0-1)}\})]$$
$$= \sigma^{\tau_j} \cap Pow^*(\{\sigma^{\alpha_1}, \ldots, \sigma^{\alpha_\ell}\}) \cap range(f^{(r_0-1)}).$$

We do this as follows. Let $u \in \overline{\tau_j}^{(r_0-1)} \cap Pow^*(\{\overline{\alpha_1}^{(r_0-1)}, \ldots, \overline{\alpha_\ell}^{(r_0-1)}\})$. Then by inductive hypotheses $(E.3)^{(r_0-1)}$-$(E.5)^{(r_0-1)}$ and $(E.8)^{(r_0-1)}$, we have $f^{(r_0-1)}(u) \in \sigma^{\tau_j} \cap Pow^*(\{\sigma^{\alpha_1}, \ldots, \sigma^{\alpha_\ell}\}) \cap range(f^{(r_0-1)})$. To prove the converse inclusion, let

$$t \in \sigma^{\tau_j} \cap Pow^*(\{\sigma^{\alpha_1}, \ldots, \sigma^{\alpha_\ell}\}) \cap range(f^{(r_0-1)}).$$

Let $u \in Dom(f^{(r_0-1)})$ such that $f^{(r_0-1)}(u) = t$. From $(E.3)^{(r_0-1)}$, $(E.8)^{(r_0-1)}$, it follows $u \in \overline{\tau_j}^{(r_0-1)} \cap Pow^*(\{\overline{\alpha_1}^{(r_0-1)}, \ldots, \overline{\alpha_\ell}^{(r_0-1)}\})$, thus proving also

$$\sigma^{\tau_j} \cap Pow^*(\{\sigma^{\alpha_1}, \ldots, \sigma^{\alpha_\ell}\}) \cap range(f^{(r_0-1)})$$
$$\subseteq f^{(r_0-1)}[\overline{\tau_j}^{(r_0-1)} \cap Pow^*(\{\overline{\alpha_1}^{(r_0-1)}, \ldots, \overline{\alpha_\ell}^{(r_0-1)}\})].$$

This proves the validity of Assertion $(C.4)^{(r_0)}$, and in turn completes the demonstration that Assertion $C^{(r_0)}$ is satisfied.

After the execution of the assignment statements in C_{r_0}, we have

$$\overline{\pi}^{(r_0)} = \overline{\pi}^{(r_0-1)} \cup \Delta_\pi^{(r_0)}, \text{ for each } \pi \in \Pi,$$

and

$$f^{(r_0)} = f^{(r_0-1)} \cup f_{\tau_1}^{(r_0)} \cup \ldots \cup f_{\tau_k}^{(r_0)},$$

where $f_{\tau_j}^{(r_0)}$ is a one-to-one correspondence between $\Delta_{\tau_j}^{(r_0)}$ and the set $\sigma^{\tau_j} \cap Pow^*(\{\sigma^{\alpha_1}, \ldots, \sigma^{\alpha_\ell}\}) \setminus (\sigma^A)^{(r_0)} \setminus \text{range}(f^{(r_0-1)})$, $j = 1, \ldots, k$.

To prove that Assertion $D^{(r_0)}$ holds, we need Assertion $(E.9)^{(r_0)}$, which can be proved exactly in the same way as in the previous case in which $|\overline{\alpha}_i^{(r_0-1)}| < \varrho$ for all $i = 1, \ldots, \ell$.

From $(E.9)^{(r_0)}$ and $(C.3)^{(r_0)}$ we have

$$|\overline{\tau}_j^{(r_0)} \cap Pow^*(\{\overline{\alpha_1}^{(r_0)}, \ldots, \overline{\alpha_\ell}^{(r_0)}\})|$$
$$= |(\overline{\tau}_j^{(r'-1)} \cup \Delta_{\tau_j}^{(r_0)}) \cap Pow^*(\{\overline{\alpha_1}^{(r_0-1)}, \ldots, \overline{\alpha_\ell}^{(r_0-1)}\})|$$
$$= |\overline{\tau}_j^{(r_0-1)} \cap Pow^*(\{\overline{\alpha_1}^{(r_0-1)}, \ldots, \overline{\alpha_\ell}^{(r_0-1)}\})| + |\Delta_{\tau_j}^{(r_0)}|$$
$$= |\sigma^{\tau_j} \cap Pow^*(\{\sigma^{\alpha_1}, \ldots, \sigma^{\alpha_\ell}\}) \setminus (\sigma^A)^{(r_0)}|$$

for each $j = 1, \ldots, k$, which proves that Assertion $D^{(r_0)}$ is satisfied.

To complete the proof of the Main Inductive Lemma, we only need to show that in the present case in which $|\overline{\alpha_{j_0}}^{(r_0-1)}| \geq \varrho$ for some $j_0 \in \{1, \ldots, \ell\}$, Assertion $E^{(r_0)}$ is satisfied.

As regards $(E.1)^{(r_0)}$, it is enough to observe that $\{Dom(f^{(r_0-1)})\} \cup \{Dom(f_{\tau_j}^{(r_0)}): j = 1, \ldots, k\}$ and $\{range(f^{(r_0-1)})\} \cup \{range(f_{\tau_j}^{(r_0)}): j = 1, \ldots, k\}$ are both families of pairwise disjoint sets, and that $f^{(r_0-1)}$, $f_{\tau_1}^{(r_0)}, \ldots, f_{\tau_k}^{(r_0)}$ are all one-to-one functions.

Concerning $(E.2)^{(r_0)}$, it suffices to notice that $Dom(f_{\tau_j}^{(r_0)}) = \Delta_{\tau_j}^{(r_0)} \subseteq \tau_j^{(r_0)}$, for all $j = 1, \ldots, k$. Moreover, as $f_{\tau_j}^{(r_0)}[\Delta_{\tau_j}^{(r_0)}] \subseteq \sigma^{\tau_j}$, Assertion $(E.3)(r_0)$ follows at once.

Next, let π be a place of P such that $|\overline{\pi}^{(r_0)}| < \varrho$. Then $|\Delta_\pi^{(r_0)}| < \varrho$, which easily implies $\Delta_\pi^{(r_0)} \subseteq Dom(f^{(r_0)})$. Therefore by inductive hypothesis $(E.4)^{(r_0-1)}$ we obtain $(E.4)^{(r_0)}$.

Concerning $(E.5)^{(r_0)}$, let π be a trapped place. If π is not a target of the P-node A, then $\overline{\pi}^{(r_0)} = \overline{\pi}^{(r_0-1)}$ and the inductive hypothesis $(E.5)^{(r_0-1)}$ implies $(E.5)(r_0)$. On the other hand, if π is a target of the P-node A, then $\overline{\pi}^{(r_0)} = \overline{\pi}^{(r_0-1)} \cup \Delta_\pi^{(r_0)}$, with $\overline{\pi}^{(r_0-1)} \subseteq Dom(f^{(r_0-1)})$ (cf. $(E.5)^{(r_0-1)}$) and $\Delta_\pi^{(r_0)} \subseteq Dom(f_\pi^{(r_0)}) \subseteq Dom(f^{(r_0)})$. Therefore, $\overline{\pi}^{(r_0)} \subseteq Dom(f^{(r_0)})$, so that the injectivity of $f^{(r_0)}$ combined with $(E.3)^{(r_0)}$ gives $|\overline{\pi}^{(r_0)}| \leq |\sigma^\pi|$, which proves $(E.5)^{(r_0)}$.

As regards $(E.6)^{(r_0)}$, let α and β be any two distinct places of P. By inductive hypothesis, $\overline{\alpha}^{(r_0)} \cap \overline{\beta}^{(r_0)} = (\overline{\alpha}^{(r_0-1)} \cap \Delta_\beta^{(r_0)}) \cup (\overline{\beta}^{(r_0-1)} \cap \Delta_\alpha^{(r_0)})$, so that in order to prove that $\overline{\alpha}^{(r_0)} \cap \overline{\beta}^{(r_0)} = \emptyset$, it is enough to show that $\overline{\gamma}^{(r_0-1)} \cap \Delta_\delta^{(r_0)} = \emptyset$, for any two distinct places γ and δ. Suppose by contradiction that there exists $u \in \overline{\gamma}^{(r_0-1)} \cap \Delta_{\delta}^{(r_0)}$. Then, by $(C.3)^{(r_0)}$, $u \in Pow^*(\{\overline{\alpha_1}^{(r_0-1)}, \ldots, \overline{\alpha_\ell}^{(r_0-1)}\})$, which by $(E.9.b)^{(r_0-1)}$ implies that $\Delta_{\delta}^{(r_0)} \cap \overline{\gamma}^{(r_0-1)} = \emptyset$. But this contradicts our assumption $\overline{\gamma}^{(r_0-1)} \cap \Delta_{\delta}^{(r_0)} \neq \emptyset$, and consequently proves Assertion $(E.6)^{(r_0)}$.

Finally, concerning Assertions $(E.7)^{(r_0)}$ and $(E.8)^{(r_0)}$, we notice that the proofs given in the analysis of the preceding case in which $|\overline{\alpha_j}^{(r_0-1)}| < \varrho$ for all $i = 1, \ldots, \ell$ can be repeated word for word in the present case too. Also, we already observed that the same is true for Assertion $(E.9)^{(r_0)}$.

This completes the analysis of the inductive step in the case in which $|\overline{\alpha_{j_0}}^{(r_0-1)}| \geq \varrho$ for some $j_0 \in \{1, \ldots, \ell\}$, thereby concluding the proof of the Main Inductive Lemma.   ∎

The following section will show that the sequence of calls $C_1, C_2, \ldots$, in K is finite and that in fact the computation K terminates.

8.5   Termination proof

Again, we denote by K an execution of the IVAA and by $C_1, C_2, C_3, \ldots$ the sequence of calls to the procedure *Distribute* arranged in the order in which they occur during the computation K. Also, we use the notation $\overline{\pi}^{(r)}$, $\Delta_{\pi}^{(r)}$, $f^{(r)}$, Assertion $X^{(r)}$ with the same meaning as in the preceding sections.

In this section we prove the following lemma, which has been already stated in Section 8.3.

Lemma 8.13 (Termination) *The number of calls $C_1, C_2, C_3, \ldots$ in the computation K is bounded by $(\varrho-2) \cdot (n-1) \cdot 2^{n-2} + 3 \cdot 2^{n-1} - 2$, where $n = |\Pi|$, and ϱ is the constant which appears in condition (C4) of Theorem 8.1. Moreover, when the last call to the procedure* Distribute *is made, all places $\pi \in \Pi$ are blocked.*

Proof. We begin by establishing an upper bound on the number of calls $C_1, C_2, C_3, \ldots$ made during the computation K. Let $A = \{\alpha_1, \ldots \alpha_\ell\}$ be a nonempty P-node, and let $C_{r_1}, C_{r_2}, C_{r_3}, \ldots$, with $1 \leq r_1 < r_2 < r_3 < \ldots$, be the subsequence of $C_1, C_2, C_3, \ldots$ consisting of all the calls *Distribute*$(\alpha_1, \ldots, \alpha_\ell)$ made from the Propagation Phase and such that $|\overline{\alpha_i}^{(r_j-1)}| < \varrho$, for all $i = 1, \ldots, \ell$ and $j = 1, 2, 3, \ldots$. It follows by $(B.2)^{(r_j)}$ that for each $j = 1, 2, 3, \ldots$

$$\bigcup_{\beta \in T(A)} \overline{\beta}^{(r_{j+1}-1)} \supseteq \bigcup_{\beta \in T(A)} \overline{\beta}^{(r_j)} \supseteq Pow^*(\{\overline{\alpha_1}^{(r_j-1)}, \ldots, \overline{\alpha_\ell}^{(r_j-1)}\}),$$

whereas $\bigcup_{\beta \in T(A)} \overline{\beta}^{(r_{j+1}-1)} \not\supseteq Pow^*(\{\overline{\alpha_1}^{(r_{j+1}-1)}, \ldots, \overline{\alpha_\ell}^{(r_{j+1}-1)}\})$. Therefore, we deduce that $\bigcup_{i=1}^{\ell} \overline{\alpha_i}^{(r_j-1)} \subsetneq \bigcup_{i=1}^{\ell} \overline{\alpha_i}^{(r_{j+1}-1)}$, $j = 1, 2, 3, \ldots$. But

$|\bigcup_{i=1}^{\ell} \overline{\alpha_i}^{(r_1-1)}| \geq \ell$ and $|\bigcup_{i=1}^{\ell} \overline{\alpha_i}^{(r_j-1)}| \leq (\varrho - 1) \cdot \ell$. Hence, if we denote by $N(A)$ the number of all the calls C_{r_j}, $Distribute(\alpha_1, \ldots, \alpha_\ell)$, made from the Propagation Phase and such that $|\overline{\alpha_i}^{(r_j-1)}| < \varrho$ for all $i = 1, \ldots, \ell$, we have

$$N(A) \leq (\varrho - 2) \cdot |A| + 1.$$

Easy calculations prove

$$\sum_{\substack{A \subseteq \Pi \setminus \{\pi^{qv}\} \\ \& \ A \neq \emptyset}} N(A)$$

$$= (2^{n-1} - 1) + (\varrho - 2) \cdot \sum_{i=1}^{n-1} \binom{n-1}{i} i$$

$$= (\varrho - 2) \cdot (n - 1) \cdot 2^{n-2} + 2^{n-1} - 1.$$

Furthermore, it is easy to see that the empty node is processed just once, and that by Lemma 8.11(b)–(c) each non-empty P-node can be processed at most two times more by the procedure *Distribute*. Therefore, the sequence of calls $C_1, C_2, C_3, \ldots$ in K is finite and if we denote by ξ its length the following inequality holds

$$\xi \leq (\varrho - 2) \cdot (n - 1) \cdot 2^{n-2} + 3 \cdot 2^{n-1} - 2,$$

proving the first half of the Termination Lemma.

Next we will prove the slightly stronger fact that the computation K necessarily terminates, i.e. reaches a point in which every place is blocked. This will be established by first proving that if this is false, the set $\bigcup_{\substack{\pi \in \Pi \\ \& \ |\overline{\pi}| < \varrho}} (\sigma^\pi \setminus f[\overline{\pi}])$

must be non-empty, and then obtaining a contradiction from this fact.

Remark 8.7 As in Section 8.4, for simplicity we will often write $\overline{\pi}$ in place of $\overline{\pi}^{(\xi)}$, for all $\pi \in \Pi$, and f in place of $f^{(\xi)}$, where ξ is the length of the sequence of calls $C_1, C_2, C_3, \ldots$ in K.   □

Suppose therefore that K does not terminate, i.e. that after the last call C_ξ to the procedure *Distribute*, K will remain permanently in a state Σ in which there are unblocked places. It is easy to see by examining the IVAA code that this implies that the following two statements must be true.

(S.1) Let ϑ_0 be the minimum unblocked place in state Σ. Then (the blocking phase never makes any additional node blocked, i.e.) either

(S.1.1) ϑ_0 is trapped and $|\overline{\vartheta_0}| \neq |\sigma^{\vartheta_0}|$; or

(S.1.2) ϑ_0 is non-trapped, $|\overline{\vartheta_0}| \geq \varrho$ and for some P-node $\{\alpha_1, \dots, \alpha_\ell\}$ having ϑ_0 among its targets, and such that $|\overline{\alpha_j}| < \varrho$ for all $j = 1, \dots, \ell$, we have

$$|Pow^*(\{\overline{\alpha_1}, \dots \overline{\alpha_\ell}\}) \cap \overline{\vartheta_0}| \neq |Pow^*(\{\sigma^{\alpha_1}, \dots, \sigma^{\alpha_\ell}\}) \cap \sigma^{\vartheta_0}|;$$

or

(S.1.3) ϑ_0 is non-trapped and $|\overline{\vartheta_0}| < \varrho$.

(S.2) (No call to *Distribute* is made on behalf of any unblocked node, i.e.) for every unblocked P-node $\{\alpha_1, \dots, \alpha_\ell\}$, either

(S.2.1) $\overline{\alpha_{j_0}} = \emptyset$, for some $j_0 \in \{1, \dots, \ell\}$; or

(S.2.2) $0 < |\overline{\alpha_j}| < \varrho$, for all $j = 1, \dots, \ell$, and

$$Pow^*(\{\overline{\alpha_1}, \dots, \overline{\alpha_\ell}\}) \subseteq \bigcup_{\gamma \in T(\{\alpha_1, \dots, \alpha_\ell\})} \overline{\gamma}.$$

To begin with, we establish the following lemma.

Lemma 8.17 *Assume that the foregoing statements (S.1) and (S.2) holds, and let $A = \{\alpha_1, \dots, \alpha_\ell\}$ be a P-node such that $|\overline{\alpha_i}| < \varrho$, for all $i = 1, \dots, \ell$. Then*

$$Pow^*(\overline{\alpha_1}, \dots, \overline{\alpha_\ell}) \subseteq (\bigcup_{\beta \in T(A)} \overline{\beta}) \cap Dom(f).$$

Proof. In view of Assertion (E.7), it is enough to show that $Pow^*(\overline{\alpha_1}, \dots, \overline{\alpha_\ell}) \subseteq \bigcup_{\beta \in T(A)} \overline{\beta}$. If the P-node $\{\alpha_1, \dots, \alpha_\ell\}$ is blocked, let C_{r_0} be the last call *Distribute*$(\alpha_1, \dots, \alpha_\ell)$. Then

$$\begin{aligned}
&Pow^*(\{\overline{\alpha_1}, \dots, \overline{\alpha_\ell}\}) \\
={} &Pow^*(\{\overline{\alpha_1}^{(r_0-1)}, \dots, \overline{\alpha_\ell}^{(r_0-1)}\}) \\
\subseteq{} &\bigcup_{\beta \in T(A)} \overline{\beta}^{(r_0)} \subseteq \bigcup_{\beta \in T(A)} \overline{\beta}.
\end{aligned}$$

On the other hand, if the P-node is unblocked, our conclusion follows from statement (S.2), and in any case the lemma holds.   ∎

Next we prove the following lemma, from which it will by easy to deduce that $\bigcup_{\substack{\pi \in \Pi \\ \& \, |\overline{\pi}| < \varrho}} (\sigma^\pi \setminus f[\overline{\pi}]) \neq \emptyset$.

Lemma 8.18 *The hypothesis that statements (S.1) and (S.2) hold implies that in state Σ there is some place γ of P such that*

$$|\overline{\gamma}| < min(\varrho, |\sigma^\gamma|).$$

Proof. We distinguish three cases, according to whether (S.1.1), or (S.1.2), or (S.1.3) holds.

Case: (S.1.1) *holds.* Suppose that ϑ_0 is a trapped place and that $|\overline{\vartheta_0}| \neq |\sigma^{\vartheta_0}|$. Therefore it follows from Assertion (E.5) that $|\overline{\vartheta_0}| < |\sigma^{\vartheta_0}| < \varrho$, which clearly yields $|\overline{\vartheta_0}| < min(\varrho, |\sigma^{\vartheta_0}|)$.

Case: (S.1.2) *holds.* Next assume that ϑ_0 is non-trapped but for some P-node $\{\alpha_1, \ldots, \alpha_\ell\}$ having ϑ_0 as a target, and such that $|\overline{\alpha_i}| < \varrho$ for all $i = 1, \ldots, \ell$, we have

$$|Pow^*(\{\overline{\alpha_1}, \ldots, \overline{\alpha_\ell}\}) \cap \overline{\vartheta_0}| \neq |Pow^*(\{\sigma^{\alpha_1}, \ldots, \sigma^{\alpha_\ell}\}) \cap \sigma^{\vartheta_0}|.$$

The preceding lemma implies that $Pow^*(\{\overline{\alpha_1}, \ldots, \overline{\alpha_\ell}\}) \subseteq Dom(f)$. But Lemmas 8.3 and 8.17 yield

$$\begin{aligned} &f[Pow^*(\{\overline{\alpha_1}, \ldots, \overline{\alpha_\ell}\}) \cap \overline{\vartheta_0}] \\ = \ &Pow^*(\{f[\overline{\alpha_1}], \ldots, f[\overline{\alpha_\ell}]\}) \cap f[\overline{\vartheta_0}] \\ \subseteq \ &Pow^*(\{\sigma^{\alpha_1}, \ldots, \sigma^{\alpha_\ell}\}) \cap \sigma^{\vartheta_0}. \end{aligned}$$

Therefore, by the injectivity of f,

$$|Pow^*(\{\overline{\alpha_1}, \ldots, \overline{\alpha_\ell}\}) \cap \overline{\vartheta_0}| < |Pow^*(\{\sigma^{\alpha_1}, \ldots, \sigma^{\alpha_\ell}\}) \cap \sigma^{\vartheta_0}|.$$

Next we show that

$$Pow^*(\{f[\overline{\alpha_1}], \ldots, f[\overline{\alpha_\ell}]\}) \cap f[\overline{\vartheta_0}] = Pow^*(\{f[\overline{\alpha_1}], \ldots, f[\overline{\alpha_\ell}]\}) \cap \sigma^{\vartheta_0}.$$

Let $t \in Pow^*(\{f[\overline{\alpha_1}], \ldots, f[\overline{\alpha_\ell}]\}) \cap \sigma^{\vartheta_0} = f[Pow^*(\{\overline{\alpha_1}, \ldots, \overline{\alpha_\ell}\})] \cap \sigma^{\vartheta_0}$. Then by Assertion (E.8), $t = f(u)$, for some $u \in Pow^*(\{\overline{\alpha_1}, \ldots, \overline{\alpha_\ell}\}) \cap \overline{\vartheta_0}$, which in particular proves $t = f(u) \in f[\overline{\vartheta_0}]$. Thus

$$Pow^*(\{f[\overline{\alpha_1}], \ldots, f[\overline{\alpha_\ell}]\}) \cap \sigma^{\vartheta_0} \subseteq Pow^*(\{f[\overline{\alpha_1}], \ldots, f[\overline{\alpha_\ell}]\}) \cap f[\overline{\vartheta_0}].$$

The converse inclusion is an immediate consequence of Assertion (E.3), and therefore the equality follows.

This in turn implies

$$|Pow^*(\{f[\overline{\alpha_1}], \ldots, f[\overline{\alpha_\ell}]\}) \cap \sigma^{\vartheta_0}| < |Pow^*(\{\sigma^{\alpha_1}, \ldots, \sigma^{\alpha_\ell}\}) \cap \sigma^{\vartheta_0}|,$$

and thus, by Assertion (E.3), $f[\overline{\alpha_{j_0}}] \subsetneq \sigma^{\alpha_{j_0}}$, for some $j_0 \in \{1, \ldots, \ell\}$. Therefore $|\overline{\alpha_{j_0}}| < min(\varrho, |\sigma^{\alpha_{j_0}}|)$, proving our lemma, at least in the case in which (S.1.2) holds.

Case: (S.1.3) *holds.* Finally suppose that ϑ_0 is non-trapped and that $|\overline{\vartheta_0}| < \varrho$. Then $|\overline{\vartheta_0}| < \varrho \leq |\sigma^{\vartheta_0}|$, yielding $|\overline{\vartheta_0}| < min(\varrho, |\sigma^{\vartheta_0}|)$ and completing the proof of the lemma in all possible cases. ∎

Corollary 8.6 *If statements (S.1) and (S.2) hold, then*

$$\bigcup_{\substack{\pi \in \Pi \\ \& \ |\overline{\pi}| < \varrho}} (\sigma^{\pi} \setminus f[\overline{\pi}]) \neq \emptyset$$

Proof. The preceding lemma implies that there exists a place γ such that $|\overline{\gamma}| < min(\varrho, |\sigma^{\gamma}|)$. Therefore $\emptyset \neq \sigma^{\gamma} \setminus f[\overline{\gamma}] \subseteq \bigcup_{\substack{\pi \in \Pi \\ \& \ |\overline{\pi}| < \varrho}} (\sigma^{\pi} \setminus f[\overline{\pi}])$ and the corollary follows. ■

Now we are ready to show how to derive a contradiction from the fact that

$$\bigcup_{\substack{\pi \in \Pi \\ \& \ |\overline{\pi}| < \varrho}} (\sigma^{\pi} \setminus f[\overline{\pi}]) \neq \emptyset$$

To this end let $t_0 \in \sigma^{\pi_0} \setminus f[\overline{\pi_0}]$ be an element of *minimal* rank in $\bigcup_{\substack{\pi \in \Pi \\ \& \ |\overline{\pi}| < \varrho}} (\sigma^{\pi} \setminus f[\overline{\pi}])$, where $|\overline{\pi_0}| < \varrho$.

From Lemma 8.7(c) it follows that $t_0 \in \sigma^{\pi_0} \cap Pow^*(\{\sigma^{\alpha_1}, \ldots, \sigma^{\alpha_\ell}\})$, for some P-node $A = \{\alpha_1, \ldots, \alpha_\ell\}$ having π_0 as a target.

There are only three possibilities: either $\overline{\alpha_{i_0}} = \emptyset$ for some $i_0 \in \{1, \ldots, \ell\}$, or $0 < |\overline{\alpha_i}| < \varrho$ for all $i = 1, \ldots, \ell$, or $\overline{\alpha_i} \neq \emptyset$ for all $i = 1, \ldots, \ell$ and $|\overline{\alpha_{j_0}}| \geq \varrho$ for some $j_0 \in \{1, \ldots, \ell\}$. Below we will show that all cases lead to a contradiction, thereby proving that at the end of the computation K all places must be blocked.

Case: $\overline{\alpha_{i_0}} = \emptyset$ *for some* $i_0 \in \{1, \ldots, \ell\}$. Since $t_0 \cap \sigma^{\alpha_{i_0}} \neq \emptyset$, for every $u_0 \in t_0 \cap \sigma^{\alpha_{i_0}}$ we have $rk\, u_0 < rk\, t_0$ and $u_0 \in \sigma^{\alpha_{i_0}} \setminus f[\overline{\alpha_{i_0}}] \subseteq \bigcup_{\substack{\pi \in \Pi \\ \& \ |\overline{\pi}| < \varrho}} (\sigma^{\pi} \setminus f[\overline{\pi}])$, contradicting the minimality of $rk\, t_0$ and ruling out this first case.

Case: $0 < |\overline{\alpha_i}| < \varrho$ *for all* $i = 1, \ldots, \ell$. As proved in Lemma 8.17, in this case we have $Pow^*(\{\overline{\alpha_1}, \ldots, \overline{\alpha_\ell}\}) \subseteq (\bigcup_{\beta \in T(A)} \overline{\beta}) \cap Dom(f)$ and therefore by Lemmas 8.3 and 8.17

$$f[Pow^*(\{\overline{\alpha_1}, \ldots \overline{\alpha_\ell}\})] = Pow^*(\{f[\overline{\alpha_1}], \ldots, f[\overline{\alpha_\ell}]\}) \subseteq Pow^*(\{\sigma^{\alpha_1}, \ldots, \sigma^{\alpha_\ell}\}).$$

Thus, since

$$t_0 \in Pow^*(\{\sigma^{\alpha_1}, \ldots, \sigma^{\alpha_\ell}\}) \setminus Pow^*(\{f[\overline{\alpha_1}], \ldots, f[\overline{\alpha_\ell}]\})$$

it follows that for some place α_{i_0} there must exist an element u_0 such that $u_0 \in t_0 \cap \sigma^{\alpha_{i_0}} \setminus f[\overline{\alpha_{i_0}}]$. But then $rk\, u_0 < rk\, t_0$ and $u_0 \in \bigcup_{\substack{\pi \in \Pi \\ \& \ |\overline{\pi}| < \varrho}} (\sigma^{\pi} \setminus f[\overline{\pi}])$, which shows that the present case is inconsistent.

Finally we prove that even the last case is contradictory.

Case: $\overline{\alpha_i} \neq \emptyset$ *for all* $i = 1, \ldots, \ell$ *and* $|\overline{\alpha_{i_0}}| \geq \varrho$ *for some* $j_0 \in \{1, \ldots, \ell\}$. In this case, statement (S.2) yields that the P-node $\{\alpha_1, \ldots, \alpha_\ell\}$ is marked either visited or blocked. In any event, as $|\overline{\pi_0}| < \varrho$ it follows that the place π_0 must be trapped. Thus by Assertion D we have

$$|\sigma^{\pi_0} \cap Pow^*(\{\sigma^{\alpha_1}, \ldots, \sigma^{\alpha_\ell}\}) \setminus \sigma^A| = |\overline{\pi} \cap Pow^*(\{\overline{\alpha_1}, \ldots, \overline{\alpha_\ell}\})|.$$

where

$$\sigma^A = \begin{cases} \emptyset & \text{if A is blocked} \\ \{\sigma^{\alpha_1} \cup \ldots \cup \sigma^{\alpha_\ell}\} & \text{otherwise} \end{cases}$$

Since $\sigma^{\pi_0} \cap Pow^*(\{\sigma^{\alpha_1}, \ldots, \sigma^{\alpha_\ell}\}) \setminus \sigma^A \subseteq range(f)$, whereas by Assertion (E.3) $t_0 \in \sigma^{\pi_0} \cap Pow^*(\{\sigma^{\alpha_1}, \ldots \sigma^{\alpha_\ell}\}) \setminus range(f)$, the only possibility is that the P-node A is unblocked and $t_0 = \sigma^{\alpha_1} \cup \ldots \cup \sigma^{\alpha_\ell}$. In particular some $\alpha_{k_0} \in A$ must be unblocked. Thus $\vartheta_0 \leq \alpha_{k_0}$ and consequently $rk\, \sigma^{\vartheta_0} \leq rk\, \sigma^{\alpha_{k_0}}$ (cf. Definition 8.6).

If $|\overline{\vartheta_0}| < \varrho$, then Assertion (E.4) and statement (S.1) yield $\emptyset \neq \sigma^{\vartheta_0} \setminus f[\overline{\vartheta_0}] \subseteq \bigcup_{\substack{\pi \in \Pi \\ \& \, |\overline{\pi}| < \varrho}} (\sigma^\pi \setminus f[\overline{\pi}])$, and therefore for every $t_1 \in \sigma^{\vartheta_0} \setminus f[\overline{\vartheta_0}]$ we have

$$\begin{aligned} rk\, t_1 \quad &< \quad rk\, \sigma^{\vartheta_0} \leq rk\, \sigma^{\alpha_{k_0}} \\ &\leq \quad rk\, (\sigma^{\alpha_1} \cup \ldots \cup \sigma^{\alpha_\ell}) = rk\, t_0. \end{aligned}$$

This contradicts the minimality of $rk\, t_0$ and rules out the possibility that $|\overline{\vartheta_0}| < \varrho$.

On the other hand, if $|\overline{\vartheta_0}| \geq \varrho$, by statement (S.1) we deduce that for some P-node $\{\beta_1 \ldots, \beta_k\}$ having ϑ_0 among its targets, and such that $|\overline{\beta_j}| < \varrho$ for all $j = 1, \ldots, k$, we have

$$|Pow^*(\{\overline{\beta_1}, \ldots, \overline{\beta_k}\}) \cap \overline{\vartheta_0}| \neq |Pow^*(\{\sigma^{\beta_1}, \ldots \sigma^{\beta_k}\}) \cap \sigma^{\vartheta_0}|.$$

This implies

$$Pow^*(\{\sigma^{\beta_1}, \ldots, \sigma^{\beta_k}\}) \cap \sigma^{\vartheta_0} \setminus f[Pow^*(\{\overline{\beta_1}, \ldots, \overline{\beta_k}\}) \cap \overline{\vartheta_0}] \neq \emptyset,$$

which by Assertion (E.4) in turn gives

$$Pow^*(\{\sigma^{\beta_1}, \ldots, \sigma^{\beta_k}\}) \cap \sigma^{\vartheta_0} \setminus f[Pow^*(\{\overline{\beta_1}, \ldots, \overline{\beta_k}\})] \neq \emptyset,$$

i.e.

$$Pow^*(\{\sigma^{\beta_1}, \ldots, \sigma^{\beta_k}\}) \cap \sigma^{\vartheta_0} \setminus Pow^*(\{f[\overline{\beta_1}], \ldots, f[\overline{\beta_k}]\}) \neq \emptyset.$$

Let $u \in Pow^*(\{\sigma^{\beta_1}, \ldots, \sigma^{\beta_k}\}) \cap \sigma^{\vartheta_0} \setminus Pow^*(\{f[\overline{\beta_1}], \ldots, f[\overline{\beta_k}]\})$. For some $i_0 \in \{1, \ldots, k\}$, $u \cap \sigma^{\beta_{i_0}} \subseteq f[\overline{\beta_{i_0}}]$. Hence, for each $u' \in u \cap \sigma^{\beta_{i_0}} \setminus f[\overline{\beta_{i_0}}]$,

$$u' \in \bigcup_{\substack{\pi \in \Pi \\ \& \, |\overline{\pi}| < \varrho}} (\sigma^\pi \setminus f[\overline{\pi}])$$

and

$$
rk\, u' < rk\, u < rk\, \sigma^{\vartheta_0} \;\;\leq\;\; rk\, \sigma^{\alpha_{k_0}}
$$
$$
\leq\;\; rk\,(\sigma^{\alpha_1} \cup \ldots \cup \sigma^{\alpha_\iota}) = rk\, t_0,
$$

contradicting again the minimality of $rk\, t_0$.

Having shown that a contradiction is derived even in this last case, it follows that our initial assumption (i.e. that there are unblocked places after the last call C_ξ) is false.

This concludes the proof of the Termination Lemma.   ■

9. MAP CONSTRUCTS

9.1 Introduction and preliminaries

In this chapter we consider the satisfiability problem for a class of unquantified formulae of set theory involving several operators and predicates of the elementary theory of relations and maps (cf. [CS88]). It will be proved that a sound and complete decision test exists for this problem.

Similar results concerning map operators in a set-theoretic framework can be found in [FOS80b], which solves the satisfiability problem for the three-sorted language consisting of set variables with the operators $\setminus, \cup$ and the predicates $=, \in$, cardinality variables with the operator $+$ and the predicates $=, <, \leq$, function variables with the operators D (domain), R (range) and the predicates $single\text{-}valued(f)$, $one\text{-}to\text{-}one(f)$, and mixed terms of type $\#(t)$ and $f[t]$, where $\#$ is the cardinality operator, f is a function variable, and t is a set term.

Notice that in the above language, constructs like $f_1 = f_2 \cup f_3$, $f_1 = f_2 \setminus f_3$, $x \in f_1$, $f_1 \in f_2$, PAIR_IN(x, y, f) (which says that the ordered pair $[x, y]$ is in f), where f_1, f_2, f_3 are function variables and x, y are set variables, are forbidden.

The language $\mathcal{L}$ considered in this chapter allows one to deal with such literals and properly extends the purely set-theoretic part (i.e. the part without cardinality constructs) of the theory considered in [FOS80b].

In detail, the class of formulae expressible in the language $\mathcal{L}$ consists of the propositional combinations (obtained by using the connectives $\neg$, $\&$, $\vee$, $\rightarrow$, $\leftrightarrow$, but with no explicit use of existential or universal quantifiers) of set-theoretic clauses of the forms

$$x = y \cap z, \, x = y \setminus z, \, x = y \cup z,$$
$$x = y, \, x = \emptyset,$$
$$x \subseteq y,$$
$$x \in y,$$
$$d = Df, \, r = Rf,$$
$$y = f[x], \, y = f^{-1}[x],$$
$$\text{PAIR\_IN}(x, y, f),$$
$$\text{RESTR}(g, f, x),$$

$$\text{INV}(f, g),$$
$$\text{SINGLEVALUED}(f),$$
$$\text{INJECTIVE}(f), \tag{9.1}$$

where x, y, z, f, g stand for variables or the constant $\emptyset$.

The intended meaning of the operators and predicates in (9.1) is as follows. Write the subset of f consisting of all ordered pairs in f as pairs(f), so that each p in pairs(f) has the form $p = [u, v]$ (below we will be more specific on the set representation of $[u, v]$). Then

- Df (the *domain* of f) designates the set

$$\{u : [u, v] \in \text{pairs}(f), \text{ for some } v\};$$

- Rf (the *range* of f) designates the set

$$\{v : [u, v] \in \text{pairs}(f), \text{ for some } u\};$$

- $f[x]$ (the *set image* of x under f) designates the set

$$\{v : [u, v] \in \text{pairs}(f), \text{ for some } u \text{ in } x\};$$

- $f^{-1}[x]$ (the *inverse set image* of x under f) designates the set

$$\{u : [u, v] \in \text{pairs}(f), \text{ for some } v \text{ in } x\};$$

- the predicate PAIR_IN(x, y, f) is true if and only if $[x, y] \in$ pairs(f);

- the predicate RESTR(g, f, x) stands for

$$(\forall u)(\forall v)([u, v] \in g \leftrightarrow ([u, v] \in f \,\&\, u \in x)),$$

  i.e. it expresses that the set of pairs in g coincides with the set of pairs in f whose first component belongs to x, that is, g is a restriction of f to x;

- the predicate INV(f, g) holds if and only if pairs(f) $= (\text{pairs}(g))^{-1}$, i.e. for each pair $[u, v]$, $[u, v]$ belongs to f if and only if the inverse pair $[v, u]$ belongs to g;

- the predicate SINGLEVALUED(f) holds whenever the relation pair(f) is single-valued, i.e. f contains no two distinct pairs of the form $[u, v]$ and $[u, v']$;

- the predicate INJECTIVE(f) stands for

$$(\forall u)(\forall u')(\forall v)([u, v] \in f \,\&\, [u', v] \in f \rightarrow u = u').$$

The remaining operators and predicates that appear in (9.1) have their standard interpretation.

Notice that the language $\mathcal{L}$ cannot express the fact that a set f is a relation in the ordinary sense, i.e. f contains only ordered pairs (this will be an immediate corollary of the construction to be given in Section 9.3). In particular, one cannot express in $\mathcal{L}$ that a set f is an 'ordinary' function. So, for example, according to our definition, each f admits an entire proper class of inverses $(\mathrm{pairs}(f))^{-1} \cup S$, as S ranges over the sets which contain no pairs, and, likewise, for each fixed x there is a class of restrictions of f to x.

It is plain that by converting formulae to disjunctive normal form one can reduce the satisfiability problem for $\mathcal{L}$ to the satisfiability problem for conjunctions of atoms or negations of atoms of type (9.1). As a matter of fact, even occurrences of literals of type $y = f[x]$, $y \neq f[x]$, $y = f^{-1}[x]$, $y \neq f^{-1}[x]$, $r = Rf$, $r \neq Rf$, $\mathrm{RESTR}(g, f, x)$, $\neg\,\mathrm{RESTR}(g, f, x)$, $\mathrm{INJECTIVE}(f)$, and $\neg\,\mathrm{INJECTIVE}(f)$, can be eliminated easily, preserving satisfiability, in the order that follows.

$y = f^{-1}[x]$ can be replaced by

$$\mathrm{INV}(h, f) \ \& \ y = h[x] \ ;$$

$y \neq f^{-1}[x]$ can be replaced by

$$\mathrm{INV}(h, f) \ \& \ y \neq h[x] \ ;$$

$y = f[x]$ can be replaced by

$$\mathrm{RESTR}(h, f, x) \ \& \ y = Rh \ ;$$

$y \neq f[x]$ can be replaced by

$$\mathrm{RESTR}(h, f, x) \ \& \ y \neq Rh \ ;$$

$r = Rf$ can be replaced by

$$\mathrm{INV}(h, f) \ \& \ r = Dh \ ;$$

$r \neq Rf$ can be replaced by

$$\mathrm{INV}(h, f) \ \& \ r \neq Dh \ ;$$

$\mathrm{RESTR}(g, f, x)$ can be replaced by

$$D(f \setminus g) \cap x = \emptyset \ \& \ D(g \setminus f) = \emptyset \ \& \ Dg \subseteq x \ ;$$

$\neg$ RESTR(g, f, x) can be replaced by

$$(\text{PAIR\_IN}(u, v, g) \,\&\, (\neg\, \text{PAIR\_IN}(u, v, f) \lor u \notin x))$$
$$\lor \quad (\neg\, \text{PAIR\_IN}(u, v, g) \,\&\, \text{PAIR\_IN}(u, v, f) \,\&\, u \in x) \,;$$

INJECTIVE(f) can be replaced by

$$\text{INV}(h, f) \,\&\, \text{SINGLEVALUED}(h) \,;$$

$\neg$ INJECTIVE(f) can be replaced by

$$\text{INV}(h, f) \,\&\, \neg\, \text{SINGLEVALUED}(h) \,;$$

where h, u, v stand for newly introduced variables.

In order to introduce additional simplifications, observe that

- the literal $\neg\, \text{PAIR\_IN}(x, y, f)$ is equisatisfiable with

$$(\text{PAIR\_IN}(x, y, g) \,\&\, g \cap f = \emptyset) \,;$$

- the literal $\neg\, \text{INV}(f, g)$ is equisatisfiable with

$$(\text{PAIR\_IN}(x, y, f) \;\leftrightarrow\; \neg\text{PAIR\_IN}(y, x, g)) \,;$$

- the literal $\neg\, \text{SINGLEVALUED}(f)$ is equisatisfiable with

$$(\text{PAIR\_IN}(x, y, f) \,\&\, \text{PAIR\_IN}(x, y', f) \,\&\, y \neq y') \,.$$

Therefore, by applying a normalization process of the kind described in Section 3.8, the satisfiability problem for $\mathcal{L}$ can further be reduced to the satisfiability problem for the subfamily of conjunctions of un-negated atoms of the following types

$$x = y \setminus z \,, x = y \cup z \,,$$
$$x \in y \,,$$
$$d = Df \,,$$
$$\text{PAIR\_IN}(x, y, f) \,,$$
$$\text{INV}(f, g) \,,$$
$$\text{SINGLEVALUED}(f) \,, \tag{9.2}$$

where x, y, z, f, g, stand for variables.

It is convenient to restrict ourselves, without loss of generality, to the problem of *injective satisfiability*. To this end, we recall (cf. Definition 3.9) that a set-theoretic formula P is said to be injectively satisfied by an interpretation M if M satisfies P and M maps distinct variables into distinct sets (in which case M is called an *injective model* of P).

In conclusion, we have the following result (cf. Lemma 3.3).

Lemma 9.1 *The satisfiability problem for propositional combinations of atoms of type (9.1) is equivalent to the injective satisfiability problem for conjunctions of* un-negated *atoms of type (9.2).* □

The ordered pair notion can of course be represented by any one of several more primitive set-theoretic constructions. To complete all details of the proof which is to follow, we need to choose one such, so for specificity we will define the ordered pair notion by

$$[x, y] = \{\{0, \{x\}\}, \{2, \{y\}\}\}, \tag{9.3}$$

where the integers 0 and 2 have definitions given by von Neumann, namely $0 = \emptyset$, $2 = \{\emptyset, \{\emptyset\}\}$. Hence a set is an ordered pair if it has exactly two elements e_0, e_2, each of which is a pair. Exactly one of these, namely e_0 (respectively, e_2), must have 0 (respectively, 2) as an element and the other element of each must be a singleton. It is then plain that the first component x and the second component y of $[x, y]$ can be recovered as the elements of these uniquely characterized singletons. Notice that we have chosen to encode ordered pairs by means of (9.3) only to slightly simplify our presentation. In fact the decidability result that will be presented in this chapter is independent of the specific ordered pair representation. Therefore the same result holds under the widespread convention (due to Kuratowski) of representing the ordered pair (x, y) by the set $\{\{x\}, \{x, y\}\}$.

Before closing the present section, we intend to recall some terminology on graphs which will be used later on in Section 9.3 (cf. [Jec78]).

By *graph* we mean a set N of *nodes* together with a set $E \subseteq N \times N$ of *edges*. Usually we write $u \Rightarrow v$ to denote the edge which connects the startpoint u with the endpoint v.

Definition 9.1 *A graph* $G = (N, E)$ *is said to be* well-founded *if it has no infinite descending chain* $\cdots \Rightarrow u_n \Rightarrow \cdots \Rightarrow u_2 \Rightarrow u_1$. □

Recall from Chapter 2 that an analogue of the arithmetic induction principle holds for WFGs. As an application, given a WFG $G = (N, E)$, a notion of height has been easily defined on N, by putting

$$\text{height}(v) = 0 \text{ if } v \text{ has no predecessors},$$

$$\text{height}(v) = \bigcup \{\text{height}(u) + 1 : u \Rightarrow v \text{ is in } E\}.$$

Let us tune to our current purposes the definition of extensionality introduced in Chapter 2.

Definition 9.2 *A graph* $G = (N, E)$ *is said to be* quasi-extensional, *if for all* $v_1, v_2 \in N$,

$$\{u \in N : u \Rightarrow v_1 \text{ is in } E\} = \{u \in N : u \Rightarrow v_2 \text{ is in } E\} \neq \emptyset$$

implies $v_1 = v_2$. □

A one-to-one realization of a WFG (see Chapter 2) is called a *representation*. That is,

Definition 9.3 *A function* R *defined on the set* N *of nodes of a WFG* $G = (N,E)$ *and with values in a class of sets is called a* representation *of* G *if for all* $v_1, v_2 \in N$

(i) $R(v_1) = R(v_2)$ *implies* $v_1 = v_2$,

(ii) $R(v_1) \in R(v_2)$ *if and only if* $v_1 \Rightarrow v_2$ *is in* E.   □

9.2   The decision algorithm

Let P be a conjunction of simple clauses of type (9.2). For technical reasons, and without any loss of generality, we assume that P contains the following clauses

$$q_0 = q_0 \setminus q_0 \,,\, U = DU \,,\, \mathrm{INV}(U,U), \, x \setminus U = q_0 \,,\, x \in U \,,\, \mathrm{PAIR\_IN}(x,y,U),$$
$$(9.4)$$

for all variables x and y occurring in P and distinct from U, where the variable U occurs in P only within the clauses (9.4). To prove that the above assumption does not affect in any way our decidability result, we only need to verify that given a set Q of clauses of type (9.2), and putting

$$Q' =_{\mathrm{Def}} Q \quad \& \quad q_0 = q_0 \setminus q_0 \,\&\, U = DU \,\&\, \mathrm{INV}(U,U) \,\&\, q_0 \in U$$
$$\& \quad \&_{\substack{x,y \text{ occur} \\ \text{in } Q}} (x \setminus U = q_0 \,\&\, x \in U \,\&\, \mathrm{PAIR\_IN}(x,y,U)) \,,$$

with q_0 and U new variables not already occurring in Q, the formulae Q and Q' are equisatisfiable.

Clearly any model M' of Q' is a model for Q. On the other hand, let M be a model for Q. In order to extend M to a model for Q', begin by putting $Mq_0 = \emptyset$, so that $q_0 = q_0 \setminus q_0$ is satisfied. Next let α be a limit ordinal such that

$$rk \, Mx < \alpha \,, \text{ for all } x \text{ occurring in } Q\,.$$

Then put

$$MU = V_\alpha \,,$$

where V_α is the α-th level of the von Neumann hierarchy of all sets, i.e. the family of all sets with rank less than α. It is plain that M so extended models correctly all conjuncts in Q'.

So, let P be a conjunction of clauses of type (9.2) which satisfies the additional hypothesis introduced at the beginning of the present section. Let M be an injective model of P. Below we will derive a collection of effectively verifiable conditions on the structure of P which are necessary for P to be injectively satisfiable. In the next section we will prove that such conditions are also sufficient for the injective satisfiability of P, therefore proving, in view of Lemma 9.1, that the class of formulae in the language $\mathcal{L}$ has a solvable satisfiability problem.

Up to the end of this section M will denote a given and fixed injective model of P.

Let $V = \{y_1, y_2, \ldots, y_m\}$ be the collection of all distinct variables occurring in P.

The presence in P of clauses (9.4) yields the following lemma:

Lemma 9.2 (a) $Mx \in U$, $[Mx, My] \in MU$, and $Mx \subseteq MU$, for all variables x, y in $V \setminus \{U\}$.

(b) If $[p, q] \in MU$ then $p \in MU$. Conversely, if $p \in MU$, then $[p, q] \in MU$, for some q.

(c) $[p, q] \in MU$ if and only if $[q, p] \in MU$, for all pairs $[p, q]$.   □

Let $\sigma_1, \sigma_2, \ldots, \sigma_n$ be the non-empty regions of the Venn diagram of the sets $My_1, My_2, \ldots, My_m$ in the universal space MU (notice that by Lemma 9.2(a) $\bigcup_{i=1}^{m} My_i = MU$). Then for every set σ_i we define

$$\pi_i(y) = \begin{cases} 0 & \text{if } \sigma_i \cap My = \emptyset \\ 1 & \text{if } \sigma_i \subseteq My, \end{cases} \tag{9.5}$$

where y ranges over V.

We put $\Pi = \{\pi_1, \pi_2, \ldots, \pi_n\}$. Notice that for each y in V we have

$$My = \bigcup_{\substack{\pi \in \Pi \\ \& \; \pi(y)=1}} \sigma^\pi, \tag{9.6}$$

where we designate by σ^π the region of the Venn diagram relative to the model M and the set of clauses P which induces the function π, according to the definition (9.5).

From (9.6) and the disjointness of the sets σ_i, we can easily deduce some properties of the functions π_i. Let $x = y \cup z$ occur in P. Then $Mx = My \cup Mz$, i.e.

$$\bigcup_{\pi(x)=1} \sigma^\pi = \bigcup_{\pi(y)=1} \sigma^\pi \cup \bigcup_{\pi(z)=1} \sigma^\pi = \bigcup_{\pi(y)=1 \vee \pi(z)=1} \sigma^\pi.$$

Hence for each $\pi \in \Pi$, $\pi(x) = \pi(y) \vee \pi(z)$. Analogously, if $x = y \setminus z$ occurs in P, then we can prove that $\pi(x) = \pi(y) \& \neg\pi(z)$, for all π in Π. Thus, clearly, Π is a set of places of P (cf. Definition 8.1).

Remark 9.1 In what follows we will often write $\pi \subseteq x$ when $\pi(x) = 1$.   □

Next, let x be in $V \setminus \{U\}$. It follows from Lemma 9.2(a) that $Mx \in MU$ and therefore $Mx \in \sigma^{\pi^x}$, for some place π^x. Notice that if $x \in y$ occurs in P, then $Mx \in My$, which in particular implies $\sigma^{\pi^x} \subseteq My$, i.e. $\pi^x(y) = 1$. Thus the function $x \mapsto \pi^x$ just defined, associates each variable in P distinct from U with a place at the same variable (cf. Definition 8.2).

In analogy to the definition of places at variables, we can set the following definition of *places at pairs* of variables.

Definition 9.4 *Given two variables x and y in $V \setminus \{U\}$, a place π is said to be at the pair $[x, y]$ if $\pi(f) = 1$ for all clauses of type $PAIR\_IN(x, y, f)$ occurring in P.* □

Therefore, for every x, y in $V \setminus \{U\}$, by letting $\pi^{x,y}$ be the place of P such that $[Mx, My] \in \sigma^{\pi^{x,y}}$ (cf. Lemma 9.2(a)), we have that $\pi^{x,y}$ is a place of P at the pair $[x, y]$.

Having proved the existence of the set of places Π and of the functions $x \mapsto \pi^x$ and $(x, y) \mapsto \pi^{x,y}$, such that π^x and $\pi^{x,y}$ are places of P respectively at the variable x and at the pair $[x, y]$, we begin by listing a collection of conditions which are necessary for P to be satisfiable.

Let x and y be any two distinct variables in P. Then, as the model M is injective, $Mx \neq My$, so that by (9.6) there must exist a place $\pi \in \Pi$ such that $\pi(x) \neq \pi(y)$. This gives us a first necessary condition.

Condition C1. *For all distinct x, y occurring in P, there exists a place of P, $\pi \in \Pi$, such that $\pi(x) \neq \pi(y)$.* □

Next we observe that the rank ordering over $\{\sigma_1, \sigma_2, \ldots, \sigma_n\}$ induces a linear ordering $\lhd$ over Π such that

$$\text{if } rk\, \sigma^\alpha < rk\, \sigma^\beta \text{ then } \alpha \lhd \beta, \text{ for all } \alpha, \beta \in \Pi. \tag{9.7}$$

Let $\alpha(x) = 1$, for some $\alpha \in \Pi$ and $x \in V \setminus \{U\}$. Then, plainly, $rk\, \sigma^\alpha \leq rk\, Mx < rk\, \sigma^{\pi^x}$, so that $\alpha \lhd \pi^x$. This yields a second necessary condition.

Condition C2. *If $\alpha(x) = 1$, with $\alpha \in \Pi$ and $x \in V \setminus \{U\}$, then $\alpha \lhd \pi^x$.* □

It is implicit in the results of Chapter 6 that, in the absence of the map constructs D, INV, PAIR_IN, and SINGLEVALUED, conditions C1 and C2 are also sufficient for the injective satisfiability of P.

All the essential complications that need to be faced are connected with the presence in P of finitely many clauses of the form $d = Df$, $INV(f, g)$, $PAIR\_IN(x, y, f)$, $SINGLEVALUED(f)$. Note, for example, that in the presence of such clauses a satisfiable set of statements may possess infinite models only, the statement

$$f \neq \emptyset \ \& \ f = Df \tag{9.8}$$

being a case in point. This has the model

$$f = \{0, [0, 1], [[0, 1], 2], [[[0, 1], 2], 3], \ldots\},$$

but since each $x \in f$ is a member of a member of a member of some $y \in f$, (9.8) cannot have finite models.

The following definitions take a step toward elucidating the logical weight of clauses in P of type $d = Df$, $\mathrm{INV}(f, g)$, and $\mathrm{SINGLEVALUED}(f)$.

For each $\alpha \in \Pi$, put

$$dom(\alpha) = \{\beta \in \Pi : \mathrm{Dom}(\sigma^\alpha) \cap \sigma^\beta \neq \emptyset\}, \qquad (9.9)$$

$$inv(\alpha) = \{\beta \in \Pi : \text{there exists } [p, q] \in \sigma^\alpha \text{ such that } [q, p] \in \sigma^\beta\}, \qquad (9.10)$$

so that dom and inv are both functions from Π into $Pow(\Pi)$.

The functions dom and inv have some remarkable properties.

Let $\beta \in dom(\alpha)$ and let $d = Df$ be a D-clause of P such that $\alpha \subseteq f$. Hence $\sigma^\alpha \subseteq Mf$, so that $\mathrm{Dom}(\sigma^\alpha) \subseteq \mathrm{Dom}(Mf) = Md$, which by (9.9) yields $\sigma^\beta \subseteq Md$, i.e. $\beta \subseteq d$.

Next, let $\beta \in inv(\alpha)$. It follows immediately from definition (9.10) that $\alpha \in inv(\beta)$. Moreover, assume that either $\mathrm{INV}(f, g)$ or $\mathrm{INV}(g, f)$ occurs in P. If $\alpha \subseteq f$, i.e. $\sigma^\alpha \subseteq Mf$, let $[p, q] \in \sigma^\alpha$ be such that $[q, p] \in \sigma^\beta$. As $[p, q] \in Mf$, we have $[q, p] \in Mg$ implying $\sigma^\beta \subseteq Mg$, i.e. $\beta \subseteq g$. Analogously, one proves that if $\beta \subseteq g$ then $\alpha \subseteq f$.

Notice also that if $\beta \in inv(\alpha)$, then $\mathrm{pairs}(\sigma^\alpha) \neq \emptyset$, $\mathrm{pairs}(\sigma^\beta) \neq \emptyset$, i.e. $\mathrm{Dom}(\sigma^\alpha) \neq \emptyset$, $\mathrm{Dom}(\sigma^\beta) \neq \emptyset$, so that by Lemma 9.2 we have $\mathrm{Dom}(\sigma^\alpha) \subseteq MU$, $\mathrm{Dom}(\sigma^\beta) \subseteq MU$, which in turn implies $dom(\alpha) \neq \emptyset$, $dom(\beta) \neq \emptyset$. In addition, if $dom(\alpha) \neq \emptyset$, then, by (9.9), $\mathrm{pairs}(\sigma^\alpha) \neq \emptyset$. But, from Lemma 9.2(c), $(\mathrm{pairs}(\sigma^\alpha))^{-1} \subseteq MU$, thereby proving that $inv(\alpha) \neq \emptyset$.

The preceding discussion gives us the following necessary conditions:

Condition C3. *If $\beta \in dom(\alpha)$, then*

(i) $inv(\alpha) \neq \emptyset$;

(ii) *for all D-clauses $d = Df$ in P, if $\alpha(f) = 1$ then $\beta(d) = 1$.*   □

Condition C4. *If $\beta \in inv(\alpha)$, then*

(i) $\alpha \in inv(\beta)$;

(ii) $dom(\alpha) \neq \emptyset$;

(iii) *for all INV-clauses $\mathrm{INV}(f, g)$ or $\mathrm{INV}(g, f)$ in P,*

$$\alpha(f) = 1 \text{ if and only if } \beta(g) = 1.$$

□

In order to state the next condition, we need to introduce the notion of *D-pair*.

Definition 9.5 *A D-pair (relative to P, Π, dom, etc.) is a pair (Γ, π) with $\Gamma \subseteq \Pi$ and $\pi \in \Pi$ such that*

(i) $\Gamma \neq \emptyset$;

(ii) $\pi \in dom(\gamma)$, *for all* $\gamma \in \Gamma$;

(iii) *for all* $\gamma_1, \gamma_2 \in \Gamma$, *if* $\gamma_1(f) = \gamma_2(f) = 1$, *for some variable* f *such that the clause* SINGLEVALUED(f) *appears in* P, *then* $\gamma_1 = \gamma_2$;

(iv) *for all D-clauses* $d = Df$ *present in* P, *if* $\pi(d) = 1$ *then there exists* $\gamma_0 \in \Gamma$ *such that* $\gamma_0(f) = 1$. □

We have the following lemma.

Lemma 9.3 *Let* $\pi \in \Pi$ *and let* $p \in \sigma^\pi$. *Put*

$$\Gamma_p = \{\gamma \in \Pi : p \in \text{Dom}(\sigma^\gamma)\}. \tag{9.11}$$

Then (Γ_p, π) *is a D-pair.*

Proof. We need to verify that conditions (i)–(iv) of Definition 9.5 are all fulfilled.

Condition (i) follows immediately from Lemma 9.2(b).

Concerning (ii), note that if $\gamma \in \Gamma_p$ then $p \in \text{Dom}(\sigma^\gamma) \cap \sigma^\pi$, i.e., by (9.9), $\pi \in dom(\gamma)$.

Next let $\gamma_1, \gamma_2 \in \Gamma$ and suppose that $\gamma_1(f) = \gamma_2(f) = 1$, where SINGLEVALUED$(f)$ occurs in P. We have $p \in \text{Dom}(\sigma^{\gamma_1}) \cap \text{Dom}(\sigma^{\gamma_2})$, that is there exist q_1 and q_2 such that $[p, q_1] \in \sigma^{\gamma_1}$ and $[p, q_2] \in \sigma^{\gamma_2}$. But $\sigma^{\gamma_1} \cup \sigma^{\gamma_2} \subseteq Mf$ and Mf is single-valued. Therefore q_1 must coincide with q_2, i.e. $[p, q_1] \in \sigma^{\gamma_1} \cap \sigma^{\gamma_2}$. As the σs are pairwise disjoint sets, it follows that $\gamma_1 = \gamma_2$, thus verifying condition (iii).

Finally, as regards (iv), let $d = Df$ be a D-clause in P such that $\pi(d) = 1$. Then $p \in \sigma^\pi \subseteq Md = \text{Dom}(Mf)$, i.e. $[p, q] \in Mf = \bigcup_{\gamma(f)=1} \sigma^\gamma$, for some q, i.e. $[p, q] \in \sigma^{\gamma_0}$ for some place γ_0 such that $\gamma_0(f) = 1$. In particular one has $p \in \text{Dom}(\sigma^{\gamma_0})$, which yields $\gamma_0 \in \Gamma_p$ and in turn completes the verification of condition (iv). ∎

Let $\alpha \in \Pi$ be such that $dom(\alpha) \neq \emptyset$. Hence there must exist $\pi \in \Pi$ such that $\text{Dom}(\sigma^\alpha) \cap \sigma^\pi \neq \emptyset$. Let $p \in \text{Dom}(\sigma^\alpha) \cap \sigma^\pi$. From the preceding lemma it follows that (Γ_p, π) is a D-pair, and since obviously $\alpha \in \Gamma_p$, we have the following necessary condition.

Condition C5. *Let* $\alpha \in \Pi$ *be such that* $dom(\alpha) \neq \emptyset$. *Then there exists a D-pair* (Γ, π) *such that* $\alpha \in \Gamma$. □

Definition 9.6 *For any two sets s, t, we write $s \in^+ t$ if there exists a finite chain of intermediate elements $s_1, s_2, \ldots, s_k$, with $k \geq 0$, such that $s \in s_1 \in s_2 \in \cdots \in s_k \in t$.* $\qquad\square$

Then for each variable x in P distinct from U we put

$$\Pi_x = \{\pi \in \Pi : Mx \in^+ \sigma^\pi\}. \tag{9.12}$$

By definition, for every $x \in V \setminus \{U\}$ we have $Mx \in \sigma^{\pi^x}$ (see the discussion following Remark 9.1). Hence we obtain another necessary condition.

Condition C6. *For all x in $V \setminus \{U\}$, $\pi^x \in \Pi_x$.* $\qquad\square$

Moreover, let $\alpha(x) = 1$ and $\pi \in \Pi_x$, where $x \in V \setminus \{U\}$. Therefore $\sigma^\alpha \subseteq Mx \in^+ \sigma^\pi$, which yields $rk\, \sigma^\alpha < rk\, \sigma^\pi$. By (9.7) the latter inequality implies $\alpha \lhd \pi$. Hence we have

Condition C7. *Let $\alpha(x) = 1$ and $\pi \in \Pi_x$ for some $x \in V \setminus \{U\}$. Then $\alpha \lhd \pi$.*

Notice that condition C2 is a consequence of conditions C6 and C7. $\qquad\square$

Recall that, by definition, $[Mx, My] \in \sigma^{\pi^{x,y}}$, for all $x, y \in V \setminus \{U\}$ (see the discussion just after Definition 9.4). Therefore $\pi^{x,y} \in \Pi_x \cap \Pi_y$. Also, as $Mx \in \text{Dom}(\sigma^{\pi^{x,y}}) \cap \pi^x$, we have $\pi^x \in dom(\pi^{x,y})$. Finally, notice that $[Mx, My] \in \sigma^{\pi^{x,y}}$ and $[My, Mx] \in \sigma^{\pi^{y,x}}$, for all $x, y \in V \setminus \{U\}$. Hence (9.10) yields $\pi^{y,x} \in inv(\pi^{x,y})$.

Summing up the preceding discussion, we obtain the following condition which is necessary for the injective satisfiability of P.

Condition C8. *For all $x, y \in V \setminus \{U\}$ we have:*

(i) $\pi^{x,y} \in \Pi_x \cap \Pi_y$;

(ii) $\pi^x \in dom(\pi^{x,y})$;

(iii) $\pi^{y,x} \in inv(\pi^{x,y})$. $\qquad\square$

Put, as in (9.11), $\Gamma_{Mx} = \{\gamma \in \Pi : Mx \in \text{Dom}(\sigma^\gamma)\}$, with $x \in V \setminus \{U\}$. Then clearly $\Gamma_{Mx} \subseteq \Pi_x$ and $\pi^{x,y} \in \Gamma_{Mx}$, for each $y \in V \setminus \{U\}$. Moreover, for each $\gamma \in \Gamma_{Mx}$, since $Mx \in \text{Dom}(\sigma^\gamma)$ then $[Mx, q] \in \sigma^\gamma$ for some q. But therefore by Lemma 9.2(c) and (9.10) $[q, Mx] \in \sigma^\delta$, for some $\delta \in inv(\gamma)$. In particular, since $Mx \in^+ \sigma^\delta$ we have also $\delta \in inv(\gamma) \cap \Pi_x$. Observe, in addition, that since $Mx \in \sigma^{\pi^x}$, from Lemma 9.3 it follows that (Γ_{Mx}, π^x) is a D-pair.

Hence we deduce the following necessary condition.

Condition C9. *Let $x \in V \setminus \{U\}$. Then there exists a D-pair (Γ, π^x) such that* *(i)* $\Gamma \subseteq \Pi_x$;

(ii) $\pi^{x,y} \in \Gamma$, *for each y in $V \setminus \{U\}$;*

(iii) $inv(\gamma) \cap \Pi_x \neq \emptyset$, *for each $\gamma \in \Gamma$.* $\qquad\square$

Next, let $\pi \in \Pi$. Since $\emptyset \neq \sigma^\pi$, Lemma 9.3 implies that there exists a D-pair (Γ, π).

If, moreover, $\pi \in \Pi_x$, for some $x \in V \setminus \{U\}$, then $Mx \in^+ \sigma^\pi$. Choose $p \in \sigma^\pi$ so that $Mx \in^+ \{p\}$. Then it follows again from Lemma 9.3 that (Γ_p, π) is a D-pair. In addition, let $\gamma \in \Gamma_p$. We have $p \in \text{Dom}(\sigma^\gamma)$, so that $Mx \in^+ \sigma^\gamma$, i.e. $\gamma \in \Pi_x$. Also, as $p \in \text{Dom}(\sigma^\gamma)$, $[p,q] \in \sigma^\gamma$ for some q. Lemma 9.2(c) implies that $[q,p] \in \sigma^\delta$ for some $\delta \in \Pi$. In fact, by (9.10), $\delta \in inv\,(\gamma)$ and since $Mx \in^+ \sigma^\delta$, we have also $\delta \in \Pi_x$.

The preceding discussion is summarized in the following condition.

Condition C10. *For each $\pi \in \Pi$ there exists a D-pair (Γ, π). In addition, for each $\pi \in \Pi_x$, with $x \in V \setminus \{U\}$, there exists a D-pair (Γ, π) such that $\Gamma \subseteq \Pi_x$ and, for each $\gamma \in \Gamma$, $inv(\gamma) \cap \Pi_x \neq \emptyset$.* $\qquad\square$

Clauses of type SINGLEVALUED(f) have already been taken into account within the definition of D-pairs. Another condition concerning such clauses, which also concludes our list of necessary conditions, is the following:

Condition C11. *If the clause SINGLEVALUED(f) appears in P, then for each $x, y, y' \in V \setminus \{U\}$ such that $\pi^{x,y}(f) = \pi^{x,y'}(f) = 1$ we have $y = y'$.* $\qquad\square$

To show that condition C11 is necessary together with conditions C1–C10 for the injective satisfiability of P we observe that if $\pi^{x,y}(f) = \pi^{x,y'}(f) = 1$, then $[Mx, My], [Mx, My'] \in Mf$. But Mf must be single-valued, hence $My = My'$. The equality $y = y'$ follows immediately from the injectivity of M.

The results proved in this section are summarized by the following lemma.

Lemma 9.4 (Soundness) *Let P be a conjunction of clauses of type $x = y \cup z$, $x = y \setminus z$, $x \in y$, $d = Df$, PAIR_IN(x, y, f), INV(f, g), SINGLEVALU-ED(f), containing also clauses (9.4). If P is injectively satisfiable, then there exist*

(i) *a set Π of places of P;*

(ii) *a function $x \mapsto \pi^x$ from $V \setminus \{U\}$ into Π such that π^x is a place at the variable x, for all $x \in V \setminus \{U\}$;*

(iii) *a function $(x, y) \mapsto \pi^{x,y}$ from $(V \setminus \{U\})^2$ into Π such that $\pi^{x,y}$ is a place at the pair (x, y), for all $x, y \in V \setminus \{U\}$;*

(iv) a linear ordering $\lhd$ of Π;

(v) two functions dom *and* inv *from Π into $Pow(\Pi)$;*

(vi) a function $x \mapsto \Pi_x$ from $V \setminus \{U\}$ into $Pow(\Pi) \setminus \{\emptyset\}$

such that conditions C1–C11 are satisfied. $\square$

Plainly, all conditions stated in the preceding lemma are effectively verifiable. Therefore, in order to establish our main result, to viz. that the class of unquantified set-theoretic formulae in the language $\mathcal{L}$ has a solvable decidability problem, it only remains to show that the conditions of Lemma 9.4 are also sufficient for the injective satisfiability of P.

This will be done in the next section.

9.3 Completeness of the decision algorithm

Again, let P be a conjunction of positive clauses of type (9.2) containing also clauses (9.4) and let V be the set of variables occurring in P. Furthermore, assume that there exist $\Pi = \{\pi_1, \pi_2, \ldots, \pi_n\}$, $x \mapsto \pi^x$, $(x,y) \mapsto \pi^{x,y}$, $\lhd$, $dom : \Pi \to Pow(\Pi)$, $inv : \Pi \to Pow(\Pi)$, $x \mapsto \Pi_x$ as from (i)–(vi) of Lemma 9.4 and satisfying conditions C1–C11 of the preceding section.

In this section we will prove that under such hypotheses P has an injective model. More specifically, we will show how to build an infinite, well-founded, quasi-extensional graph (cf. Definitions 9.1 and 9.2). This graph, called the *skeletal model* of P, will satisfy certain closure properties which depend on P and $\Pi, x \mapsto \pi^x, \ldots, x \mapsto \Pi_x$. Subsequently we will show that this graph admits a representation (cf. Definition 9.3) from which an injective model M^* of P can easily be extracted.

In the presentation of the below Construction Process, we will make use of the following definition.

Definition 9.7 *For any two sets $\Gamma_1, \Gamma_2 \subseteq \Pi$, we put*

$\Gamma_1 \prec \Gamma_2$ *if there exist $\gamma_2 \in \Gamma_2$ such that $\gamma_1 \lhd \gamma_2$, for all $\gamma_1 \in \Gamma_1$.*

In addition, for all $x, y \in V$, we put

$$x \prec y \quad \text{if} \quad \{\pi \in \Pi : \pi \subseteq x\} \prec \{\pi \in \Pi : \pi \subseteq y\}.$$

$\square$

Notice that the relation $\prec$ is acyclic and therefore extendible to a linear ordering, which throughout the present section will be denoted by the same symbol $\prec$.

In detail, the skeletal model is constructed by the following infinite process:

Algorithm 9.1 (Construction Process)

[INITIALIZATION PHASE]

Assume $\Pi = \{\pi_1, \ldots, \pi_n\}$, with $n = |\Pi|$. Let $N_0, N_1, \ldots, N_{n+3}$ be disjoint sets with cardinalities:

$$|N_0| = 2^n;$$
$$|N_i| = \omega, \text{ for } i = 1, \ldots, n;$$
$$|N_{n+1}| = (|V| - 1)^2;$$
$$|N_{n+2}| = |N_{n+3}| = \omega$$

(for definiteness one might put, e.g.,

$$N_0 = \{(p_0)^j : j = 1, \ldots, 2^n\},$$
$$N_1 = \{(p_1)^j : j \in \omega\},$$

etc., where p_i denotes the $(1 + i)$-th prime natural number). The members of $\mathcal{N} = \bigcup_{i=0}^{n+3} N_i$ will be called *nodes*. Arbitrarily choose bijections

$$\text{node} : Pow(\Pi) \to N_0,$$
$$\text{node}_2 : (V \setminus \{U\})^2 \to N_{n+1},$$

and denote for brevity

$$\text{node}(\pi) =_{\text{Def}} \text{node}(\{\pi\}) \text{ for all } \pi \text{ in } \Pi,$$
$$\text{node}(x) =_{\text{Def}} \text{node}(\{\pi \in \Pi : \pi(x) = 1\}).$$

Define:

$$N_I = \bigcup_{i=0}^{n+1} N_i,$$

$$
\begin{aligned}
E_I = \;& \{v \Rightarrow \text{node}(\Sigma) : i \in \{1, \ldots, n\}, v \in N_i, \Sigma \subseteq \Pi | \pi_i \in \Sigma\} \cup \\
& \{\text{node}(x) \Rightarrow \text{node}(\Sigma) : x \in V \setminus \{U\}, \Sigma \subseteq \Pi | \pi^x \in \Sigma\} \cup \\
& \{\text{node}_2(x, y) \Rightarrow \text{node}(\Sigma) : x, y \in V \setminus \{U\}, \Sigma \subseteq \Pi | \pi^{x,y} \in \Sigma\} \cup \\
& \{\text{node}(x) \overset{L}{\Rightarrow} \text{node}_2(x, y) : x, y \in V \setminus \{U\}\} \cup \\
& \{\text{node}(y) \overset{R}{\Rightarrow} \text{node}_2(x, y) : x, y \in V \setminus \{U\}\},
\end{aligned}
$$

$$G_I = (N_I, E_I).$$

Thus G_I is a graph (coarsely schematized by Figure 9.1), of which N_I and E_I are the set of nodes and the set of edges. Some edges (written in the form $v \Rightarrow w$) are untagged, while others (written in the form $v \overset{L}{\Rightarrow} w$ or in the form $v \overset{R}{\Rightarrow} w$) bear the tag 'L' or the tag 'R'. Notice that any edge leading from a node of the form $\text{node}(x)$ to the corresponding node $\text{node}_2(x, x)$ occurs twice in the graph G_I: once with tag 'L', once with tag 'R'. Notice also that every node in N_I is an endpoint of some untagged edge of G_I; more specifically:

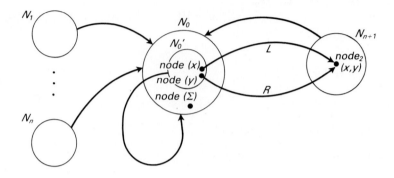

Fig. 9.1. Schematic picture of G_I

- exactly 2^{n-1} edges (among which the edge $v \Rightarrow \{\pi_i\}$) leave each $v \in N_i$, for $i = 1, \ldots, n$;

- infinitely many edges lead from inside $\bigcup_{i=1}^n N_i$ to node(Σ) for every $\Sigma \subseteq \Pi$, $\Sigma \neq \emptyset$, and hence

- infinitely many edges lead to each node, other than node(q_0), belonging to N_0 (cf. condition C1);

- exactly $(|V| - 1) \cdot 2^{n-1}$ edges lead from inside N_0 into N_0;

- exactly $2 \cdot (|V| - 1)$ tagged edges lead from each node in N_0 of the form node(x) into N_{n+1};

- every node v in N_{n+1} can be written in the form node$_2(x, y)$, so that exactly 2^{n-1} edges (among which the edge $v \Rightarrow \text{node}(\pi^{x,y})$) leave v, and exactly one edge tagged 'L' (respectively, tagged 'R') leads from inside N_0 to v.

Nodes in $N_{n+2} \cup N_{n+3}$ have not been used so far; as a matter of fact, these nodes have been introduced in sight of the subsequent phases (Stabilization Phase and Completion Phase).

The Initialization Phase ends with the assignment of proper initial values to a few program variables. For clarity, we decompose N_0 as the disjoint union $N_0 = N_0' \cup N_0''$, where

$$N_0' = \{\text{node}(x) : x \in V \setminus \{U\}\},$$

and set:

(I.1) $E \leftarrow E_I$;

(I.2) $L_0 \leftarrow [v : v \in \bigcup\limits_{i=1}^{n} N_i]$;

(I.3) $L_1 \leftarrow [v : v \in N_0' \cup N_{n+1}]$;

(I.4) $S_0 \leftarrow [v : v \in N_{n+2}]$;

(I.5) $S_1 \leftarrow [v : v \in N_{n+3}]$;

(I.6) $m \leftarrow \{[\text{node}(x), \{\pi^{x,y} : y \in V \setminus \{U\}\}] : x \in V \setminus \{U\}\} \cup$
$\{[v, \emptyset] : v \in \bigcup\limits_{i=1}^{n+2} N_i\}$;

so that $m(v)$ is initially non-empty only for nodes $v \in N_0'$ (indeed, it follows from $v = \text{node}(x)$ that $\pi^{x,q_0} \in m(v)$). In particular, L_0 results from imposing an arbitrary well-ordering of ordinal ω to $\bigcup_{i=1}^{n} N_i$. In a similar manner, S_0 and S_1 result from N_{n+2} and from N_{n+3} respectively. The invariants

$N = \{ v : v$ coincides with the tail x or with the head y of some
(tagged or untagged) edge $x \Rightarrow y$ in $E \}$,

$G = (N, E)$,

will be tacitly maintained throughout the subsequent phases of this Construction Process; in our intended meaning, G, N and E are the skeletal model, its set of nodes and its edges.

Let us to begin to clarify the purpose of the above initialization. The Construction Process that we are describing is to force certain closure properties, which will be discussed at length below, on the skeletal model G. To this end, each node v for which the edge $v \Rightarrow \text{node}(\pi_i)$ is in E, for some $i = 1, 2, \ldots, n$, will be processed suitably. Precisely in order to set up a convenient data structure for this processing, all nodes in $\bigcup_{i=1}^{n} N_i$ have been arranged in the ordered list L_0. Also, we will maintain a second ordered list L_1, of finite length, into which each new node produced during the processing of old nodes will be inserted. The '*mark*' function m from nodes of L_0 and L_1 into subsets of Π will also have to be maintained: as a matter of fact, $m(v)$, which initially equals $\emptyset$, will eventually become a singleton set, for each v in $\bigcup_{i=1}^{n} N_i$.

[END INITIALIZATION PHASE]

[STABILIZATION PHASE]

(S.1) $r \leftarrow 0$;

(S.2) **DO FOREVER**

    (S.3) $r \leftarrow 1 - r$;

    (S.4) Let v be the node contained in the first location of L_r.

    (S.5) Discard v from the beginning of L_r.

    (S.6) Let π be the unique place in Π such that $v \Rightarrow \text{node}(\pi)$ is in E.

    (S.7) Let $T_v = \{v\} \cup \{w \in N : \text{there is a path leading from } w \text{ to } v\}$.

    (S.8) **CASE** $T_v \cap \{\text{node}(x) : x \in V \setminus \{U\}\} = \emptyset$:

      (S.9) Let (Γ, π) be a D-pair such that $m(v) \subseteq \Gamma$.

      (S.10) **FOR EACH** $\gamma \in \Gamma \setminus m(v)$ **DO**

        (S.11) Let $\iota_\gamma \in inv(\gamma)$, and let (I_γ, β_γ) be a D-pair such that $\iota_\gamma \in I_\gamma$.

        (S.12) *Stabilize*$(v, \gamma, \iota_\gamma, \beta_\gamma)$

        [Note: The code for the procedure *Stabilize* is shown below.]

      END FOR EACH.

    (S.13) **CASE** $T_v \cap \{\text{node}(x) : x \in V \setminus \{U\}\} \neq \emptyset$:

      (S.14) Let x_0 be a $\prec$-maximal variable such that $\text{node}(x_0) \in T_v$ (cf. Definition 9.7).

      (S.15) Notice that $\pi \in \Pi_{x_0}$.

      (S.16) Let (Γ, π) be a D-pair such that $m(v) \subseteq \Gamma$, $\Gamma \subseteq \Pi_{x_0}$, and for each $\gamma \in \Gamma$ $inv(\gamma) \cap \Pi_{x_0} \neq \emptyset$.

      (S.17) **FOR EACH** $\gamma \in \Gamma \setminus m(v)$ **DO**

        (S.18) Let $\iota_\gamma \in inv(\gamma) \cap \Pi_{x_0}$, and let (I_γ, β_γ) be a D-pair such that $\iota_\gamma \in I_\gamma$.

        (S.19) *Stabilize*$(v, \gamma, \iota_\gamma, \beta_\gamma)$.

      END FOR EACH.

    END CASE.

END DO FOREVER.

PROCEDURE *Stabilize*(v, λ, μ, ν)

[Note: When *Stabilize*(v, λ, μ, ν) is called, v is a node in N and λ, μ, ν are places of P in Π.]

(PS.1) Let w be the first node in the list L_0 such that $m(w) = \emptyset$ and $w \Rightarrow \text{node}(\nu)$ is in E.

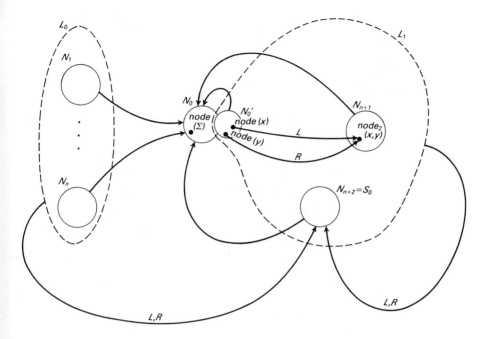

Fig. 9.2. Schematic picture of G_S

(PS.2) Let $u_{v,w}$ and $u_{w,v}$ be the nodes contained in the first two locations of S_0. Discard them from the beginning of S_0.

Put

(PS.3) $E \leftarrow E \cup \{v \overset{L}{\Rightarrow} u_{v,w}, w \overset{R}{\Rightarrow} u_{v,w}, w \overset{L}{\Rightarrow} u_{w,v}, v \overset{R}{\Rightarrow} u_{w,v}\}.$

[Note: Thus the new nodes $u_{v,w}$, $u_{w,v}$ have been implicitly added to N.]

(PS.4) Add $u_{v,w}$ and $u_{w,v}$ at the end of the list L_1.

(PS.5) $E \leftarrow E \cup \{u_{v,w} \Rightarrow \text{node}(\Sigma) : \Sigma \subseteq \Pi | \lambda \in \Sigma\};$

(PS.6) $E \leftarrow E \cup \{u_{w,v} \Rightarrow \text{node}(\Sigma) : \Sigma \subseteq \Pi | \mu \in \Sigma\};$

(PS.7) $m(w) \leftarrow \{\mu\};$

END PROCEDURE *Stabilize.*

[END STABILIZATION PHASE]

[COMPLETION PHASE]

(C.1) Arrange all startpoints (or 'tails') of tagged edges contained in E in an ordered list L_2 such that node(q_0) is the first element in L_2.

(C.2) **FOR EACH** v in L_2 **DO**

(C.3) Let v', v_L, and v_R be the nodes contained in the first three locations of S_1. Discard them from S_1.

[Notice: In the following $\underline{0}$ and $\underline{2}$ will stand respectively for node(q_0) and (node(q_0))$_L$.]

Put

(C.4) $E \leftarrow E \cup \{v \Rightarrow v', v' \Rightarrow v_L, \underline{0} \Rightarrow v_L, v' \Rightarrow v_R, \underline{2} \Rightarrow v_R\}$.

[Note: Thus the new nodes $\{v', v_L, v_R\}$ have been implicitly added to N.]

END FOR EACH.

(C.5) **FOR EACH** tagged edge $v \overset{L}{\Rightarrow} w$ (respectively, $v \overset{R}{\Rightarrow} w$) in E **DO**

(C.6) Substitute in E the tagged edge $v \overset{L}{\Rightarrow} w$ (respectively, $v \overset{R}{\Rightarrow} w$) with the untagged edge $v_L \Rightarrow w$ (respectively, $v_R \Rightarrow w$).

END FOR EACH.

[END COMPLETION PHASE]

END CONSTRUCTION PROCESS. □

Remark 9.2 In what follows, we will denote by $G_I = (N_I, E_I)$, $G_S = (N_S, E_S)$, and $G_C = (N_C, E_C)$ the skeletal model as constructed at the end of the Initialization Phase, Stabilization Phase, and Completion Phase of the above Construction Process (see Figures 9.1,2,3). □

The Construction Process (CP for brevity) contains some implicit assertions (cf. for example lines (S.4), (S.6), (S.9), (S.11), (S.14), (S.15), (S.16), (S.18), (PS.1), etc.). Below we will show that under the hypotheses stated at the beginning of the present section, all assertions are true in any computation K of the CP. Subsequently we will prove that the graph $G_C = (N_C, E_C)$ is well-founded and quasi-extensional (cf. Definitions 9.1 and 9.2). Also we will show that G_C enjoys several closure properties which will allow us to define a suitable representation $Repr$ of G_C for which the assignment $M^*x = Repr(\text{node}(x))$, for each x occurring in our conjunction P, is injective and satisfies P.

Let K be a computation of the CP above. We begin by stating some simple properties of the skeletal model G_C:

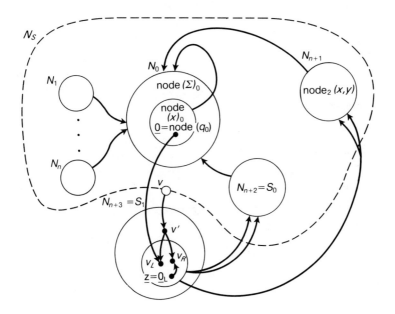

Fig. 9.3. Schematic picture of G_C

Lemma 9.5 *(a) Let v be a node such that $v \Rightarrow \text{node}(\pi)$ is in E_C for some $\pi \in \Pi$. Then, during the computation K, v is inserted either in L_0 or in L_1, and subsequently processed by the Stabilization loop (S.2).*

(b) Conversely, if during the computation K a node is introduced in one of the lists L_0 and L_1, then there is a unique place $\pi \in \Pi$ such that the edge $v \Rightarrow \text{node}(\pi)$ is in E_C (for specificity, we will indicate this uniquely characterized place by ${}^v\pi$).

(c) No $u \in N_1 \cup N_2 \cup \cdots \cup N_n$ has any predecessors.

(d) $\text{node}(q_0)$ has no predecessors.

(e) For all $\Sigma \subseteq \Pi$ such that $\text{node}(\Sigma) \notin \{\text{node }(x) : x \in V \setminus \{U\}\}$, $\text{node}(\Sigma)$ has no outgoing edges.

(f) After execution of (C.1), the elements of the list L_2 are exactly all nodes v such that $v \Rightarrow \text{node}(\pi)$ is in E_C for some $\pi \in \Pi$.

Proof. Concerning (d), it is enough to notice that by (9.4) and Definition 8.1, $\{\pi \in \Pi : \pi(q_0) = 1\} = \emptyset$. The remaining points of the lemma can be easily proved by inspecting the code of the Construction Process. ∎

Lemma 9.5(b) implies that assertion (S.6) is satisfied. By inducting on K we will show that assertions (S.4), (S.9), (S.11), (S.15), (S.16), and (S.18) are also satisfied.

Notice that the assertion $L_r \neq [\ ]$ implicit in (S.4) holds the $(2 \cdot i)$-th (respectively, the first) time the body of (S.2) is executed, because then $L_r = L_0$ has infinitely many components (respectively, $L_r = L_1$ has $\text{node}(q_0)$ as one of its components). Furthermore, (S.4) will continue to hold if some node v with $m(v) = \emptyset$ is ever placed in L_1. In fact, even if this v is removed afterwards —clearly at (S.5)— from L_1, short later *Stabilize* will be invoked from (S.13) or (S.20), because $\Gamma \setminus m(v) \neq \emptyset$ at line (S.11) or (S.18), by Definition 9.5(i). *Stabilize*, in turn, will cause new nodes u with $m(u) = \emptyset$ to be put into L_1, so that L_1 can never become empty during the Stabilization Phase. In order to validate (S.4) it will hence suffice to prove that *Stabilize* is invoked at least once. As a matter of fact, unless *Stabilize* is invoked during the first execution of the body of (S.2), $m(v)$ will be $\emptyset$, and consequently *Stabilize* will be invoked, during the second execution of the body of (S.2). Incidentally, this argument shows that *Stabilize* is invoked infinitely many times by the Construction Process, so that $N_C \supseteq N_{n+2}$ and, as a consequence, $N_C \supseteq N_{n+3}$, $N_C = \bigcup_{i=0}^{n+3} N_i$.

Let now v be a node selected at line (S.4), let ${}^v\pi$ be the unique place in Π such that $v \Rightarrow \text{node}({}^v\pi)$ is in E_C, and assume that at the time (S.8) is executed, $T_v \cap \{\text{node}(x) : x \in V \setminus \{U\}\} = \emptyset$, i.e. v has no ancestor, nor is it of type $\text{node}(x)$ with $x \in V \setminus \{U\}$. Since, in particular, $v \neq \text{node}(x)$ for any $x \in V \setminus \{U\}$, it follows that either $m(v) = \emptyset$, or the value $m(v)$ has been set during the execution of line (PS.7) within a call $Stabilize(v', \lambda, \mu, \nu)$. In the latter case, by induction, $\mu \in inv(\lambda)$ and $\mu \in I_\mu$ for some D-pair (I_μ, ν). Hence (PS.7) yields $m(v) = \{\mu\} \subseteq I_\mu$. Moreover, since by (PS.1) $\nu = {}^v\pi$, it follows that the D-pair (I_μ, ν) satisfies all requirements in (S.9). If, on the other hand, $m(v) = \emptyset$, then the validity of (S.9) follows from condition C10.

So, let $(\Gamma, {}^v\pi)$ be a D-pair such that $m(v) \subseteq \Gamma$ (as from (S.9)) and let $\gamma \in \Gamma \setminus m(v)$ (as from (S.10)). Definition 9.5(ii) of a D-pair implies that $dom(\gamma) \neq \emptyset$. Hence, conditions C3(i), C4(i) and (ii), and C5 yield that $inv(\gamma) \neq \emptyset$, and that for each $\iota_\gamma \in inv(\gamma)$ there exists a D-pair (I_γ, β_γ) such that $\iota_\gamma \in I_\gamma$, thus completing the verification of assertion (S.11).

Next suppose that at the time v was selected from L_0 or L_1, $T_v \cap \{\mathrm{node}(x) : x \in V \setminus \{U\}\} \neq \emptyset$. We will distinguish two cases, according to whether $v = \mathrm{node}(x)$, for some $x \in V \setminus \{U\}$, or not.

Case: $v = \mathrm{node}(x)$, with $x \in V \setminus \{U\}$. Let x_0 be the $\prec$-maximal variable such that $\mathrm{node}(x_0) \in T_v$. Then $x_0 = x$. In fact, let u be an immediate predecessor of v. If $u \in N_1 \cup N_2 \cup \cdots \cup N_n$, then $T_u \cap \{\mathrm{node}(y) : y \in V \setminus \{U\}\} = \emptyset$. On the other hand, if $u \in S_0$, then the edge connecting u to v was inserted into E during the execution of a call $Stabilize(v', \lambda, \mu, \nu)$. Plainly, then

$$T_u \cap \{\mathrm{node}(y) : y \in V \setminus \{U\}\} = T_{v'} \cap \{\mathrm{node}(y) : y \in V \setminus \{U\}\}. \qquad (9.13)$$

It is then enough to show that if $\mathrm{node}(y) \in T_u$ for some $y \in V \setminus \{U\}$ and $u \Rightarrow v$ is in E_C, then $y \prec x$. Let $u \Rightarrow v$ and let y_0 be the $\prec$-maximal variable such that $\mathrm{node}(y_0) \in T_u$. From (9.13), $\mathrm{node}(y_0) \in T_{v'}$ and y_0 is also $\prec$-maximal among the variables y such that $\mathrm{node}(y) \in T_{v'}$. Therefore, by induction and by (S.15)–(S.18), it follows that $\lambda, \mu \in \Pi_{y_0}$ and $\mathrm{node}(x) \in \{\mathrm{node}(\lambda), \mathrm{node}(\mu)\}$. This in particular yields $|\{\alpha \in \Pi : \alpha(x) = 1\}| = 1$ and the unique place α such that $\alpha(x) = 1$ is either λ or μ. If $\pi(y_0) = 1$, then, by condition C7, $\pi \lhd \lambda$ and $\pi \lhd \mu$. Hence $y_0 \prec x$ (cf. Definition 9.7), thereby showing that under the hypothesis that $v = \mathrm{node}(x)$, if $\mathrm{node}(y) \in T_v$, for some $y \in V \setminus \{U\}$, then $y \preceq x$.

Since the untagged edges of G_I are edges of G_C too, $\mathrm{node}(x) \Rightarrow \mathrm{node}(\pi^x)$ is in E_C. Therefore in the present case ${}^v\pi = \pi^x$ which, by condition C6, is a member of Π_x. Thus assertion (S.15) is completely verified. Since $v = \mathrm{node}(x)$, line (I.6) of the Construction Process implies that $m(v) = \{\pi^{x,y} \in \Pi : y \in V \setminus \{U\}\}$. Condition C9 guarantees that there exists a D-pair (Γ, π) such that $m(v) \subseteq \Gamma \subseteq \Pi_x$ and such that $inv(\gamma) \cap \Pi_x \neq \emptyset$, for every $\gamma \in \Gamma$. Hence, plainly, assertions (S.16) and (S.18) hold, at least in the case in which $v = \mathrm{node}(x)$, for some $x \in V \setminus \{U\}$.

Case: $v \notin \{\mathrm{node}(x) : x \in V \setminus \{U\}\}$. In this case the node v has been inserted into N during the execution of a call $Stabilize(v', \lambda, \mu, \nu)$, and since obviously v has at least one predecessor, from (PS.1), (PS.4), and (I.6) it follows that $m(v) = \emptyset$. Notice that since $v \neq \mathrm{node}(x)$, for all $x \in V \setminus \{U\}$, then $T_v \cap \{\mathrm{node}(x) : x \in V \setminus \{U\}\} = T_{v'} \cap \{\mathrm{node}(x) : x \in V \setminus \{U\}\}$. Therefore, if x_0 is the $\prec$-maximal variable in $V \setminus \{U\}$ such that $\mathrm{node}(x_0) \in T_v$, then as in the previous case, by inductive hypothesis we can prove that $\lambda, \mu \in \Pi_{x_0}$ and ${}^v\pi \in \{\lambda, \mu\}$, i.e. (S.15) holds. Also, much as in the preceding case, it can be proved that assertions (S.16) and (S.18) are true.

Finally, we conclude the verification that all instructions in the Construction Process are executable by observing that at initialization the sets S_0, S_1 are infinite, and that for each $\pi \in \Pi$ the list L_0 contains infinitely many nodes w such that $m(w) = \emptyset$ and $w \Rightarrow \text{node}(\pi)$ is in E_C. Hence, in particular, instructions (PS.1), (PS.2), (C.1) and (C.3) are always executable, when encountered.

The above discussion can be summarized as follows:

Lemma 9.6 *Under the hypotheses stated at the beginning of the present section, each computation of the Construction Process is successful, in the sense that all assertions (respectively, instructions) encountered are valid (respectively, executable).* □

Next we prove that the graph $G_C = (N_C, E_C)$ is well-founded.

We will begin by proving that $G_I = (N_I, E_I)$ is well-founded. Observe that N_I consists of $N_1 \cup N_2 \cup \cdots \cup N_n$ plus a finite number of additional nodes. Therefore, if G_I were not well-founded, by Lemma 9.5(c) it would contain a finite cycle

$$v_0 \Rightarrow v_1 \Rightarrow \cdots \Rightarrow v_{k-1} \Rightarrow v_0 \,,$$

where the v_is are either nodes of type $\text{node}(\Sigma)$, with $\Sigma \subseteq \Pi$, or nodes of type $\text{node}_2(x, y)$ (cf. Figure 9.1). If $v_i = \text{node}_2(x, y)$, $i \in \{0, 1, \ldots, k-1\}$, then $v_{i-1} \in \{\text{node}(x), \text{node}(y)\}$ and $v_{i+1} = \text{node}(\Sigma_{x,y})$, for some $\Sigma_{x,y} \subseteq \Pi$ such that $\pi^{x,y} \in \Sigma_{x,y}$ (($i-1$) and ($i+1$) are to be taken modulo k). For specificity, suppose that $v_{i-1} = \text{node}(x) = \text{node}(\{\pi \in \Pi : \pi(x) = 1\})$. Then, from conditions C8(i) and C7 it follows that for each $\pi \subseteq x$, $\pi \lhd \pi^{x,y}$. On the other hand, if v_i is of type $\text{node}(x)$, for some $x \in V \setminus \{U\}$, then $v_{i+1} = \text{node}(\Sigma_x)$, for some $\Sigma_x \subseteq \Pi$ such that $\pi^x \in \Sigma_x$. By C2, then, for every $\pi \subseteq x$ we have $\pi \lhd \pi^x$. In any case we have shown that by deleting from the cycle $v_0 \Rightarrow v_1 \Rightarrow \cdots \Rightarrow v_{k-1} \Rightarrow v_0$ all nodes of type $\text{node}_2(x, y)$, we obtain a subcycle of $\Rightarrow^*$ (i.e. the transitive closure of $\Rightarrow$), $v_{i_0} \Rightarrow^* v_{i_1} \Rightarrow^* \cdots \Rightarrow^* v_{i_{k'-1}} \Rightarrow^* v_{i_0}$, such that $v_{i_j} = \text{node}(\Sigma_j)$, $j = 0, 1, \ldots, k'-1$, and for each $j = 0, 1, \ldots, k'-1$ there is an $\alpha_j \in \Sigma_j$ such that for all $\pi \in \Sigma_{j-1}$ we have $\pi \lhd \alpha_j$. In particular we would have $\alpha_0 \lhd \alpha_1 \lhd \cdots \lhd \alpha_{k'-1} \lhd \alpha_0$, which is a contradiction. Thus the graph $G_I = (N_I, E_I)$ is well-founded.

During the Stabilization Phase, new edges can be added to G only by the procedure *Stabilize*. Notice that each call to *Stabilize* can cause only finitely many new edges to be added to G. Therefore to show that the Stabilization Phase preserves the well-foundedness of G, it is enough to prove that the acyclicity of G cannot be disrupted by calls to the procedure *Stabilize*.

So assume that G is acyclic prior to the execution of a call $Stabilize(v, \lambda, \mu, \nu)$. The effect on G of the call $Stabilize(v, \lambda, \mu, \nu)$ is that two new nodes, $u_{v,w}$ and $u_{w,v}$, are added to G together with the edges $v \overset{L}{\Rightarrow} u_{v,w}$, $w \overset{R}{\Rightarrow} u_{v,w}$, $v \overset{R}{\Rightarrow} u_{w,v}$, $w \overset{L}{\Rightarrow} u_{w,v}$, $u_{v,w} \Rightarrow \text{node}(\Sigma_\lambda)$, for all $\Sigma_\lambda \subseteq \Pi$ such that $\lambda \in \Sigma_\lambda$, and $u_{w,v} \Rightarrow \text{node}(\Sigma_\mu)$, for all $\Sigma_\mu \subseteq \Pi$ such that $\mu \in \Sigma_\mu$ (observe that w is a node with no incoming edges and $w \Rightarrow \text{node}(\nu)$ is in E; cf. (PS.1), Lemma 9.5(b),(c) and (I.2)). Plainly, the only way a cycle could be introduced into G is that before the call $Stabilize(v, \lambda, \mu, \nu)$ is made, there was a path leading from $\text{node}(\Sigma^*)$ to v, for some $\Sigma^* \subseteq \Pi$ such that $\Sigma^* \cap \{\lambda, \mu\} \neq \emptyset$. But then, by Lemma 9.5(e), $\text{node}(\Sigma^*) = \text{node}(x^*)$, for some $x^* \in V \setminus \{U\}$. Hence we would have $T_v \cap \{\text{node}(x) : x \in V \setminus \{U\}\} \neq \emptyset$, thus precluding, by (S.8), the possibility that the call $Stabilize(v, \lambda, \mu, \nu)$ can be made from (S.12). But even the assumption that the call $Stabilize(v, \lambda, \mu, \nu)$ is made from (S.19) is contradictory. Indeed, by letting x_0 be the $\prec$-maximal variable such that $\text{node}(x_0) \in T_v$, from (S.14), (S.16), (S.18) it would follow that $\lambda, \mu \in \Pi_{x_0}$. Thus, by condition C7, we would have $\pi \triangleleft \lambda$ and $\pi \triangleleft \mu$, for each $\pi \subseteq x_0$, which, recalling that $\Sigma^* \cap \{\lambda, \mu\} \neq \emptyset$, in turn would imply $x_0 \prec x^*$. This, plainly, contradicts the $\prec$-maximality of x_0, and consequently proves that no cycle can be introduced by the call $Stabilize(v, \lambda, \mu, \nu)$.

Hence, at the end of the Stabilization Phase the graph G is still well-founded.

Finally, by observing that, essentially, the effect of the Completion Phase on G is to substitute tagged edges $v \overset{L}{\Rightarrow} w$ (respectively, $v \overset{R}{\Rightarrow} w$) by untagged paths $v \Rightarrow v' \Rightarrow v_L \Rightarrow w$ (respectively, $v \Rightarrow v' \Rightarrow v_R \Rightarrow w$) of length 3, it follows plainly that the well-foundedness of G is also preserved during the Completion Phase.

Hence we have:

Lemma 9.7 *The graph $G_C = (N_C, E_C)$ produced by the computation K is well-founded.* $\square$

In order to prove that the graph $G_C = (N_C, I_C)$ is also quasi-extensional, we partition the set of nodes N_C as follows. In view of Lemma 9.5(f), put

$$
\begin{aligned}
Cl_1 &= \{v' : v \Rightarrow \text{node}(\pi), \text{ for some } \pi \in \Pi\}, \\
Cl_2 &= \{v_L : v \Rightarrow \text{node}(\pi), \text{ for some } \pi \in \Pi\}, \\
Cl_3 &= \{v_R : v \Rightarrow \text{node}(\pi), \text{ for some } \pi \in \Pi\}, \\
Cl_4 &= N_0 = \{\text{node}(\Sigma) : \Sigma \subseteq \Pi\}, \\
Cl_5 &= N_1 \cup N_2 \cup \cdots \cup N_n, \\
Cl_6 &= N_C \setminus \bigcup_{i=1}^{5} Cl_i = N_{n+1} \cup N_{n+2},
\end{aligned}
$$

so that $Cl_1 \cup Cl_2 \cup Cl_3 = N_{n+3}$, where v', v_L, v_R are the nodes introduced during the FOR loop (C.2) of the Completion Phase, and $N_0, N_1, \ldots, N_{n+3}$ are the sets of nodes defined in the Initialization Phase.

The following lemma, which can be proved by a direct inspection of the Construction Process code (cf. also Figures 9.2,3, Lemma 9.5(a),(b), and the remarks made inside the Initialization Phase), lists some useful facts concerning the above partition.

Lemma 9.8   (a) *The functions $v \mapsto v', v \mapsto v_L$, and $v \mapsto v_R$, from $\{v \in N_C : v \Rightarrow \mathrm{node}(\pi)$, for some $\pi \in \Pi\}$ into Cl_1, Cl_2, and Cl_3, respectively, defined during the FOR-loop (C.2) of the CP Completion Phase, are injective.*

(b) *For each v' in Cl_1, v is the unique immediate predecessor of v'.*

(c) *For each v_L in Cl_2, v' and $\underline{0}$ are the only immediate predecessors of v_L.*

(d) *For each v_R in Cl_3, v' and $\underline{2}$ are the only immediate predecessors of v_R.*

(e) *For each $\Sigma \subseteq \Pi$,*

  (e_1) *if $\Sigma = \emptyset$, then $\mathrm{node}(\Sigma) = \mathrm{node}(q_0)$ has no immediate predecessor;*

  (e_2) *if $\Sigma \neq \emptyset$, then $\mathrm{node}(\Sigma)$ has infinitely many predecessors.*

(f) *For each u in Cl_6 there exist two uniquely determined nodes v and w such that v_L and w_R are the only immediate predecessors of u. To stress on this dependence, we will index the node u with the pair v, w (thus agreeing with the notation used in (PS.2), which is here extended so that $u_{x,y} = u_{\mathrm{node}(x),\mathrm{node}(y)} = \mathrm{node}_2(x, y)$ for every member $\mathrm{node}_2(x, y)$ of N_{n+1}).*

(g) *The partial function $(v, w) \mapsto u_{v,w}$ defined in (f) above is injective.*   □

An immediate consequence of the preceding lemma is that for all $v_1, v_2 \in Cl_i, i \neq 4$, if $\{u \in N_C : u \Rightarrow v_1$ is in $E_C\} = \{u \in N_C : u \Rightarrow v_2$ is in $E_C\} \neq \emptyset$, then $v_1 = v_2$. On the other hand, if $v_1, v_2 \in Cl_4$, then $v_1 = \mathrm{node}(\Sigma_1)$ and $v_2 = \mathrm{node}(\Sigma_2)$ for some $\Sigma_1, \Sigma_2 \subseteq \Pi$. If $v_1 \neq v_2$, then $\Sigma_1 \neq \Sigma_2$, so that there exist $\pi_j \in (\Sigma_1 \setminus \Sigma_2) \cup (\Sigma_2 \setminus \Sigma_1)$. Let $u^* \in N_j$. Therefore, from the definition of E_I in the Initialization Phase, we obtain that $u^* \Rightarrow v_1$ is in E_C if and only if $u^* \Rightarrow v_2$ is not in E_C, thus showing that $\{u \in N_C : u \Rightarrow v_1$ is in $E_C\} \neq \{u \in N_C : u \Rightarrow v_2$ is in $E_C\}$.

To complete the proof of the quasi-extensionality of G_C, it only remains to show that if $v_1 \in Cl_i$ and $v_2 \in Cl_j$, with $i \neq j$ and $i, j \neq 5$, then there is a node u such that the edge $u \Rightarrow v_1$ is in E_C if and only if the edge $u \Rightarrow v_2$ is not in E_C. Since every node in Cl_1 has only one immediate predecessor, every node in $Cl_2 \cup Cl_3 \cap Cl_6$ has only two immediate predecessors, whereas every node in $Cl_4 \setminus \{\mathrm{node}(q_0)\}$ has infinitely many immediate predecessors, we can further limit ourselves to verify the above property only for $i, j \in \{2, 3, 6\}$.

If $v_1 \in Cl_2 \cap Cl_3$ and $v_2 \in Cl_6$, then either $v_1 = w_L$ or $v_1 = w_R$, for some node w such that the edge $w \Rightarrow \text{node}(\pi)$ is in E_C, with $\pi \in \Pi$. But, in any case, by (b) and (c) of Lemma 9.8, $w' \Rightarrow v_1$ is in E_C, with $w' \in Cl_1$, whereas $w' \Rightarrow v_2$ is not in E_C, since by (f) of the same lemma the immediate predecessors of v_2 are contained in $Cl_2 \cup Cl_3$. On the other hand, let $v_L \in Cl_2$ and $w_R \in Cl_3$. Points (c) and (d) of Lemma 9.8 imply respectively that $\underline{0} \Rightarrow v_L$ and $\{u \in N_C : u \Rightarrow w_R \text{ is in } E_C\} \subseteq Cl_1 \cup Cl_2$. Hence $\underline{0} \Rightarrow w_R$ is not in E_C, since $\underline{0} \in Cl_4$. This concludes the proof that the graph G_C is quasi-extensional. Thus we have:

Lemma 9.9 *At the termination of the Construction Process, the graph $G_C = (N_C, E_C)$ is quasi-extensional.* □

The following lemma collects some closure properties enjoyed by the graph G_C.

Lemma 9.10 (a) *For each $x \in V \setminus \{U\}$, $\text{node}(x) \Rightarrow \text{node}(\pi^x)$ is in E_C.*

(b) *For each $x, y \in V \setminus \{U\}$, there exists a unique node $u_{x,y} \in Cl_6$ such that $(\text{node}(x))_L \Rightarrow u_{x,y}$, $(\text{node}(y))_R \Rightarrow u_{x,y}$, and $u_{x,y} \Rightarrow \text{node}(\pi^{x,y})$ are in E_C.*
(Note: $u_{x,y}$ is an abbreviation for $u_{\text{node}(x),\, \text{node}(y)}$; cf. Lemma 9.8(f).)

(c) *Let v be a node such that $v \Rightarrow \text{node}(^v\pi)$ is in E_C, with $^v\pi \in \Pi$. Then there exists a D-pair $(\Gamma, {}^v\pi)$ such that for each $\gamma \in \Gamma$, there is a node w for which $u_{v,w} \Rightarrow \text{node}(\gamma)$ is in E_C, and, vice versa, if for some node w and place $\pi \in \Pi$ the edge $u_{v,w} \Rightarrow \text{node}(\pi)$ is in E_C, then $\pi \in \Gamma$.*

(d) *Let $u_{v,w_1} \Rightarrow \text{node}(\pi)$, $u_{v,w_2} \Rightarrow \text{node}(\pi)$ be in E_C, for some $v, w_1, w_2 \in N_C$, $\pi \in \Pi$ such that $w_1 \neq w_2$. Then $v, w_1, w_2 \in \{\text{node}(x) : x \in V \setminus \{U\}\}$.*

(e) *If $u_{v,w} \Rightarrow \text{node}(\alpha)$ is in E_C, for some $u_{v,w} \in Cl_6$ and $\alpha \in \Pi$, then $u_{w,v} \in Cl_6$ and $u_{w,v} \Rightarrow \text{node}(\beta)$ is in E_C for some $\beta \in \text{inv}(\alpha)$.*

(f) *For all $v \in N_C$ and $x \in V$, $v \Rightarrow \text{node}(x)$ is in E_C if and only if $v \Rightarrow \text{node}(\pi)$ is in E_C for some $\pi \subseteq x$.*

Proof. (a) and (b) are easily verified by inspection of the CP code.

Concerning (c), let v be a node in N_C such that $v \Rightarrow \text{node}(^v\pi)$ is in E_C. Lemma 9.5(a) implies that v is put either in L_0 or in L_1 during the computation K. Hence eventually it will be selected during the Stabilization Phase at line (S.4). From (S.9), (S.10) and (S.16), (S.17), it follows that there is a D-pair $(\Gamma^*, {}^v\pi)$ such that, for each $\gamma \in \Gamma^* \setminus m(v)$, $v_L \Rightarrow u_{v,w}$ and $u_{v,w} \Rightarrow \text{node}(\gamma)$ are in E_C, for some node w. On the other hand, if $\gamma \in m(v)$, then the same conclusion follows either from (I.1) and (I.6) or from (PS.3), (PS.6) and (PS.7), according to whether v is of type $\text{node}(x)$, $x \in V \setminus \{U\}$, or v is the element selected at (PS.1) during the execution of a call to the procedure *Stabilize*.

Conversely, if for some node w and some place $\pi^* \in \Pi$ the edge $u_{v,w} \Rightarrow$ node(π^*) is in E_C, then $v \overset{L}{\Rightarrow} u_{v,w}$ is in E_S. If $v \overset{L}{\Rightarrow} u_{v,w}$ is introduced into E_S during the Initialization Phase, then $v = \text{node}(x)$, $w = \text{node}(y)$, for some $x, y \in V \setminus \{U\}$, and $\pi^* = \pi^{x,y} \in m(\text{node}(x))$ (cf. (I.6)). Therefore, by (S.16), $m(v) \subseteq \Gamma^*$, and in particular $\pi^* \in \Gamma^*$. On the other hand, if $v \overset{L}{\Rightarrow} u_{v,w}$ is introduced during the execution of a call $Stabilize(v, \pi^*, \mu, \nu)$, then by (S.9) or (S.16) $\pi^* \in \Gamma^*$. Finally, if $v \overset{L}{\Rightarrow} u_{v,w}$ is introduced during the execution of a call $Stabilize(w, \lambda, \pi^*, \nu)$, then $m(v) = \{\pi^*\}$, so that when eventually v is selected at (S.4), (S.9), or (S.16) will imply that $\pi^* \in \Gamma^*$.

Concerning (d), let $v, w_1, w_2 \in N_C$, $\pi \in \Pi$ be such that $w_1 \neq w_2$ and $u_{v,w_1} \Rightarrow \text{node}(\pi)$, $u_{v,w_2} \Rightarrow \text{node}(\pi)$ are in E_C. By (c) above there exists a D-pair $(\Gamma, {}^v\pi)$ such that $\pi \in \Gamma$ and the edge $v \Rightarrow \text{node}({}^v\pi)$ is in E_C. If either $\pi \in \Gamma \setminus m(v)$ or $\pi \in m(v)$, where the value $m(v)$ was set at line (PS.7), then inspection of the FOR loops (S.10) and (S.17), as well as of the procedure $Stabilize$, yields $w_1 = w_2$, a contradiction. Therefore we must have $\pi \in m(v)$, where the value $m(v)$ was set at line (I.6), in which case inspection of (I.1) and (I.6) yields the conclusion.

Next, as regards (e), let $u_{v,w} \Rightarrow \text{node}(\alpha)$ be in E_C, for some $u_{v,w} \in Cl_6$. If $u_{v,w}$ was introduced into N_C during the Initialization Phase, then $v = \text{node}(x)$ and $w = \text{node}(y)$, for some $x, y \in V \setminus \{U\}$, and $\alpha = \pi^{x,y}$. Therefore, $u_{w,v} \Rightarrow \text{node}(\pi^{y,x})$ is introduced in E_C at (I.1), and it is enough to observe that by condition C8(iii), $\pi^{y,x} \in inv(\pi^{x,y}) = inv(\alpha)$. On the other hand, if $u_{v,w}$ was introduced into N_C during the execution of a call $Stabilize(v_1, \lambda, \mu, \nu)$, with $\alpha \in \{\lambda, \mu\}$, then $u_{w,v}$ is also introduced during the same call and $u_{w,v} \Rightarrow \text{node}(\beta)$, where $\beta \in \{\lambda, \mu\} \setminus \{\alpha\}$. But from (S.11) and (S.18) it follows that $\mu \in inv(\lambda)$, which by condition C4(i) yields $\lambda \in inv(\mu)$. Thus in any case $\beta \in inv(\alpha)$, and the proof of (e) is completed.

Finally, concerning (f) it is enough to observe that as soon as a new edge $v \Rightarrow \text{node}(\pi)$ is introduced in E_C, the call $Propagate(v, \pi)$ is made, with the effect that edges $v \Rightarrow \text{node}(\Sigma)$, for all $\Sigma \subseteq \Pi$ such that $\pi \in \Sigma$, are introduced into E_C. This completes the proof of the lemma. ∎

Having proved that the graph G_C is well-founded and quasi-extensional and that it possesses the closure properties stated in the preceding lemma, we next define a suitable representation of G_C from which subsequently an injective model of P will be extracted.

Put

$$A_i = \{\{3i, 3i+1, 3i+2\}\}, \quad i = 1, 2, \ldots, \tag{9.14}$$

and let I be a biunivoque correspondence from $N_1 \cup N_2 \cup \cdots \cup N_n$ into $\{A_i : i = 1, 2, \ldots\}$ ($N_1 \cup N_2 \cup \cdots \cup N_n$ is the set of all nodes in $N_C \setminus \{\text{node}(q_0)\}$ that have no immediate predecessors). Extend I to the whole N_C by putting $I(v) = \emptyset$, for all $v \in N_C \setminus (N_1 \cup N_2 \cup \cdots \cup N_n)$. Then, by induction on height(v), for all

v put:

$$Repr(v) = \{Repr(u) : u \Rightarrow v \text{ is in } E_C\} \cup I(v). \qquad (9.15)$$

The function $Repr$ is a representation of G_C (in the sense of Definition 9.3), as proved in the following lemma:

Lemma 9.11 *For all* $u, v \in N_C$

(a) *if* $Repr(u) = Repr(v)$ *then* $u = v$;

(b) $Repr(u) \in Repr(v)$ *if and only if* $u \Rightarrow v$ *is in* E_C.

Proof. (a) We will proceed by induction on $M = \max(\text{height}(u), \text{height}(v))$. If $M = 0$ then $\text{height}(u) = \text{height}(v) = 0$, i.e.

$$u, v \in N_1 \cup N_2 \cup \cdots \cup N_n \cup \{node(q_0)\}.$$

Therefore by (9.15) $Repr(u) = I(u)$ and $Repr(v) = I(v)$. Thus by the injectivity of I over $N_1 \cup N_2 \cup \cdots \cup N_n \cup \{node(q_0)\}$, if $I(u) = I(v)$, then $u = v$. Concerning the inductive step, suppose that $Repr(u) = Repr(v)$. For specificity, assume that $\text{height}(u) > 0$, so that $I(u) = \emptyset$. Let $w_1 \Rightarrow u$ be in E_C and let $t = Repr(w_1)$. Notice that by Lemma 9.8, no node in N_C has exactly three immediate predecessors. Therefore, by induction, $|Repr(w_1)| \neq 3$, that is $Repr(w_1) \in Repr(v) \setminus I(v)$ (this in particular shows that $I(v) = \emptyset$). By (9.15) there must exist $w_2 \in N_C$ such that $w_2 \Rightarrow v$ is in G_C and $Repr(w_2) = Repr(w_1)$. But $\max(\text{height}(w_1), \text{height}(w_2)) < \max(\text{height}(u), \text{height}(v))$; hence by induction $w_1 = w_2$, i.e. $w_1 \Rightarrow v$ is in G_C. Likewise, we can prove that if $w_1 \Rightarrow v$ is in E_C then so is $w_1 \Rightarrow u$. Thus by quasi-extensionality it follows $u = v$, completing the proof of (a).

(b) By definition (9.14), if $u \Rightarrow v$ is in E_C, then $Repr(u) \in Repr(v)$. Next we prove that if $Repr(u) \in Repr(v)$, $u, v \in N_C$, then the edge $u \Rightarrow v$ is in E_C. If $\text{height}(u) = 0$, then $Repr(u) = I(u)$, and by (9.14) $|Repr(u)| \leq 1$. On the other hand, as observed above, if $\text{height}(u) \neq 0$, then $|Repr(u)| \neq 3$. In any case $|Repr(u)| \neq 3$, for all $u \in N_C$. Thus $Repr(u) \notin I(v)$ and by (9.15) there must exist w such that $w \Rightarrow v$ is in E_C and $Repr(u) = Repr(w)$. But by (a) above $u = w$, hence $u \Rightarrow v$ is in E_C, proving (b) and in turn completing the proof of the lemma. ∎

Next we state some properties of the function $Repr$:

Lemma 9.12   (a) $Repr(v') = \{Repr(v)\}$, *for all* $v' \in Cl_1$;

(b) $Repr(node(q_0)) = \emptyset$

(c) $Repr(v_L) = \{\emptyset, \{Repr(v)\}\}$, *for all* $v_L \in Cl_2$;

(d) $Repr(2) = \{\emptyset, \{\emptyset\}\} = 2$;

(e) $Repr(v_R) = \{2, \{Repr(v)\}\}$, *for all* $v_R \in Cl_3$;

(f) $|Repr(v)| \geq \omega$, for all $v \in Cl_4 \setminus \{node(q_0)\}$;

(g) $|Repr(v)| = 1$, for all $c \in Cl_5$;

(h) $Repr(u_{v,w}) = [Repr(v), Repr(w)]$, for all $u_{v,w} \in Cl_6$.

Proof. The lemma follows from Lemmas 9.8, 9.9, 9.11 and the definition (9.15) of $Repr$.   ∎

An immediate consequence of the preceding lemma is the following result:

Corollary 9.1 If $[s,t] \in Repr(u)$, with $u \in N_C$, then there exist $u_{v,w} \in Cl_6$ such that the edge $u_{v,w} \Rightarrow u$ is in E_C and $Repr(u_{v,w}) = [s,t]$.   □

For each variable x occurring in P, put

$$M^*x = Repr(node(x)). \qquad (9.16)$$

Also, for each $\pi \in \Pi$ we put

$$\overline{\pi} = Repr(node(\pi)). \qquad (9.17)$$

We will prove below that M^* is an injective model of the conjunction P. We need the following lemma:

Lemma 9.13   (a) The sets $\overline{\pi}$ are non-empty and pairwise disjoint.

(b) For each variable x occurring in P, $M^*x = \bigcup_{\pi(x)=1} \overline{\pi}$.

(c) For every variable $x \in V \setminus \{U\}$, $M^*x \in \overline{\pi^x}$.

(d) For all variables $x, y \in V \setminus \{U\}$, $[M^*x, M^*y] \in \overline{\pi^{x,y}}$.

(e) If $[s,t] \in \overline{\alpha}$, with $\alpha \in \Pi$, then for some $\beta \in dom(\alpha)$, $s \in \overline{\beta}$.

(f) If $s \in \overline{\beta}$, with $\beta \in \Pi$, then there is a D-pair (Γ, β) such that for each $\gamma \in \Gamma$, $s \in Dom(\overline{\gamma})$.

(g) If $[s,t] \in \overline{\alpha}$, with $\alpha \in \Pi$, then for some $\beta \in inv(\alpha)$, $[t,s] \in \overline{\beta}$.

(h) If $[s,t_1], [s,t_2] \in M^*f$, for some variable f such that the conjunct SINGLEVALUED(f) is in P, then $t_1 = t_2$.

Proof. (a) By Lemma 9.12(f), $\overline{\pi} = Repr(node(\pi))$ is infinite. The pairwise disjointness of sets $\overline{\pi}$, $\pi \in \Pi$, follows from Lemma 9.5(a),(b), the quasi-extensionality of G_C (cf. Lemma 9.9), and Lemma 9.11.

(b) This is an immediate consequence of Lemma 9.10(f) and the definition of $Repr$ (cf. (9.15)).

(c) Lemma 9.10(a) yields that for each $x \in V \setminus \{U\}$ the edge $node(x) \Rightarrow node(\pi^x)$ is in E_C. Therefore

$$M^*x = Repr(node(x)) \in Repr(node(\pi^x)) = \overline{\pi^x}.$$

(d) Let $x, y \in V \setminus \{U\}$. Then by Lemma 9.10(b) and Lemma 9.12(h) we have

$$[M^*x, M^*y] = [Repr(\text{node}(x)), Repr(\text{node}(y))] \in \overline{\pi^{x,y}}.$$

(e) Assume that $[s, t] \in \overline{\alpha}$, for some $\alpha \in \Pi$. Then by Corollary 9.1 there exists $u_{v,w} \in Cl_6$ such that $u_{v,w} \Rightarrow \text{node}(\alpha)$ is in E_C, and $Repr(u_{v,w}) = [s, t]$. Let β be the place in Π such that $v \Rightarrow \text{node}(\beta)$ is in E_C. By Lemma 9.10(c) there exists a D-pair (Γ, β) such that in particular $\alpha \in \Gamma$. Hence by the definition itself of a D-pair (cf. Definition 9.5(ii)), $\beta \in dom(\alpha)$. In addition, since v is connected to $\text{node}(\beta)$ in G_C, it follows that $Repr(v) \in \overline{\beta}$. But $[s, t] = Repr(u_{v,w}) = [Repr(v), Repr(w)]$ (cf. Lemma 9.12(h)), hence $s \in \overline{\beta}$ with $\beta \in dom(\alpha)$, proving (e).

(f) Next, suppose that $s \in \overline{\beta}$, for some place $\beta \in \Pi$. Hence $s = Repr(v)$ with $v \Rightarrow \text{node}(\beta)$ in E_C. Again by Lemma 9.10(c) there exists a D-pair (Γ, β) such that for each γ there is a node w_γ for which $u_{v,w_\gamma} \Rightarrow \text{node}(\gamma)$ is in E_C. Hence, by Lemma 9.12(h), $[s, Repr(w_\gamma)] \in \overline{\gamma}$, which in turn implies $s \in \text{Dom}(\overline{\gamma})$.

(g) Let $[s, t] \in \overline{\alpha}$, with $\alpha \in \Pi$. Following the proof of (e) above, we have that there is a node $u_{v,w}$ in Cl_6 such that $u_{v,w} \Rightarrow \text{node}(\alpha)$ is in E_C and $Repr(u_{v,w}) = [s, t]$ (which by Lemma 9.12(h) implies $s = Repr(v)$ and $t = Repr(w)$). From Lemma 9.10(e), it follows that $u_{w,v} \Rightarrow \text{node}(\beta)$ is in E_C for some $\beta \in inv(\alpha)$. Hence $[t, s] = [Repr(w), Repr(v)] \in \overline{\beta}$, with $\beta \in inv(\alpha)$.

(h) Finally, let $[s, t_1], [s, t_2] \in M^*f$, where the clause SINGLEVALUED (f) is in P. Let $[s, t_1] \in \overline{\alpha}_1$ and $[s, t_2] \in \overline{\alpha}_2$, with $\alpha_1, \alpha_2 \subseteq f$. Corollary 9.1 and Lemma 9.11(a) give that $[s, t_1] = Repr(u_{v,w_1})$ and $[s, t_2] = Repr(u_{v,w_2})$, for some nodes v, w_1, w_2 such that $u_{v,w_1} \Rightarrow \text{node}(\alpha_1)$ and $u_{v,w_2} \Rightarrow \text{node}(\alpha_2)$ are in E_C. Hence by Lemma 9.10(c), there is a D-pair $(\Gamma, {}^v\pi)$ such that $\alpha_1, \alpha_2 \in \Gamma$ and $v \Rightarrow \text{node}({}^v\pi)$ is in E_C. But then the definition itself of a D-pair (cf. Definition 9.5(iii)) implies $\alpha_1 = \alpha_2$, which by Lemma 9.10(d) in turn gives $v, w_1, w_2 \in \{\text{node}(x) : x \in V \setminus \{U\}\}$. Let $v = \text{node}(x)$, $w_1 = \text{node}(y)$, and $w_2 = \text{node}(z)$. Then, by Lemma 9.10(b), $\pi^{x,y} = \alpha_1 = \alpha_2 = \pi^{x,z}$, which, by condition C11, yields $y = z$. Hence $t_1 = Repr(w_1) = Repr(\text{node}(y)) = Repr(w_2) = t_2$, which concludes the proof of (h) and in turn of the lemma. ∎

Now we are ready to prove that the assignment M^* is an injective model of all conjuncts in P.

As concerns the injectivity, notice that by condition C1 the function $x \mapsto \{\pi \in \Pi : \pi \subseteq x\}$ is one-to-one. Thus Lemma 9.13(a),(b) yields that $M^*x \neq M^*y$ for any two distinct variables x and y in P, i.e. M^* is injective.

Next we prove that all conjuncts of P are satisfied by the assignment M^*.

Let $x = y \cup z$ be in P. Then in view of Lemma 9.13(a),(b) and the properties of places (cf. Definition 8.1), we have

$$M^*x = \bigcup_{\pi(x)=1} \overline{\pi} = \bigcup_{\substack{\pi(y)=1 \\ \vee \pi(z)=1}} \overline{x} = \bigcup_{\pi(y)=1} \overline{\pi} \cup \bigcup_{\pi(z)=1} \overline{\pi} = M^*y \cup M^*z,$$

i.e. $x = y \cup z$ is modelled correctly by M^*.

Next, let $x = y \setminus z$ be in P. Then we have

$$M^*x = \bigcup_{\pi(x)=1} \overline{\pi} = \bigcup_{\substack{\pi(y)=1 \\ \& \ \pi(z)=0}} \overline{\pi} = \bigcup_{\pi(y)=1} \overline{\pi} \setminus \left(\bigcup_{\pi(z)=1} \overline{\pi} \right) = M^*y \setminus M^*z,$$

which shows that the conjunct $x = y \setminus z$ is also satisfied by M^*.

If $x \in y$ (respectively, PAIR_IN(x, y, f)) occurs in P, then since π^x (respectively, $\pi^{x,y}$) is a place at the variable x (respectively, at the pair $[x, y]$), then (cf. Definition 8.2 (respectively, Definition 9.4)) $\pi^x(y) = 1$ (respectively, $\pi^{x,y}(f) = 1$). Therefore Lemma 9.13(c) (respectively, Lemma 9.13(d)) yields $M^*x \in \overline{\pi}^* \subseteq M^*y$, i.e. $M^*x \in M^*y$ (respectively, $[M^*x, M^*y] \in M^*f$). Hence literals of type $x \in y$ (respectively, PAIR_IN(x, y, f)) are correctly modelled by M^*.

Next, let $d = Df$ be a clause in P. In this case we need to verify that $M^*d = \text{Dom}(M^*f)$, which we do as follows. Let $s \in M^*d$, and in particular let $\beta \subseteq d$ be such that $s \in \overline{\beta}$. From Lemma 9.13(f) it follows that there is a D-pair (Γ, β) such that for each $\gamma \in \Gamma$, $s \in \text{Dom}(\overline{\gamma})$. Also, from the definition itself of a D-pair (cf. Definition 9.5(iv)), since $\beta(d) = 1$, there is a $\gamma_0 \in \Gamma$ such that $\gamma_0(f) = 1$, where the clause $d = Df$ is in P. Hence $s \in \text{Dom}(\overline{\gamma}_0) \subseteq \text{Dom}(M^*f)$, which in turn implies $M^*d \subseteq \text{Dom}(M^*f)$. To prove the converse inclusion, let $s \in \text{Dom}(M^*f)$. Thus for some set t, $[s, t] \in M^*f$. Let $\alpha \subseteq f$ such that $[s, t] \in \overline{\alpha}$. Lemma 9.13(e) yields the existence of a place $\beta \in dom(\alpha)$ such that $s \in \overline{\beta}$. But then from Condition C3(ii), we have $\beta(d) = 1$, which implies $s \in M^*d$. Thus $\text{Dom}(M^*f) \subseteq M^*d$, and in conclusion $M^*d = \text{Dom}(M^*f)$, proving that clauses $d = Df$ are correctly modelled too.

As regards clauses of type INV, let INV(f, g) be one such. Assume that $[s, t] \in M^*f$ and let $\alpha \subseteq f$ be such that $[s, t] \in \overline{\alpha}$. Then, from Lemma 9.13(g), there exists a place $\beta \in inv(\alpha)$ such that $[s, t] \in \overline{\beta}$. To prove that $[t, s] \in M^*g$, it only remains to be shown that $\beta(g) = 1$. But this follows immediately from Condition C4(iii). Symmetrically one can prove that if $[s, t] \in M^*g$, then $[t, s] \in M^*f$. Hence M^*f is an inverse of M^*g, in the sense specified at the beginning of Section 9.1.

Finally, from Lemma 9.13(h) it follows that M^* satisfies all conjuncts in P of the form SINGLEVALUED(f).

This concludes the proof that M^* is an injective model of P. Thus we have

Lemma 9.14 (Completeness) *The conditions stated in Lemma 9.4, necessary for the injective satisfiability of a conjunction P of positive clauses of type $x = y \cup z$, $x = y \setminus z$, $x \in y$, PAIR_IN(x, y, f), $d = Df$, INV(f, g), SINGLEVALUED(f), containing clauses (9.4), are also sufficient.* □

In view of Lemma 9.1 and by observing that the conditions stated in Lemma 9.4 are effectively verifiable we have

Theorem 9.1 (Decidability) *The class of propositional combinations of set-theoretic literals of the form (9.1) has a solvable satisfiability problem, and Lemma 9.14 suggests a specific decision procedure for $\mathcal{L}$.*   □

By taking Corollary 3.1 into account, we obtain from this theorem the following proposition:

Corollary 9.2 (Decidability) *The family of $\in$-left formulae (cf. Definition 3.10) over the language $\mathcal{L}$ has a solvable satisfiability problem.*   □

10. THE UNIONSET OPERATOR

10.1 Preliminaries

In this chapter, which is a rewrite of [CFS87], we consider the satisfiability problem for the class MLSU of unquantified formulae obtained as the propositional closure of atoms of the kinds

$$x = y \cup z, \quad x = y \setminus z, \quad x \in y, \quad u = Un(y), \tag{10.1}$$

where we recall that $Un(x) = \{y : y \in z \text{ for some } z \in x\}$.

Pioneering results by Breban on this subject (see [Bre82] and [BF84]) will be generalized, at least in one respect; as a matter of fact, we will prove that the problem at hand, with no restrictions on the number of occurrences of literals of type $u = Un(y)$, admits a sound and complete decision test, while Breban solved the case in which just one occurrence of the unionset operator appears. It is worthwhile recalling, however, that in [Bre82] an unlimited number of occurrences of the singleton operator were allowed.

As observed in the preceding chapters, by applying a normalization process of the type described in Section 3.8, we can limit ourselves to considering only conjunctions of atoms of type (10.1), where each x, y, z, u is a variable. In addition, in view of Lemma 3.3, we can further limit ourselves to the *injective* satisfiability problem for the class MLSU.

Let P be a conjunction of atoms of type (10.1). As shown in Chapters 8 and 9, any model M of the statements of P defines a set of places for P (cf. Definition 8.1), and the structure of this set of places goes a long way towards describing the structure of the model M. Specifically, if p is any point appearing in the model, then the function π defined by $\pi(x) = 1$ if $p \in Mx$, $\pi(x) = 0$ if $p \notin Mx$, is clearly a place of P. Moreover, as in the preceding chapters, if we are given any model M and any place π of P, then we can consider the set

$$\sigma^\pi = \{p : p \in Mx \leftrightarrow \pi(x) = 1, \text{ for all variables } x \text{ in } P\}, \tag{10.2}$$

which is the region of the Venn diagram (of the universal space of the model M) associated with the place π. It is convenient to consider only places π for which $\sigma^\pi \neq \emptyset$ as places of the model M and to exclude the others. This will be done in what follows. With this understanding, the subsets σ^π are clearly pairwise disjoint and $\sigma^\pi \subseteq Mx$ if and only if $\pi(x) = 1$. Each set σ^π is either

wholly contained in Mx or wholly disjoint from it, and $Mx = \bigcup_{\pi(x)=1} \sigma^\pi$.
We also recall from Definition 8.2 that given a variable x, the place α is said
to be a *place at the variable x* if $\alpha(y) = 1$ whenever the literal $x \in y$ is in P;
if literals of the form $x \in y$ *do* appear in P, then the only place at x is the
one—to be denoted π^x—for which $Mx \in \sigma^{\pi^x}$.

Remark 10.1 For any place α and variable x, we will often use the notation
$\alpha \subseteq x$ to mean $\alpha(x) = 1$. $\qquad\qquad\qquad\qquad\qquad\qquad\qquad\qquad\qquad\qquad$ □

The set Π_1 of all possible places associated with the set P of clauses is
clearly finite and easily calculated. We aim at stating the condition that P
should be satisfiable using only combinatorial conditions on the clauses of P
and on the set of places which actually appears in a model M of P. This
is clearly some subset Π of Π_1, which we suppose to have been chosen in
advance.

Remark 10.2 Given a model M of P, we will denote by Π_M the set of places
π associated with the non-empty regions σ^π of the Venn diagram relative to
M (cf. (10.2)). $\qquad\qquad\qquad\qquad\qquad\qquad\qquad\qquad\qquad\qquad\qquad\qquad$ □

Without loss of generality, we will assume that P contains the clause

$$y_0 = y_0 \setminus y_0. \qquad\qquad (10.3)$$

All essential complications that need to be faced are connected with the
presence in P of finitely many clauses of the form $u_i = Un(y_i)$, which will be
referred to as the *Uclauses* of P. The variables y_i appearing on the right of
clauses of this form will be called *Uvariables*.

Since we are concerned with injective models only, and since $u = Un(y)$
and $u' = Un(y)$ implies $u = u'$, we can clearly suppose that each Uvariable y_i
appears in just one Uclause.

The following definitions take a first step toward elucidating the logical
weight of the Uclauses in P.

Definition 10.1 *Given a conjunction P of literals of type (10.1) and a set
Π of places of P, the Ugraph G of P and Π is the graph whose set of nodes
is Π, plus an additional node Ω, and whose edges are as follows:*

(i) *A directed edge connects the place π to the node Ω if and only if $\pi(y_i) = 0$
for every Uvariable y_i.*

(ii) *Otherwise, a directed edge connects the place α to the place β if and only
if $\beta(u_i) = 1$ for all Uclauses $u_i = Un(y_i)$ such that $\alpha(y_i) = 1$. (If there
are no such β, then α is not the source node of any edge of G.)*

We write $\pi \Rightarrow \psi$ when an edge connects π to ψ. $\qquad\qquad\qquad\qquad\qquad$ □

Remark 10.3 Notice that *self-loops*, i.e. edges $\alpha \Rightarrow \alpha$, can be present. □

Definition 10.2 *Let G be the Ugraph of P and Π.*

(i) *A node π of G is called* safe *if there is a directed path through G starting at π which reaches Ω.*

(ii) *A node π of G is called* null *if there is no $\beta \in \Pi \cup \{\Omega\}$ such that $\pi \Rightarrow \beta$.*

(iii) *A node π of G is called* trapped *if every sufficiently long path forward from π eventually reaches a null node.*

(iv) *A node π of G which is neither safe nor trapped is said to be* cyclic *(in this case some path forward from such a node can be extended indefinitely, but must then traverse certain other nodes repeatedly).* □

Remarks 10.4 Note that if α is trapped and $\alpha \Rightarrow \tau$, then τ is also trapped. Moreover, if α is safe, so is every β such that $\beta \Rightarrow \alpha$; hence if α is cyclic and $\alpha \Rightarrow \psi$, then ψ is either trapped or cyclic. Moreover, the null node is trapped and no trapped node is safe. □

The following lemma clarifies the intuition behind Definitions 10.1 and 10.2.

Lemma 10.1 *Let M be a model for P and let G be the Ugraph of P and Π_M.*

(a) *If $\alpha(y) = 1$ for some Uvariable y and Uclause $u = Un(y)$, then*

   (a$_1$) *if $Un(\sigma^\alpha) \cap \sigma^\beta \neq \emptyset$, then the edge $\alpha \Rightarrow \beta$ is in the Ugraph G;*

   (a$_2$) *$Un(\sigma^\alpha) \subseteq \bigcup_{\alpha \Rightarrow \beta} \sigma^\beta$.*

(b) *If $\beta(u) = 1$ for some Uclause $u = Un(y)$, then $\sigma^\beta \subseteq \bigcup_{\alpha \in A} Un(\sigma^\alpha)$, where A is the set of all places α such that $\alpha \subseteq y$ and $\alpha \Rightarrow \beta$.*

(c) *If γ is a null place, then $\gamma = \pi^{y_0}$ (we recall that π^{y_0} is the place at the variable y_0, where we are assuming that the literal $y_0 = y_0 \setminus y_0$ occurs in P).*

(Thus in particular if $\alpha \Rightarrow \Omega$, intuitively this means that the Uclauses of P tell us nothing about the set $Un(\sigma^\alpha)$.)

Proof. (a) Let $Un(\sigma^\alpha) \cap \sigma^\beta \neq \emptyset$, where $\alpha(y) = 1$ for some Uvariable y and some Uclause $u = Un(y)$. To show that the edge $\alpha \Rightarrow \beta$ is in the Ugraph G, we need to prove that $\beta(u_i) = 1$ for all Uclauses $u_i = Un(y_i)$ in P such that $\alpha(y_i) = 1$. So, let $\alpha(y_i) = 1$. Since $\sigma^\alpha \subseteq My_i$, we have $Un(\sigma^\alpha) \subseteq Un(My_i) = Mu_i$, which in turn implies $\sigma^\beta \cap Mu_i \neq \emptyset$, i.e. $\sigma^\beta \subseteq Mu_i$, as the set σ^β is a region of the Venn diagram of the sets Mx, with x ranging over the variables of P. This shows that $\beta(u_i) = 1$, and (a$_1$) is proved.

Concerning (a$_2$), since $Un(\sigma^\alpha) \subseteq Mu$, it follows that $Un(\sigma^\alpha) \subseteq \bigcup_{\beta \in \Pi} \sigma^\beta$. Thus, (a$_2$) follows at once from (a$_1$).

(b) Let $p \in \sigma^\beta$. Since

$$\sigma^\beta \subseteq Mu = Un(My) = Un\left(\bigcup_{\alpha(y)=1} \sigma^\alpha\right) = \bigcup_{\alpha(y)=1} Un(\sigma^\alpha),$$

then for some place α such that $\alpha(y) = 1$ we have $p \in Un(\sigma^\alpha)$. But, from (a_1) above, the edge $\alpha \Rightarrow \beta$ is in the Ugraph G. Therefore, $p \in \bigcup_{\alpha \in A} Un(\sigma^\alpha)$, which in turn implies $\sigma^\beta \subseteq \bigcup_{\alpha \in A} Un(\sigma^\alpha)$, where A is the set of all places α such that $\alpha \subseteq y$ and $\alpha \Rightarrow \beta$. Hence (b) is proved.

(c) Let γ be null. Then, by (a), $Un(\sigma^\gamma) \subseteq \bigcup_{\gamma \Rightarrow \beta} \sigma^\beta = \emptyset$, i.e. $\sigma^\gamma = \{\emptyset\} = \{My_0\}$. Thus $\gamma = \pi^{y_0}$. $\blacksquare$

It is very easy to see that complications greater than those encountered when no clauses $u_i = Un(y_i)$ are present must be expected in the case before us. For example, the conjunction

$$u = Un(u) \ \& \ v \in u \tag{10.4}$$

can be satisfied, but only by an infinite model. Indeed, it is immediate to verify that (10.4) is satisfied by any assignment I such that

$$Iv = \emptyset,$$
$$Iu = \{\emptyset, \{\emptyset\}, \{\{\emptyset\}\}, \ldots\}.$$

Moreover, if M is a model of (10.4), then Mu must be non-empty. So, let s_0 be any element of Mu. Since $Mu = Un(Mu)$, it follows that $s_0 \in s_1$ for some s_1 in Mu. Analogously, there must exist an s_2 in Mu such that $s_1 \in s_2$. By repeating this argument, we obtain a countably infinite sequence of distinct elements s_i belonging to Mu, which proves that Mu must be infinite in any model M of (10.4).

The arguments which follow show that it is not hard to deal with these infinities. However, worse combinatorial difficulties are connected with the possible existence of trapped places. To see why this should be so, define the *height* of a trapped place τ as one more than the length of the longest path forward from τ to a null place. Suppose that there is a model for our set of clauses, which therefore associates a set Mx with every variable x and a set σ^π with every place π. If τ is of height 1, i.e. null, we have $Un(\sigma^\tau) = \emptyset$, so $\sigma^\tau = \{\emptyset\}$; hence there can be only one such place, which must be a place at y_0 (recall that we are assuming that $y_0 = y_0 \setminus y_0$ is in P). Then it follows by an easy inductive argument exploiting Lemma 10.1(a_2) (see Lemma 10.6) that if τ is a trapped place, the rank of σ^τ is at most the height of τ. This restricts σ^τ to one of a finite collection of possible values, namely if H is the maximum height of any trapped place and, for any positive integer h, V_h is the (finite) collection of all sets of rank less than h, σ^τ must have some value in V_{H+1}. We will see in the next section that if there are no trapped places, restrictions of this kind, which prevent σ^π from being infinite and cause the combinatorial complications alluded to above, do not occur.

For ease of presentation, we first consider the simpler case in which no places are trapped. We will derive some conditions (namely C1–C4 in the next section) which are necessary for P to be satisfiable and will prove that in the absence of trapped places these conditions are also sufficient. In Section 10.3 conditions C2, C3, and C4 will be strengthened into conditions C2′, C3′, C4′ that work in the general case. Section 10.2 is hence preparatory to Section 10.3, which subsumes it. By contrasting these two sections, one can single out the combinatorial difficulties involved with trapped places.

It is conjectured that MLSU extended by the singleton operator $\{\bullet\}$ has a decidable satisfiability problem and can be successfully treated by resorting to a further generalization of the method developed in Section 10.3.

10.2 The decision algorithm in the absence of trapped places

In this section we deduce some conditions which are necessary for P to be satisfiable, regardless of the presence or absence of trapped places. Moreover we show that if trapped places are absent, then these conditions are also sufficient for the satisfiability of P.

The conditions with which we work assert that the Ugraph G of P and Π has certain connectivity properties. They imply that sets $\overline{\pi}$ associated to $\pi \in \Pi$ can be initialized in a manner assuring that the initial interpretation $Mx = \bigcup_{\pi(x)=1} \overline{\pi}$ satisfies all equalities in P and allows a subsequent 'stabilization' phase to force all remaining clauses of P of type $x \in y$ to be satisfied without disrupting any other clause already modelled correctly.

Our first condition follows at once from injectivity. Suppose that an injective model M of P exists. Note that two variables x, y have the same representation in the model M if and only if $\pi(x) = \pi(y)$ for all places π of the model. Therefore, since M is injective, for any two variables x, y occurring in P there exists a place π in Π_M such that

$$\pi(x) = 1 \leftrightarrow \pi(y) = 0 \ ,$$

or, stated in new terms,

Condition C1. *Let P be a conjunction of atoms of type (10.1), and let Π be a set of places of P relative to an injective model of P. Then no two distinct variables of P are Π-equivalent.*   □

To deduce our second condition we argue as follows. Form the union S of σ^π, π running over all trapped and cyclic places. Then

$$Un(S) \subseteq S \ . \tag{10.5}$$

Indeed, let $p \in Un(S)$. Then $p \in Un(\sigma^\pi)$ for some trapped or cyclic place. Lemma 10.1(c),(a_2) then implies that $p \in \sigma^\beta$, for some place β such that the edge $\pi \Rightarrow \beta$ is in G. Since π is not safe, β cannot be safe either, i.e. β must be trapped or cyclic. Thus $\sigma^\beta \subseteq S$ which in turn implies (10.5).

Next, take any element $p_1 \in \sigma^{\alpha_1} \subseteq S$. If $p_1 \neq \emptyset$, it has an element p_2 belonging to some σ^{α_2} such that $\alpha_1 \Rightarrow \alpha_2$ (cf. Lemma 10.1); if $p_2 \neq \emptyset$, we can repeat this argument to produce p_3 and α_3 such that $p_3 \in \sigma^{\alpha_3}$ and $\alpha_2 \Rightarrow \alpha_3$. Thus we can find a sequence $\cdots p_3 \in p_2 \in p_1$ (and a sequence $\alpha_1 \Rightarrow \alpha_2 \Rightarrow \alpha_3 \Rightarrow \cdots$, such that $p_i \in \sigma^{\alpha_i}$), which, by the set-theoretic postulate of well-foundness (cf. [Jec78]), cannot be infinite, i.e., it must end with $p_s = \emptyset = My_0$. It follows that there must be a path through G to the node π^{y_0} which is the place at y_0. This gives a second necessary condition for (injective) satisfiability.

Condition C2. *Let the set P of clauses be satisfiable by a model whose set of places is Π, and define the Ugraph G corresponding to P, Π as above. Then if there are any non-safe places in Π, there must exist a non-safe place γ which lies along a path through G from every non-safe node. Moreover, γ must be the place π^{y_0} at y_0.* □

If condition C2 is satisfied, we can define a useful auxiliary map ψ of places to places as follows:

Definition 10.3 *Let condition C2 be satisfied. Given a safe (respectively, trapped or cyclic) place π, we let $\psi(\pi)$ be any node β which is one step closer to Ω (respectively, π^{y_0}) along a path of minimum length leading from π to Ω (respectively, π^{y_0}). If $\pi \Rightarrow \Omega$, we put $\psi(\pi) = \Omega$. If π^{y_0} is null we leave $\psi(\pi^{y_0})$ undefined; otherwise if π^{y_0} is not null (which by Lemma 10.1 implies that no π is null) choose any β such that $\pi^{y_0} \Rightarrow \beta$ and put $\psi(\pi^{y_0}) = \beta$.* □

Remark 10.5 Summarizing, we note that the status of the place $\gamma = \pi^{y_0}$ conveys much information about the general structure of the Ugraph G:

(1) If γ is safe (as revealed by the fact that $\psi(\cdots(\psi(\gamma))\cdots) = \Omega$ for an adequate number of applications of ψ), then every node is safe.

(2) If γ is null (as revealed by the fact that $\psi(\gamma)$ is undefined), then γ is the sole null place, and it is also trapped. A non-safe node can be trapped or cyclic.

(3) If γ is neither safe nor null (as revealed by the fact that $\psi(\cdots(\psi(\gamma))\cdots)$ $= \gamma$ for an adequate number of applications of ψ), then every non-safe node—including γ—is cyclic.

In cases (2) and (3), for every non-safe node π other than γ, an adequate number of applications of ψ can be chosen so that $\psi(\cdots(\psi(\pi))\cdots) = \gamma$. □

The map ψ will be used later when we construct a model for P. Before this, however, we need to state additional satisfiability conditions.

Suppose once more that we have a model M for P, and derive the sets σ^π and the Ugraph G from this model as in the preceding section.

We recall that for any two sets s, t, we write $s \in^+ t$ if and only if there is a finite chain of intermediate elements s_i such that $s \in s_1 \in \cdots \in s_k \in t$ (cf. Definition 9.6).

Since from the well-foundedness axiom a circular sequence of membership relations $s_i \in^+ s_j$ is impossible, any finite collection C of sets can be enumerated in such a way as to ensure that no set s of C can satisfy $s \in^+ t$ or $s = t$ for a set t coming earlier in the sequence. In the following discussion it is supposed that the variables appearing in P are arranged in a sequence derived from such an enumeration of the sets Mx.

For each variable x, consider the set Π_x of all places π such that $Mx \in^+ \sigma^\pi$. Then plainly we must have $\pi(y) = 0$ for all $\pi \in \Pi_x$ and variables y preceding x in sequence (or equal to x). Moreover, if $Mx \in^+ \sigma^\alpha$ and $\sigma^\alpha \subseteq Mu_i = Un(My_i)$ for some Uvariable y_i and Uclause $u_i = Un(y_i)$, then by Lemma 10.1 there must exist a place $\beta \subseteq y_i$ such that $\beta \Rightarrow \alpha$, and such that $Mx \in^+ \sigma^\beta$. For each α such that $Mx \in^+ \sigma^\alpha$ for any variable x, and for each Uvariable y_i such that $\alpha \subseteq u_i$, choose any $\beta \subseteq y_i$ such that $\beta \Rightarrow \alpha$ and $Mx \in^+ \sigma^\beta$ and call it $\varphi_x(\alpha, y_i)$. Finally, define $\varphi(\alpha, y_i)$ for all Uvariables y_i such that $\alpha \subseteq u_i$ as any $\beta \subseteq y_i$ such that $\beta \Rightarrow \alpha$. This gives us an ordering of variables, a collection of maps φ, φ_x and a collection of sets Π_x of places, one for each variable x appearing in P, having the following properties:

(i) $\varphi(\alpha, y_i)$ is defined for all places α in Π and Uvariables y_i such that $\alpha \subseteq u_i$, where $u_i = Un(y_i)$ is in P, and the value $\beta = \varphi(\alpha, y_i)$ is a place such that $\beta \subseteq y_i$ and $\beta \Rightarrow \alpha$.

(ii) For each variable x, the place π^x at x, defined by $\pi^x(y) = 1$ if and only if $Mx \in My$ (i.e., the place at x), belongs to Π_x, and moreover if $\pi \in \Pi_x$ and $\pi(u_i) = 1$ for some Uclause $u_i = Un(y_i)$ in P, then $\varphi_x(\pi, y_i)$ is defined and the place $\beta = \varphi_x(\pi, y_i)$ is such that

(ii)$_1$ $\beta \in \Pi_x$;

(ii)$_2$ $\beta \Rightarrow \pi$ is in G;

(ii)$_3$ $\beta(y_i) = 1$.

(iii) For each variable x, none of the places $\pi \in \Pi_x$ satisfies $\pi(y) = 1$ for any variable y which is either equal to x or comes before x in the enumeration of variables defined above.

Remark 10.6 In what follows, it will be convenient to call an enumeration of variables and maps φ and φ_x having properties (i)–(iii) a *good Uorder* (of variables) and *good Umaps* respectively; we will not bother to introduce a corresponding term for the set Π_x of places, though, of course, such sets of places must be defined in connection with any purportedly good Umap φ_x. $\square$

The preceding discussion allows us to state a third condition necessary for (injective) satisfiability:

Condition C3. *Let P, Π, G, etc., be as in condition C2 above. Then (if P is satisfiable) a place $\pi^x \in \Pi$ such that $\pi^x(y) = 1$ if $x \in y$ occurs in P must be defined for each variable x appearing in P and there must exist sets $\Pi_x \subseteq \Pi$ for each variable x, a good Uorder of variables, and good Umaps φ and φ_x, which by definition will have the above properties (i), (ii), (iii).* $\square$

Still one more necessary condition remains to be stated. To see what this is, let M, σ^π, π^x, etc. be as above. Then if $u_i = Un(y_i)$ is a Uclause of P and $Mx \in My_i$, we must have $Mx \subseteq Mu_i$. Therefore if $\pi(x) = 1$, i.e. $\sigma^\pi \subseteq Mx$, then $\sigma^\pi \subseteq Mu_i$, i.e. $\pi(u_i) = 1$. Hence the following condition must obviously be satisfied.

Condition C4. *If $u_i = Un(y_i)$ is a Uclause of P, $\pi^x(y_i) = 1$ and $\pi(x) = 1$, then $\pi(u_i) = 1$.* $\square$

This completes the statement of all conditions for injective satisfiability, at least *in the absence of trapped places*. Summing up, we have established the following result.

Lemma 10.2 (Soundness) *Let M be an injective model of P. Then conditions C1–C4 hold.* $\square$

We can now go on to show that if there are no trapped places in the Ugraph G of P, and if conditions C1–C4 are all satisfied, then an injective model for the clauses of P can be constructed. The construction of this model is easy once a sufficient supply of 'auxiliary' and 'secondary' elements is assured (see definition below); accordingly, we will begin by assuming that such elements with the needed properties have been constructed, and will show how these can be used to build a model M. After this, the narrower technical problem of constructing the auxiliary and secondary elements will be addressed.

Definition 10.4 *Assume that condition C2 is satisfied, and let the set Π of places, the Ugraph G, the map ψ, etc., be as in condition C2. A good assignment of auxiliary and secondary elements to the nodes of G is any pair of maps Aux, Sec which associate sets to the nodes of the Ugraph G in such a way that, if we*

- *denote by Aux_π the set $Aux(\pi)$ and by Sec_π the set $Sec(\pi)$, for every node π of G;*

- call auxiliary *(respectively,* secondary*)* elements resident at π *the members of* Aux_π *(respectively, of* Sec_π*);*

- *for all* $p \in Aux_\pi \cup Sec_\pi$*, call* residence *of p the place* π*;*

then

(a) Aux_π *and* Sec_π *are countably infinite for each node* π *in* G*; moreover all auxiliary elements have cardinality 1, whereas all secondary elements have cardinality at most 2;*

(b) $\emptyset \in Sec_{\pi^{v_0}}$ *;*

(c) *every secondary element b must satisfy* $b \in^+ a$ *(cf. Definition 9.6), for some auxiliary element a not necessarily sharing its residence with b;*

(d) *no two auxiliary elements (not necessarily resident at the same node)* a, a' *can satisfy* $a \in^+ a'$*;*

(e) *if* $\psi(\alpha) \neq \Omega$*, then*

$$Un\left(Aux_\alpha \cup Sec_\alpha\right) \subseteq Sec_{\psi(\alpha)},$$

i.e. every member of an auxiliary or secondary element resident at the place α *is a secondary element resident at the place* $\psi(\alpha)$*;*

(f) *the sets in* $\{Aux_\pi : \pi \in \Pi\} \cup \{Sec_\pi : \pi \in \Pi\}$ *are pairwise disjoint.* □

We then have the following lemma.

Lemma 10.3 *Assume that conditions C1–C4 hold, that the Ugraph G has no trapped nodes and that there exist maps Aux, Sec which are a good assignment of auxiliary and secondary elements to the nodes of G. Then P is injectively satisfiable.*

Proof: Let $\Pi, G, \psi, \varphi, \varphi_x, \Pi_x$, etc., be as above and such that conditions C1–C4 hold. Assume also that Aux, Sec are a good assignment of auxiliary and secondary elements to the nodes of G and that the Ugraph G has no trapped nodes. We now exhibit a process which will construct an injective model for P.

INITIALIZATION PHASE

1. For each node π in G, arrange the set Aux_π of auxiliary elements resident at π into infinite pairwise disjoint lists $L_{\pi,x}$, one for each variable x in P.

2. For all $\pi \in \Pi$, put

$$\overline{\pi} \leftarrow Sec_\pi.$$

3. Arrange the set of all secondary elements, i.e. the set $\bigcup_{\pi \in G} Sec_\pi$, into an infinite list L_0.

4. Initialize L_1 (*accessory list*) to the empty list.

   (Comment: Note that at the end of step 4, by (a) and (f) of Definition 10.4, all the $\overline{\pi}$ are non-empty and pairwise disjoint. Therefore all clauses in P of type $x = y \cup z$, $x = y \setminus z$ are correctly modelled by the initial injective interpretation $Mx = \bigcup_{\pi(x)=1} \overline{\pi}$.)

END INITIALIZATION PHASE

φ-STABILIZATION PHASE

$r \leftarrow 1$

DO FOREVER

$r \leftarrow 1 - r$

IF L_r is empty **THEN**

(Comment: $r = 1$)

$r \leftarrow 0$

END IF;

Let p be the first component of the list L_r.

Discard it from L_r.

Let π be the place such that $p \in \overline{\pi}$.

(Comment: If p is a secondary or auxiliary element, then π is the residence of p.)

FOR ALL Uclauses $u_i = Un(y_i)$, $i = 1, \ldots, m$, such that $\pi(u_i) = 1$ **DO**

- pick the first component a of the list L_{π,y_0} and discard it from L_{π,y_0};
- let $\beta_i = \varphi(\pi, y_i)$; put

$$\overline{\beta_i} \leftarrow \overline{\beta_i} \cup \{\{p, a\}\};$$
$$\overline{\pi} \leftarrow \overline{\pi} \cup \{a\};$$

- add the sets $\{p, a\}$ and a to the end of the list L_1;

END FOR ALL;

END DO FOREVER;

END φ-STABILIZATION PHASE.

FOR ALL x in $P \setminus \{y_0\}$, following the good Uorder of variables **DO**

Put
$$\overline{\pi^x} \leftarrow \overline{\pi^x} \cup \{ \bigcup_{\pi(x)=1} \overline{\pi} \}.$$

Initialize L_1 to the list consisting of the single component $\bigcup_{\pi(x)=1} \overline{\pi}$.

φ_x-STABILIZATION PHASE

> **DO FOREVER**
>
> > **IF** L_1 is empty **THEN** *QUIT FOREVER* **END IF**;
> > Let p be the first component of the list L_1.
> > Discard it from L_1.
> > Let π be the 'residence' of p, viz. the place such that $p \in \overline{\pi}$.
> > **FOR ALL** Uclauses $u_i = Un(y_i)$, $i = 1, \ldots, m$, such that $\pi(u_i) = 1$ **DO**
> >
> > - pick the first component a of the list $L_{\pi,x}$ and discard it from $L_{\pi,x}$;
> > - let $\beta_i = \varphi_x(\pi, y_i)$; put
> >
> > $$\begin{aligned} \overline{\beta_i} &\leftarrow \overline{\beta_i} \cup \{\{p, a\}\}; \\ \overline{\pi} &\leftarrow \overline{\pi} \cup \{a\}; \end{aligned}$$
> >
> > - add $\{p, a\}$ and a to the end of the list L_1;
> > **END FOR ALL**;
>
> **END DO FOREVER**;

END φ_x-STABILIZATION PHASE

END FOR ALL.

We claim that if for all variables x in P we put
$$M^*x = \bigcup_{\pi(x)=1} \overline{\pi},$$

where the sets $\overline{\pi}$ are those produced by the above process, then M^* is an injective model for P. We begin by showing that after the φ-stabilization phase, the corresponding temporary assignment M^* is an injective model of the equalities in P, i.e. the clauses in P of type

$$x = y \cup z , \quad x = y \setminus z , \quad u = Un(y).$$

In order to prove that M^* is an injective model of the equalities in P of the first two types, it is enough to show that the sets $\overline{\pi}$ remain mutually disjoint during the whole φ-stabilization phase.

Notice that any element p which is added to the list L_1 during the φ-stabilization phase is either

(i) a secondary element, or

(ii) a pair $\{p', a'\}$, with a' an auxiliary element, or

(iii) an auxiliary element.

In any case, it follows that when the element p is processed with an auxiliary element a, then $p \neq a$. Indeed, in case (i), from Definition 10.4, $p \in^+ a'$, for some auxiliary a'. Therefore $p \neq a$, since otherwise $a \in^+ a'$, contradicting Definition 10.4(d). If $p = \{p', a'\}$, i.e. case (ii) holds, then from $p = a$ it would follow $a' \in^+ a$, which contradicts again Definition 10.4(d). Finally, to show that $p \neq a$ even in the last case (iii) in which p is an auxiliary element, it is enough to observe that while auxiliary elements are picked from the list L_{π, y_0} they are promptly discarded from L_{π, y_0}.

Having proved that no set $\{p, a\}$ generated during the φ-stabilization phase can collapse into a singleton, we next show that when it is formed, the pair $\{p, a\}$ is distinct from all elements previously inserted into any of the sets $\overline{\pi}$, and so must be the auxiliary element a. Indeed, a, which is a singleton, cannot be a previously formed pair; we will also see below that it cannot equal any of the sets $\bigcup_{\pi(x)=1} \overline{\pi}$ that we form, because such sets are either empty (i.e., $\bigcup_{\pi(y_0)=1} \overline{\pi}$) or infinite. For the same reason, $\{p, a\}$ can never equal a set $\bigcup_{\pi(x)=1} \overline{\pi}$ or a previously used auxiliary element, nor can it equal any secondary element b, since then there would exist an auxiliary a' such that $a \in^+ a'$, which is impossible. Finally, $\{p, a\}$ can never equal any previously formed pair $\{q, a'\}$, since this could only happen if $p = a'$ and $q = a$; but when p is processed, the element a has already been used and therefore it is no longer present in the list of auxiliary elements L_{π, y_0}.

The argument just given shows that the sets $\overline{\pi}$ remain disjoint throughout the φ-stabilization phase and therefore, by the definition of places, the intermediate assignment M^* models correctly all equality clauses in P of type $x = y \cup z$ and $x = y \setminus z$. Furthermore, since the sets $\overline{\pi}$ are also non-empty, condition C1 implies that the assignment M^* is injective.

To prove also that clauses of type $u = Un(y)$ are modelled correctly even at this intermediate stage, we argue as follows. The φ-stabilization process continues until a pair $\{p, a\}$ has been formed for every p inserted into any one of the sets $\overline{\pi}$. Moreover, whenever $\{p, a\}$ is inserted into $\overline{\beta}_i$, with $\beta_i = \varphi(\pi, y_i)$, we have $\pi(u_i) = 1$ and p is already in $\overline{\pi}$. The auxiliary element a is put in $\overline{\pi}$, but all members of a are secondary elements which will have been put into the set $\overline{\psi(\pi)}$, if $\psi(\pi) \neq \Omega$, i.e. if $\pi(y_i) = 1$ for any Uvariable y_i. Hence, since the condition

$$Un\left(\bigcup_{\alpha(y_i)=1} \overline{\alpha} \right) \subseteq \bigcup_{\beta(u_i)=1} \overline{\beta}$$

holds initially for every Uclause $u_i = Un(y_i)$, it holds throughout the stabilization process. Thus if M^*x denotes the value $\bigcup_{\pi(x)=1} \overline{\pi}$ for every variable x in P, then $Un(M^*y_i) \subseteq M^*u_i$ must hold when the stabilization process ceases to generate new pairs. But because of all the pairs $\{p, a\}$ inserted, we must also have $M^*u_i \subseteq Un(M^*y_i)$, and therefore we must have $M^*u_i = Un(M^*y_i)$, for every Uclause $u_i = Un(y_i)$ in P.

Next we show that after completion of the φ_x-stabilization phases the assignment M^* still satisfies all equality clauses in P and in addition it satisfies clauses of type $x \in y$ too. (Notice that the φ_{y_0}-stabilization phase can be bypassed, since $\emptyset \in \overline{\pi^{y_0}}$ is true after the initialization phase (cf. Definition 10.4(b)).

In this connection it is enough to notice the following.

• M^*x cannot be identical to any previously generated element. To see this, note that, for reasons already explained, M^*x cannot be identical to any auxiliary or secondary element, or any pair $\{p, a\}$. Moreover, no two sets M^*x, M^*y can be equal, since at the end of the initialization phase $M^*x \cap (\bigcup_{\pi \in \Pi} Sec_\pi) = \bigcup_{\pi(x)=1} Sec_\pi$, and this relationship is never disrupted by a subsequent insertion of an element of $\bigcup_{\pi \in \Pi} Sec_\pi$ into any of the sets $\overline{\pi}$.

• If $\pi^x(y_i) = 1$ and $\pi(x) = 1$, then $\pi(u_i) = 1$ by condition C4. Hence if M^*x is inserted into M^*y_i, all elements of M^*x must already belong to M^*u_i, proving that the relationship $Un(M^*y_i) \subseteq M^*u_i$ is not disrupted by insertion of M^*x into $\overline{\pi^x}$. Thus application of the φ_x-stabilization process restores all relationships $M^*u_i = Un(M^*y_i)$.

• By condition C3, no α which is included either in x (i.e., $\alpha(x) = 1$) or in a variable y which comes before x in the good Uorder of variables can be component of a chain α_i of places satisfying $\pi^x = \alpha_1$, $\alpha_{i+1} = \varphi_x(\alpha_i, y_{j_i})$. However, it is only such places that are affected either by insertion of M^*x into π^x or by the subsequent stabilization process. It follows that no already established relationship $M^*y = \bigcup_{\pi(y)=1} \overline{\pi}$ is disrupted by the said insertion and stabilization operations. This guarantees that literals of type $x \in y$ are correctly modelled by M^*. Therefore at the end of the process described above, M^* will be a model for all the clauses of P (provided that a good assignment of auxiliary and secondary elements to the nodes of G exists). Thus the lemma is proved.   ∎

We therefore will have proved that conditions C1–C4 are necessary and sufficient for the injective satisfiability of P (at least in the situation in which there are no trapped places) as soon as we show how to construct a *good assignment of auxiliary and secondary elements to the nodes of G*. This will be done in the following lemma.

Lemma 10.4 *Assume that conditions C1–C4 hold and that the Ugraph G has no trapped nodes. Then there exists a good assignment of auxiliary and secondary elements to the nodes of G.*

Proof. We can proceed as follows.

Begin with all places π such that $\psi(\pi) = \Omega$. Assign disjoint infinite sets of integers $n \geq 3$ to these places, and for each integer n assigned to a place π build the singleton $\{n\}$. Divide the set of all singletons assigned to π into two infinite subsets and define the singletons in one of the two subsets to be *auxiliary* elements resident at π, and the other singletons to be *secondary* elements resident at π.

Next suppose that there are cyclic places π, but continue to suppose that there are no trapped places. Then, as has been shown earlier, there is a place $\gamma = \pi^{y_0}$ at y_0 and a path through the Ugraph G (see condition C2) to γ from any other cyclic node.

Hence, by definition of the map ψ, there is some cycle $\gamma_1, \ldots, \gamma_{m+1}$ of length m at least 2, such that $\gamma_1 = \gamma_{m+1} = \gamma$, and $\gamma_{i+1} = \psi(\gamma_i), i = 1, \ldots, m$. (Note that this cycle is allowed to contain repetitions.) Define the set $\emptyset_n$ for all $n \geq 0$ by $\emptyset_0 = \emptyset$, $\emptyset_{i+1} = \{\emptyset_i\}$, and let all the sets $\emptyset_n$ of this form with $n \equiv 1 - j$ (mod m) be secondary elements resident at γ_j. (Notice that $\emptyset$ is a secondary element resident at $\gamma_1 = \pi^{y_0}$ and therefore condition (b) of Definition 10.4 is satisfied.) Then form all pairs $\{\emptyset_n, \emptyset_{n+m}\}$ and let all such pairs with $n \equiv -j$ (mod m) be additional secondary elements resident at γ_j. Finally, form all singletons $s_n = \{\{\emptyset_n, \emptyset_{n+m}\}\}$ and let all those with $n \equiv -1 - j$ (mod m) be resident at γ_j. Take the infinite set of the singletons of this last form resident at γ_j and divide this set, in any convenient way, into disjoint parts, both infinite; define the singletons belonging to one of these parts to be *auxiliary* elements resident at γ_j, while the singletons of the other part are defined to be *secondary* elements resident at γ_j.

Next define further singletons $s_{n,j}$ by $s_{n,1} = s_n$, $s_{n,j+1} = \{s_{n,j}\}$. It is easy to see that $s_{n,j} \in^+ s_{\ell,k}$ if and only if $n = \ell$ and $j < k$. Indeed, $s_{n,j} \in^+ s_{\ell,k}$ implies that $\{\emptyset_n, \emptyset_{n+m}\} \in^+ s_{\ell,k}$, and then clearly $\{\emptyset_n, \emptyset_{n+m}\} \in^+ \{\{\emptyset_\ell, \emptyset_{\ell+m}\}\}$, so either $\{\emptyset_n, \emptyset_{n+m}\} = \{\emptyset_\ell, \emptyset_{\ell+m}\}$, implying $n = \ell$, or $\{\emptyset_n, \emptyset_{n+m}\} \in^+ \emptyset_{\ell+m}$, which is impossible. But once we know that $s_{n,j} \in^+ s_{\ell,k}$ implies $n = \ell$, it follows trivially that it must also imply $k > j$.

We have associated infinitely many auxiliary and secondary elements of the form $\{n\}$, $n \geq 3$, with each place π such that $\psi(\pi) = \Omega$. Much as previously, define $s^*_{n,\ell}$, by $s^*_{n,1} = \{n\}$, $s^*_{n,j+1} = \{s^*_{n,j}\}$. Then, $s_{n,j} \in^+ s^*_{\ell,k}$ would imply that $\{\emptyset_n, \emptyset_{n+m}\} \in^+ s^*_{\ell,k}$, and hence $\{\emptyset_n, \emptyset_{n+m}\} \in^+ \ell$, which is impossible since all the elements of an integer are themselves integers. For the same reason, $s^*_{n,j} \in^+ s_{\ell,k}$ is impossible, and $s^*_{n,j} \in^+ s^*_{\ell,k}$ implies that $n = \ell$ and $j < k$.

At this point we have associated infinitely many auxiliary and secondary elements $s_{n,\ell}$ or $s^*_{n,\ell}$ with each place γ' of the cycle $\gamma_1, \ldots, \gamma_{m+1}$, and with each π such that $\psi(\pi) = \Omega$, and it only remains to extend this association to the remaining cyclic and safe places. For this, a simple iterative construction can be used. Regard a place as having been *treated* if secondary and auxiliary elements $s_{n,j}$ of $s^*_{n,j}$ have already been associated with it. If any untreated places remain, choose some α which has already been treated, but for which there remain untreated $\beta_1, \ldots, \beta_k$ such that $\psi(\beta_1) = \cdots = \psi(\beta_k) = \alpha$. Divide the infinitely many *secondary* elements $s_{n,j}$ or $s^*_{n,j}$ resident at α into k subsequences, all infinite, and define the elements $s_{n,j+1}$ (or $s^*_{n,j+1}$) such that $s_{n,j}$ (or $s^*_{n,j}$) belongs to the ith of these subsequences to be resident at β_i, $i = 1, \ldots, k$. Divide the infinite set of resident items thereby associated with each of the β_is into two infinite subsequences, and define the elements of one of these subsequences to be *auxiliary* elements resident at β_i, while the elements of the other subsequence are defined to be *secondary* elements resident at β_i. Continue in this way as long as any untreated places remain. Finally, in order to ensure that every secondary element b is a member of some auxiliary element, we adopt the technical convention of forming $\{b\}$ as an auxiliary element with residence at Ω whenever b is a singleton secondary element for which $\{b\}$ is not otherwise introduced.

It is clear that the collection of auxiliary and secondary elements constructed in this way defines naturally a *good assignment of auxiliary and secondary elements to the nodes of the graph G*, as from Definition 10.4. ∎

In view of the preceding lemma, Lemma 10.3 can now be restated as follows.

Lemma 10.5 (Completeness) *Assume that conditions C1–C4 hold and that the Ugraph G has no trapped nodes. Then P is injectively satisfiable.* □

This completes our treatment of the case in which no trapped places exist, i.e. shows that if the Ugraph G appearing in condition C2 has no trapped nodes, then conditions C1–C4 are necessary and sufficient for the injective satisfiability of P by a model having Π as its set of places. The case in which trapped places may exist is considered in the next section.

10.3   The decision algorithm when trapped places are present

The construction of an injective model of P in the presence of trapped places is a bit subtler than that applicable in the case considered in the preceding section. The main differences stem from the fact that in this case the role of the single place $\gamma = \pi^{y_0}$ must be played by a finite set of places, called $\gamma_1, \gamma_2, \ldots, \gamma_k$ in the discussion which follows; moreover, sets associated with trapped places can only range over a finite family of finite sets known *a priori*. This last limitation makes the stabilization phase more complicated.

Suppose again that M is an injective model for a conjunction P of clauses of type (10.1). Let Π, G, σ^π, etc. be as in the preceding section. Also, suppose that the Ugraph G may contain trapped places (in which case it follows from Definition 10.1 and Lemma 10.1 that the place π^{y_0} at y_0 is null).

Remark 10.7 In this section we will use the convention of subscribing the edge relation with the name of the graph containing it. So, for example, we will write $\alpha \Rightarrow_G \beta$ to indicate the edge $\alpha \Rightarrow \beta$ of the Ugraph G. □

The first necessary (and sufficient) condition for P to be injectively satisfiable is equal to condition C1 stated in the previous section.

Condition C1′. *Let P be a conjunction of atoms of type (10.1), and let Π be a set of places of P relative to an injective model of P. Then no two distinct variables of P are Π-equivalent.* □

The following definition introduces the concept of *height* of a trapped place.

Definition 10.5 *Let τ be a trapped place of the Ugraph G. Then the* height *of τ, denoted $\mathrm{height}(\tau)$, is defined as one more than the length of the longest path forward from τ to π^{y_0}.* □

We recall that for each natural number h, V_h designates the finite family of all sets of rank less than h (cf. Section 2.5). Note (for implicit use in what follows) that as an immediate consequence of Lemma 2.4(iv), the union of subsets of V_h is itself a subset of V_h.

Lemma 10.6 *If τ is trapped, then $\sigma^\tau \subseteq V_{\mathrm{height}(\tau)}$.*

Proof. We proceed by induction on $\mathrm{height}(\tau)$. From Lemma 10.1(a_2), it follows that

$$Un(\sigma^\tau) \subseteq \bigcup_{\tau \Rightarrow_G \beta} \sigma^\beta,$$

where for all places β such that $\tau \Rightarrow_G \beta$, we have $\mathrm{height}(\beta) < \mathrm{height}(\tau)$. Thus, by induction,

$$Un(\sigma^\tau) \subseteq V_{\mathrm{height}(\tau)-1},$$

which implies the lemma. ■

Let H be the maximum height of any trapped place τ in G (if there are no trapped places, then $H = 0$.

For all places π in Π, we put

$$\Sigma^\pi = \sigma^\pi \cap V_{H+1}. \tag{10.6}$$

The following lemma refines Lemma 10.1(a_2),(b) for the case of trapped places.

Lemma 10.7 *If the place τ is trapped, then*

(a) $\sigma^\tau = \Sigma^\tau \subseteq V_H$;

(b) $Un(\Sigma^\tau) \subseteq \bigcup_{\tau \Rightarrow_G \beta} \Sigma^\beta$;

(c) *if $\tau(u) = 1$ for some Uclause $u = Un(y)$ in P, and all places α such that $\alpha \subseteq y$ and $\alpha \Rightarrow_G \tau$ are trapped, then $\Sigma^\tau \subseteq \bigcup_{\alpha \in A} Un(\Sigma^\alpha)$, where A is the set of all places α such that $\alpha \subseteq y$ and $\alpha \Rightarrow_G \tau$.*

Proof. (a) follows at once from (10.6) and Lemma 10.6.

Concerning (b), we notice that from Lemma 10.1(a_2), we have $Un(\sigma^\tau) \subseteq \bigcup_{\tau \Rightarrow_G \beta} \sigma^\beta$. But, from (a) above, $Un(\Sigma^\tau) = Un(\sigma^\tau) \subseteq V_{H+1}$. Therefore

$$Un(\Sigma^\tau) \subseteq \left(\bigcup_{\tau \Rightarrow_G \beta} \sigma^\beta \right) \cap V_{H+1} = \bigcup_{\tau \Rightarrow_G \beta} (\sigma^\beta \cap V_{H+1}) = \bigcup_{\tau \Rightarrow_G \beta} \Sigma^\beta,$$

which proves (b).

Finally, concerning (c), we only need to observe that from Lemma 10.1(b), we have $\sigma^\tau \subseteq \bigcup_{\alpha \in A} Un(\sigma^\alpha)$, with A the set of places α such that $\alpha \subseteq y$ and $\alpha \Rightarrow_G \tau$, and that, moreover, from (a) above, $\sigma^\tau = \Sigma^\tau \subseteq V_H$, and similarly for all places α such that $\alpha \Rightarrow_G \tau$, $\sigma^\alpha = \Sigma^\alpha \subseteq V_H$. ∎

Next we introduce the concept of *trapped variables*.

Definition 10.6 *A variable x of P is said to be trapped if every π in Π such that $\pi \subseteq x$ is trapped.* □

In the lemma below we prove that the variable y_0 (where we are assuming that $y_0 = y_0 \setminus y_0$ occurs in P) is trapped.

Lemma 10.8 *The variable y_0 is trapped.*

Proof. Since $y_0 = y_0 \setminus y_0$ is in P, for all $\pi \in \Pi$, $\pi(y_0) = 0$. Therefore y_0 is vacuously trapped. ∎

Another useful fact is stated in the following lemma.

Lemma 10.9 *Let x be a trapped variable of P. Then*

$$Mx = \bigcup_{\pi(x)=1} \Sigma^\pi \in \Sigma^{\pi^x}.$$

Proof. Indeed, from Lemma 10.7,

$$Mx = \bigcup_{\pi(x)=1} \sigma^\pi = \bigcup_{\pi(x)=1} \Sigma^\pi \subseteq V_H.$$

Therefore, since $V_{H+1} = Pow(V_H)$, $Mx \in V_{H+1}$, proving the lemma. ∎

We have also the following lemma.

Lemma 10.10 *Let* x *be a variable of* P *such that the place at* x, π^x, *is trapped. Then, the variable* x *is trapped (and, by the preceding lemma,* $\bigcup_{\pi(x)=1} \Sigma^\pi \in \Sigma^{\pi^x}$ *).*

Proof. Since $Mx \in \sigma^{\pi^x}$, it follows that $Mx \subseteq Un(\sigma^{\pi^x})$. Therefore, if $\pi \subseteq x$, then $\sigma^\pi \cap Un(\sigma^{\pi^x}) \neq \emptyset$, which by Lemma 10.1($a_1$) implies $\pi^x \Rightarrow_G \pi$. But π^x is trapped, therefore so is the place π, proving that all places $\pi \in \Pi$ such that $\pi \subseteq x$ are trapped, i.e. the variable x is trapped. ∎

Definition 10.7 *For* $\alpha, \beta \in \Pi \cup \{\Omega\}$, *we put* $\alpha \sim \beta$ *if* $\alpha \Rightarrow^*_G \beta$ *and* $\beta \Rightarrow^*_G \alpha$, *where* $\gamma \Rightarrow^*_G \delta$ *means that there is a directed path in* G, *possibly null (i.e.* $\gamma = \delta$), *from* γ *to* δ. □

The following lemma states some immediate properties of the relation $\sim$ (cf. Remarks 10.4).

Lemma 10.11   *(a)* $\sim$ *is an equivalence relation;*

(b) the partition of Π *induced by* $\sim$ *refines the crude partition of the set of all places into safe, trapped, and cyclic places.* □

By $[\alpha]$ we will denote the equivalence class (relative to the relation $\sim$) containing α. (These are the strongly connected components of the Ugraph G. See [AHU76].)

Define the auxiliary directed graph $\overline{G}$ induced by the Ugraph G as the graph whose nodes are the equivalence classes of $\sim$ and whose edges are the following:

$[\alpha] \Rightarrow_{\overline{G}} [\beta]$ is an edge of $\overline{G}$ if $[\alpha] \neq [\beta]$ and there are $\alpha' \in [\alpha]$, $\beta' \in [\beta]$ such that $\alpha' \Rightarrow_G \beta'$ is an edge of the Ugraph G.

It is obvious that $\overline{G}$ has no self-loops, and that $\overline{G}$ is acyclic. Next suppose that there are cyclic places. Consider the subgraph

$$K = \{[\alpha] \in \overline{G} : \alpha \text{ is cyclic}\}$$

induced by $\overline{G}$. Obviously K is acyclic too, and hence, since K is finite, there exist elements of K with no outgoing edges to any other element of K. Therefore we have the following lemma.

Lemma 10.12 *The set* $K_0 = \{[\alpha] \in K : [\alpha] \text{ has no outgoing edge in } K\}$ *is non-empty.* □

We have also the following properties.

Lemma 10.13 *Let* $[\alpha] \in K_0$. *Then*

(a) if $[\alpha] \Rightarrow_{\overline{G}} [\beta]$, β *must be trapped;*

(b) there must be at least one trapped place β *such that* $[\alpha] \Rightarrow_{\overline{G}} [\beta]$;

(c) every element in $[\alpha]$ lies in a cycle of G.

Proof. *(a)* By the definition of K_0, β cannot be cyclic; and clearly β cannot be safe because otherwise α would also be safe. Therefore β is trapped.

(b) As observed earlier (see condition C2 and Remark 10.5), there is a path from every non-safe place to the place π^{y_0} at y_0, and the place π^{y_0} cannot be an element of $[\alpha]$ since it is not cyclic (cf. Lemma 10.1(c)). Then *(b)* follows from *(a)* above.

(c) Take any element α' in $[\alpha]$. Then there must be a path from α' to a cycle of places. But no edges along this path can exit $[\alpha]$, since if any did it would have to terminate at a trapped place, which is clearly impossible. It follows that $[\alpha]$ must contain at least one cycle; but then, since all the elements of $[\alpha]$ are equivalent, it follows that every element of $[\alpha]$ lies in a cycle. ∎

Let

$$[\gamma_1], [\gamma_2], \ldots, [\gamma_k] \tag{10.7}$$

be the elements of K_0.

Lemma 10.14 *There is a path from every cyclic place of P to at least one of the places γ_i.*

Proof. The lemma follows immediately from the very definition of K_0 and Lemma 10.13(c). ∎

For each $[\gamma_i]$, $i = 1, \ldots, k$, consider the set

$$S_i = \bigcup_{\beta \in [\gamma_i]} \sigma^\beta.$$

Let s_i be an element of S_i having minimal rank. Without loss of generality, we can assume that $s_i \in \sigma^{\gamma_i}$.

Various useful properties of the elements s_i now follow easily.

Lemma 10.15   *(a) $s_i \in \Sigma^{\gamma_i}$;*

(b) $s_i \notin^+ s_j$, for any $i, j \in \{1, 2, \ldots, k\}$;

(c) $s_i \notin^+ \Sigma^\tau$, for each $i \in \{1, 2, \ldots, k\}$ and each trapped place τ.

Proof. From Lemma 10.1(a_2), it plainly follows

$$s_i \subseteq \bigcup_{\gamma_i \Rightarrow_G \beta} \sigma^\beta.$$

Moreover, by the minimality of the rank of s_i, no element of s_i belongs to S_i; from which it follows that every such element lies outside the union $\bigcup \sigma^\alpha$ extended over all cyclic places α. Thus

$$s_i \subseteq \bigcup_{\substack{\gamma_i \Rightarrow_G \beta \\ \&\ \beta\ is\ trapped}} \sigma^\beta = \bigcup_{\substack{\gamma_i \Rightarrow_G \beta \\ \&\ \beta\ is\ trapped}} \Sigma^\beta.$$

Hence $s_i \in \sigma^{\gamma_i} \cap V_{H+1} = \Sigma^{\gamma_i}$, proving (a).

Concerning (b), suppose by contradiction that there exist i, j such that $s_i \in^+ s_j$. Then $i \neq j$ and for some $u_1, \ldots, u_\ell$ we must have $s_i \in u_1 \in \cdots \in u_\ell \in s_j$. That, by Lemma 10.1($a_1$) implies the existence of places $\beta_1, \ldots, \beta_\ell$ such that $\gamma_j \Rightarrow_G \beta_\ell \Rightarrow_G \cdots \Rightarrow_G \beta_1 \Rightarrow_G \gamma_i$, and then plainly all the β_ts must be cyclic places. From this it follows at once that $[\gamma_j] \Rightarrow^*_K [\gamma_i]$ which, by the definition of the $[\gamma_i]$s, is a contradiction. This completes the proof of (b).

Finally, concerning (c), assume by contradiction that $s_i \in^+ \Sigma^\tau$, for some s_i and some trapped place τ. Then, by reasoning as above, there would exists a chain $\tau \Rightarrow_G \beta_\ell \Rightarrow_G \cdots \Rightarrow_G \beta_1 \Rightarrow_G \gamma_i$. But this is impossible, since τ is trapped while γ_i is cyclic. Thus (c) is proved, and the proof of the lemma is completed. ∎

The preceding results show that if there exists an injective model M of P with places Π, Ugraph G, etc., involving trapped places τ, the following combinatorial condition must be satisfied.

Condition C2′. *Let H be the maximum height, in the Ugraph G, of any trapped place τ, and, as above, let V_{H+1} be the set of all sets of rank less than $H + 1$. Then there must exist a map $\pi \mapsto \Sigma^\pi$ of places in Π to pairwise disjoint subsets of V_{H+1} such that $\Sigma^\pi \neq \emptyset$ whenever π is trapped, and there must exist a map $x \mapsto \pi^x$ from the set of all variables occurring in P to the set of all places, such that:*

(i) If $x \in y$ occurs in P then $\pi^x(y) = 1$.

(ii) If the variable x is trapped (cf. Definition 10.6), then $\bigcup_{\pi(x)=1} \Sigma^\pi \in \Sigma^{\pi^x}$.

(iii) If τ is a trapped place, then $Un(\Sigma^\tau) \subseteq \bigcup_{\tau \Rightarrow_G \beta} \Sigma^\beta$.

(iv) If there are any cyclic places, then there exists a non-empty set $\{\gamma_1, \ldots, \gamma_k\}$ of such places, each lying in some cycle of the Ugraph G, and for each $i = 1, \ldots, k$ a non-empty element $s_i \in \Sigma^{\gamma_i}$, such that

(iv_1) $s_i \notin^+ s_j$, for any $i, j = 1, \ldots, k$;

(iv_2) $s_i \notin^+ \Sigma^\tau$, for $i = 1, \ldots, k$ and any trapped place τ.

(iv_3) $s_i \subseteq \bigcup_{\substack{\gamma_i \Rightarrow_G \beta \\ \& \ \beta \ \text{is trapped}}} \Sigma^\beta$, for every $i = 1, \ldots, k$.

(iv_4) There must exist a path through the Ugraph G forward from every cyclic place α to some γ_i, $i = 1, 2, \ldots, k$. □

Next we define various maps which are useful for the construction of a model for P. Let $u_i = Un(y_i)$ be a Uclause of P. From Lemma 10.1(a_1), for each place $\alpha \subseteq u_i$, there exists a place $\beta \subseteq y_i$ such that $\beta \Rightarrow_G \alpha$. If any such β is non-trapped, choose such a β and call it $\varphi(\alpha, y_i)$; otherwise, let $\varphi(\alpha, y_i)$ be any trapped place $\beta \subseteq y_i$ such that $\beta \Rightarrow_G \alpha$. Notice that if $\varphi(\alpha, y_i)$ is trapped, then the place α must be trapped too, and therefore, by Lemma 10.7(c), $\Sigma^\alpha \subseteq \bigcup_{\beta \in B} Un(\Sigma^\beta)$, where B is the set of all places β such that $\beta \subseteq y_i$ and $\beta \Rightarrow_G \alpha$.

As in the case, treated previously, in which no trapped places exist, we can define sets Π_x, a good Uorder of the variables occurring in P and good Umaps $\varphi_x(\alpha, y_i)$, one for each variable x in P (cf. Remark 10.6), having properties (ii), (iii) of condition C3 as well as the following property:

(iv) *If π^x is non-trapped, then all places in Π_x are non-trapped.*

To show that if a model M of P exists condition (iv) can always be satisfied along with the other conditions (ii) and (iii), we reason as follows. Let π^x be non-trapped and $\pi \in \Pi_x$. We recall that $\Pi_x = \{\pi \in \Pi : Mx \in^+ \sigma^\pi\}$. Therefore $Mx \in^+ \sigma^\pi$. It follows that there are elements $u_1, \dots, u_\ell$ such that $Mx \in u_1 \in \cdots \in u_\ell \in \sigma^\pi$. Suppose that π is a trapped place. Let $\beta_1, \dots, \beta_\ell$ be places such that $u_j \in \sigma^{\beta_j}$, for $j = 1, \dots, \ell$. The preceding chain of memberships implies (inductively) that $\pi = \beta_\ell \Rightarrow_G \cdots \Rightarrow_G \beta_1 \Rightarrow_G \pi^x$. Therefore, since we are assuming that π is trapped, all places in the sequence must be trapped, contradicting our assumption that π^x is not trapped.

Summing up, the preceding discussion shows that if there exists a model M of P with places Π, Ugraph G, etc., involving trapped places, then the following combinatorial condition must be satisfied.

Condition C3'. *There must exist maps $\varphi(\alpha, y_i)$ and $\varphi_x(\alpha, y_i)$ defined for places $\alpha \in \Pi$, variables x of P, and Uclauses $u_i = Un(y_i)$ in P, both having values (when defined) which are places $\beta \in \Pi$ such that $\beta \Rightarrow_G \alpha$ and $\beta(y_i) = 1$. Moreover, there must exist a good Uorder of the variables and a set of places Π_x, for each variable x, such that if the place π^x at x is non-trapped, then all places in Π_x are non-trapped, and φ_x must be a good Umap in the sense of the condition C3 stated in the preceding section, i.e. $\varphi_x(\pi, y_i)$ must be defined for every $\pi \in \Pi_x$ and every Uclause $u_i = Un(y_i)$ such that $\pi \subseteq u_i$. In addition, $\varphi(\pi, y_i)$ must be defined whenever $\pi \subseteq u_i$ and must be non-trapped if there is any non-trapped $\beta \subseteq y_i$ such that $\beta \Rightarrow_G \pi$. Moreover, if $\varphi(\pi, y_i)$ is trapped, then we must have $\Sigma^\pi \subseteq \bigcup_{\beta \in B} Un(\Sigma^\beta)$, where B is the set of all places β such that $\beta \subseteq y_i$ and $\beta \Rightarrow_G \pi$.* $\square$

Finally, the next condition is identical to condition C4 stated in the preceding section.

Condition C4'. *If $u_i = Un(y_i)$ is a Uclause of P, $\pi^x(y_i) = 1$ and $\pi(x) = 1$, then $\pi(u_i) = 1$.* $\square$

So far we have proved the following lemma.

Lemma 10.16 (Soundness) *Let M be an injective model of P, where the Ugraph G relative to P and Π_M may contain trapped nodes. Then conditions $C1'$–$C4'$ hold.* □

We shall now complete our analysis by showing that conditions $C1'$–$C4'$ are also sufficient for injective satisfiability of the clauses P by a model having the set of places Π. Suppose therefore that these conditions are satisfied. To construct a model M^* for the clauses P, we will use much the same method as in the easier case in which no trapped places exist, but with the difference that during stabilization phases no elements are ever added to a set $\bar{\tau}$ if the place τ is trapped.

We begin by considering the case in which cyclic places do occur. (However, if there are no cyclic places the proof is much the same; in this case the reader has only to ignore what is said about cyclic places in the following paragraphs.)

Our first step is to define an auxiliary map ψ from non-trapped places to non-trapped places. Clause (iv) of condition $C2'$ implies the existence of cyclic places $\gamma_1, \ldots, \gamma_k$, each lying in some simple cycle which we designate by

$$\gamma_i = \beta_{i,1} \Rightarrow_G \beta_{i,2} \Rightarrow_G \cdots \Rightarrow_G \beta_{i,m_i+1} = \gamma_i, \tag{10.8}$$

where $m_i \geq 1$, for $i = 1, \ldots, k$.

Definition 10.8 *For each of the places $\beta_{i,j}$, we put $\psi(\beta_{i,j}) = \beta_{i,j+1}$, $1 \leq j \leq m_i$, with the understanding that j and $j+1$ are taken modulo m_i. (Note that no two of the cycles $\beta_{i,1}, \ldots, \beta_{i,m_i+1}$ intersect.) For all remaining cyclic places α we put $\psi(\alpha) = \beta$, where β lies one step closer than α along some shortest path through the Ugraph G to an element in one of these cycles. If α is safe, we put $\psi(\alpha) = \Omega$ if $\alpha \Rightarrow_G \Omega$; otherwise we put $\psi(\alpha) = \beta$, where β lies one step closer to Ω than does α (again, along some shortest path through G to Ω).* □

Condition (iv) of $C2'$ ensures that ψ is well-defined for all cyclic places. Moreover, it is obvious that if α is cyclic then $\psi(\alpha)$ is cyclic too.

Having defined the auxiliary map ψ, we can now adapt the definition of good assignment of auxiliary and secondary elements to the nodes of the Ugraph G in the case in which trapped places are present.

Definition 10.9 *Assume that condition $C2'$ is satisfied, and let the set Π of places, the Ugraph G, the map ψ, etc., be as above. A good assignment of auxiliary and secondary elements to the nodes of G (when trapped places may be present, and relative to the sets Σ^π of condition $C2'$) is any pair of maps Aux, Sec which associate sets to the nodes of the Ugraph G in such a way that, if we*

- *denote by Aux_π the set $Aux(\pi)$ and by Sec_π the set $Sec(\pi)$, for every node π of G;*

- *call* auxiliary *(respectively, secondary) elements resident at π the members of Aux_π (respectively, of Sec_π);*

- *for all $p \in Aux_\pi \cup Sec_\pi \cup \Sigma^\pi$, call* residence *of p the place π;*

then

- *(a) Aux_π and Sec_π are countably infinite for each non-trapped node π in G; moreover all auxiliary elements have cardinality 1, whereas all secondary elements have cardinality at most 2;*

- *(b) every secondary element must satisfy $b \in^+ a$ (cf. Definition 9.6), for some auxiliary element a not necessarily sharing its residence with b;*

- *(c) no two auxiliary elements a, a' (not necessarily resident at the same node) can satisfy $a \in^+ a'$;*

- *(d) if $\psi(\alpha) \neq \Omega$, every member of an auxiliary or secondary element resident at a non-trapped place α is either a secondary element resident at the place $\psi(\alpha)$, or a member of Σ^τ, for some trapped place τ such that $\alpha \Rightarrow_G \tau$, the latter possibility only arising for members of secondary elements;*

- *(e) no auxiliary or secondary element a resident at a non-trapped place α satisfies $a \in^+ \Sigma^\tau$, for any trapped place τ;*

- *(f) the sets in*

$$\{Aux_\pi : \pi \in \Pi \ and \ \pi \ is \ non\text{-}trapped\}$$
$$\cup\{Sec_\pi : \pi \in \Pi \ and \ \pi \ is \ non\text{-}trapped\}$$

  are all pairwise disjoint;

- *(g) $(Aux_\pi \cup Sec_\pi) \cap \Sigma^\tau = \emptyset$, for all non-trapped $\pi \in \Pi$ and trapped $\tau \in \Pi$;*

- *(h) any auxiliary or secondary element resident at a safe place has rank greater than H, where H is the maximum height of any trapped place τ in G;*

- *(i) any auxiliary or secondary element a resident at a cyclic place and different from the sets s_i, $i = 1, \ldots, k$, satisfies $s_j \in^+ a$, for some $j \in \{1, \ldots, k\}$.* $\square$

In analogy with the preceding section, we will prove that if conditions C1′–C4′ hold and there exist a good assignment of auxiliary and secondary elements to the nodes of G, then P is injectively satisfiable even in the presence of trapped variables. Subsequently, we will show that the existence of a good assignment follows from conditions C1′–C4′, thus proving that such conditions are also sufficient for P to be injectively satisfiable.

Lemma 10.17 *Assume that conditions C1'–C4' hold, that the Ugraph G may have trapped nodes and that there exists maps Aux, Sec which are a good assignment of auxiliary and secondary elements to the nodes of G according to Definition 10.9. Then P is injectively satisfiable.*

Proof: Let $\Pi, G, \psi, \varphi, \varphi_x, \Pi_x$, etc., be as above and such that conditions C1'–C4' hold. Assume also that Aux, Sec are a good assignment of auxiliary and secondary elements to the nodes of G (according to Definition 10.9) and that the Ugraph G has trapped nodes. We will first exhibit a process which constructs an injective model for P when non-trapped places are present. Later we will observe that the case in which all places are trapped is much more easily treated.

INITIALIZATION PHASE

1. Arrange the set

$$
\left(\bigcup_{\substack{\pi \in G \\ \& \ \pi \ non-trapped}} Sec_\pi \right) \cup \left(\bigcup_{\substack{\tau \in G \\ \& \ \tau \ trapped}} \Sigma^\tau \right)
$$

   into an infinite list L_0 (note that this can be done since we have assumed that there are non-trapped places, so that the above set is countably infinite by Definition 10.9(a)).

2. For each non-trapped node π in G, arrange the set Aux_π of auxiliary elements resident at π into infinite pairwise disjoint lists $L_{\pi,x}$, one for each variable x in P.

3. Initialize L_1 (*accessory list*) to the empty list.

4. For all non-trapped places $\pi \in \Pi$, put

$$
\overline{\pi} \leftarrow Sec_\pi.
$$

5. For all trapped places $\tau \in \Pi$, put

$$
\overline{\tau} \leftarrow \Sigma^\tau.
$$

   (Comment: Note that at the end of step 5, by (a), (f), and (g) of Definition 10.9, all the $\overline{\pi}$ are non-empty and pairwise disjoint. Therefore all clauses in P of type $x = y \cup z$, $x = y \setminus z$ are correctly modelled by the initial injective interpretation $Mx = \bigcup_{\pi(x)=1} \overline{\pi}.$)

END INITIALIZATION PHASE

φ-ψ-STABILIZATION PHASE

$r \leftarrow 1$

DO FOREVER

$r \leftarrow 1 - r$

IF L_r is empty **THEN** *CONTINUE* **END IF**;

Let p be the first component of the list L_r.

Discard it from L_r.

Let π be the residence of p, viz. the place such that $p \in \overline{\pi}$.

FOR ALL Uclauses $u_i = Un(y_i)$, $i = 1, \ldots, m$, such that $\pi(u_i) = 1$ and such that the place $\varphi(\pi, y_i)$ is non-trapped **DO**

- let $\beta_i = \varphi(\pi, y_i)$;
- pick the first component a of the list $L_{\psi(\beta_i), y_0}$ and discard it from $L_{\psi(\beta_i), y_0}$ (observe that since β_i is non-trapped, then $\psi(\beta_i)$ is a non-trapped place and therefore $\psi(\beta_i)$ has associated auxiliary and secondary elements);
- put

$$\overline{\beta_i} \quad \leftarrow \quad \overline{\beta_i} \cup \{\{p, a\}\};$$

$$\overline{\psi(\beta_i)} \quad \leftarrow \quad \overline{\psi(\beta_i)} \cup \{a\};$$

- add the sets $\{p, a\}$ and a to the end of the list L_1;

END FOR ALL;

END DO FOREVER;

END φ-ψ-STABILIZATION PHASE.

FOR ALL x in $P \setminus \{y_0\}$ such that the place π^x is non-trapped and such that $\bigcup_{\pi(x)=1} \overline{\pi} \not\subseteq \{s_1, \ldots, s_k\}$, following the good Uorder of variables **DO**

Put

$$\overline{\pi^x} \leftarrow \overline{\pi^x} \cup \{ \bigcup_{\pi(x)=1} \overline{\pi} \}.$$

Intialize L_1 to the list consisting of the single component $\bigcup_{\pi(x)=1} \overline{\pi}$.

φ_x-STABILIZATION PHASE

(Comment: Note that since π^x is non-trapped, then by condition C3' all places β for which $\overline{\beta}$ is affected by the φ_x-stabilization phase are non-trapped and therefore there is no need to make use of the map ψ during the present stabilization loop.)

DO FOREVER

    IF L_1 is empty **THEN** *QUIT* **END IF**;

    Let p be the first component of the list L_1.

    Discard it from L_1.

    Let π be the residence of p.

    FOR ALL Uclauses $u_i = Un(y_i)$, $i = 1, \ldots, m$, such that $\pi(u_i) = 1$ **DO**

        • pick the first component a of the list $L_{\pi,x}$ and discard it from $L_{\pi,x}$;

        • let $\beta_i = \varphi_x(\pi, y_i)$; put

$$\overline{\beta_i} \;\leftarrow\; \overline{\beta_i} \cup \{\{p, a\}\};$$
$$\overline{\pi} \;\leftarrow\; \overline{\pi} \cup \{a\};$$

        • add $\{p, a\}$ and a to the end of the list L_1;

    END FOR ALL;

END DO FOREVER;

END φ_x-STABILIZATION PHASE

END FOR ALL.

Notice that after the initialization phase we will always have $\overline{\tau} = \Sigma^\tau$, for all trapped places $\tau \in \Pi$.

Much as in the proof of Lemma 10.3, we will prove that if for all variables x in P we put

$$M^*x = \bigcup_{\pi(x)=1} \overline{\pi},$$

where the sets $\overline{\pi}$ are those produced by the above process, then M^* is an injective model of P. We first show that the temporary assignment obtained just after the end of the φ-ψ-stabilization phase is an injective model of the equalities of P, i.e., of the clause in P of type

$$x = y \cup z, \quad x = y \setminus z, \quad u = Un(y).$$

To show that M^* is an injective model of the equalities in P of the first two forms, we only need to prove that the sets $\overline{\pi}$ remain mutually disjoint during the whole φ-ψ-stabilization phase.

Note that when it is generated, the pair $\{p, a\}$ must be distinct from all elements previously inserted into any of the $\overline{\pi}$, and so must the auxiliary element a. Indeed, the singleton a cannot be a previously formed pair, nor can it equal any element of any Σ^{τ}, τ trapped, or any secondary element resident at any non-trapped place or any previously used auxiliary element a. Moreover, $\{p, a\}$ can never equal any previously used auxiliary element, nor can it equal any secondary element b, since then there would exist an auxiliary a' such that $a \in^{+} a'$, which is impossible. Finally, for the same reason as in the simpler case considered earlier, in which there exist no trapped places, $\{p, a\}$ can never equal any previously formed pair $\{q, a'\}$.

It follows that the sets $\overline{\pi}$ remain disjoint throughout the φ-ψ-stabilization process, which continues until a pair $\{p, a\}$ has been formed for every p inserted into any set $\overline{\pi}$ such that there is a Uclause $u_i = Un(y_i)$ for which $\pi \subseteq u_i$ and $\varphi(\pi, y_i)$ is non-trapped. Moreover, before the φ-ψ-stabilization process begins, we have $Un(\overline{\alpha}) \subseteq \bigcup_{\beta(u_i)=1} \overline{\beta}$ for each Uclause $u_i = Un(y_i)$ and $\alpha \subseteq y_i$. Indeed, for α trapped $Un(\Sigma^{\alpha}) \subseteq \bigcup_{\alpha \Rightarrow_G \beta} \Sigma^{\beta}$ by (iii) of condition C2'. Moreover by Definition 10.9(d) above, every element of a secondary element p inserted into $\overline{\alpha}$ is either a secondary element inserted into $\psi(\beta) \subseteq u_i$, or an element of Σ^{τ} for some trapped τ such that $\tau \subseteq u_i$ and $\alpha \Rightarrow_G \tau$. On the other hand, the φ-ψ-stabilization process does not disturb this condition, since a pair $\{p, a\}$ is only inserted into $\overline{\beta}$, where $\beta \subseteq y_i$, when p is already in some $\overline{\alpha}$, with $\alpha \subseteq u_i$; moreover a is then inserted into $\overline{\psi(\beta)}$, which must also satisfy $\psi(\beta) \subseteq u_i$. (Note also that by Definition 10.9(d) above, when a is inserted into $\overline{\psi(\beta)}$, all the elements of a are already present in $\bigcup_{\beta(u_i)} \overline{\beta}$.)

Thus, for each Uclause $u_i = Un(y_i)$ and $\alpha \subseteq y_i$, we continue to have $Un(\overline{\alpha}) \subseteq \bigcup_{\beta(u_i)=1} \overline{\beta}$ at the end of the φ-ψ-stabilization process. However, we also have $\overline{\beta} \subseteq Un(\overline{\varphi(\beta, y_i)})$ if $\beta \subseteq u_i$ and $\varphi(\beta, y_i)$ is non-trapped. Moreover, if $\varphi(\beta, y_i)$ is trapped, then β is trapped too and it follows by condition C3' that $\overline{\beta} = \Sigma^{\beta} \subseteq \bigcup_{\gamma \in B} Un(\Sigma^{\gamma})$, where B is the set of all places γ such that $\gamma \subseteq y_i$ and $\gamma \Rightarrow_G \beta$ (all these places are trapped). But in this case $\bigcup_{\gamma \in B} Un(\Sigma^{\gamma}) = \bigcup_{\gamma \in B} Un(\overline{\gamma})$, and hence $\overline{\beta} \subseteq \bigcup_{\gamma \in B} Un(\overline{\gamma})$ in every case; i.e., at the end of the φ-ψ-stabilization process all Uclauses are correctly modelled. Moreover, since the non-empty sets $\overline{\beta}$ remain disjoint throughout the φ-ψ-stabilization process, the temporary assignment M^* is injective and it satisfies all clauses $x = y \cup z$ and $x = y \setminus z$ too. In addition, since the value M^*x assigned to a variable x is always understood to be $\bigcup_{\pi(x)=1} \overline{\pi}$, it follows from (i) and (ii) of condition C2' that the clauses $x \in y$ containing a given variable x are correctly modelled whenever the place π^x is trapped (indeed if π^x is trapped, by condition C4' the variable x is also trapped).

Next we show that after completion of the φ_x-stabilization phases the assignment M^* satisfies all clauses in P of type $x \in y$ also. It is enough to establish that no set M^*x inserted into a set π^x at the start of a phase of the construction described above is equal to a previously constructed auxiliary or secondary element a, or a pair $\{p, a\}$, or an element of a set Σ^τ with τ trapped, or a previously constructed model M^*y. This can be shown as follows. Suppose, first, that x is non-trapped, so that M^*x contains countably infinite elements. Thus, M^*x clearly cannot equal any auxiliary element a or pair $\{p, a\}$, nor can it equal any secondary element b, since every such element is $b \in^+ a$ for some auxiliary a, and thus we would have $a' \in^+ a$, where $a' \in M^*x$. Moreover, $M^*x = M^*y$ cannot hold if the variables x and y are distinct because at every stage of our construction the sets $\overline{\pi}$ remain pairwise disjoint. Finally, $M^*x \notin \bigcup_{\tau \; trapped} \Sigma^\tau$, because, as observed above, M^*x has infinitely many elements. This shows that if x is non-trapped, neither the insertion of M^*x into $\overline{\pi}$, nor the subsequent stabilization process disrupts the disjointness of sets $\overline{\pi}$. Next consider the case in which the variable x is trapped, but in which the place π^x is non-trapped (since otherwise we would not have to insert M^*x into $\overline{\pi^x}$; but notice that by (ii) of condition C2', if π^x were trapped, M^*x would already be there). Since x is trapped, M^*x has rank at most H (we recall that we are assuming that H is the maximum height of any trapped place τ in G). Hence M^*x is different from any auxiliary or secondary element resident at any safe place, since these elements have height greater than H (cf. Definition 10.9(h)). For the same reason M^*x is different from any pair $\{p, a\}$ with a resident at a safe place. On the other hand, M^*x cannot equal any auxiliary a or pair $\{p, a\}$ with a resident at a cyclic place, nor can it equal any secondary element b resident at a cyclic place and different from the s_is, $i = 1, \ldots, k$, because for each such pair or element c, by (i) of Definition 10.9 we have $s_j \in^+ c$ for some $j \in \{1, \ldots, k\}$, whereas by (iv) of C2', it follows from $M^*x \subseteq \bigcup_{\tau \; trapped} \Sigma^\tau$ that $s_j \notin^+ M^*x$, for all $j = 1, \ldots, k$. Moreover, by (ii) of C2', we have $M^*x \in \Sigma^{\pi^x}$, and hence, since π^x is non-trapped, it follows by the disjointness of the Σ^τ, τ trapped, that $M^*x \notin \bigcup_{\tau \; trapped} \Sigma^\tau$. For the same reason, M^*x can neither equal any s_i which does not belong to Σ^{π^x} nor equal an element $s_i \in \Sigma^{\pi^x}$ since we suppose that M^*x is not in $\overline{\pi^x}$ before processing of the variable x (whereas during the initialization phase, every s_j is inserted into the set $\overline{\pi}$ for which $s_j \in \Sigma^\pi$). Finally M^*x cannot equal any M^*y with y distinct from x since at every stage of our construction the sets $\overline{\pi}$ are non-empty and pairwise disjoint.

As in the absence of trapped places, the insertion of M^*x into $\overline{\pi^x}$ does not upset any relationship $Un(M^*y_i) \subseteq M^*u_i$, since whenever M^*x is inserted into M^*y_i we have $\pi^x \subseteq y_i$, and then $\alpha \subseteq x$ implies $\alpha \subseteq u_i$ by condition C4', so $Un(M^*y_i) \subseteq M^*u_i$ remains valid.

Taken all in all it follows that, just as in the simpler case considered previously (i.e. in the absence of trapped places), all the clauses of P will be modelled correctly at the end of the series of steps described. This shows that when non-trapped places exist, P is satisfiable by an injective model having Π as its set of places if and only if conditions C1'–C4' are satisfied (at least under the assumption that a good assignment of auxiliary and secondary elements to the non-trapped places of P can be found). Note that even though the wording of the preceding occasionally assumes that cyclic places are present, no real use is made of the existence of cyclic places; i.e. by simply ignoring what is said about such places one can still build an injective model of P.

If all places are trapped, it turns out as in the preceding case that by putting $\bar{\pi} = \Sigma^\pi$ for all places $\pi \in \Pi$ (in agreement with Step 5 of the Initialization Phase), the assignment $M^*x = \bigcup_{\pi(x)=1} \bar{\pi}$ is an injective model of P. Thus the lemma is completely proved.   ∎

To complete the proof that conditions C1′–C4′ are also sufficient for P to be injectively satisfiable, it only remains to verify that these imply the existence of a good assignment of auxiliary and secondary elements to the non-trapped places of P (cf. Definition 10.9).

Lemma 10.18 *Assume that conditions C1′–C4′ hold and that the Ugraph G may contain trapped nodes. Then there exists a good assignment of auxiliary and secondary elements to the non-trapped nodes of G.*

Proof. We proceed as follows. Let $\gamma_1, \ldots, \gamma_k$ be as from (iv) of condition C2′, as well as the elements $s_i \in \Sigma^{\gamma_i}$, $i = 1, \ldots, k$. Recall that all places γ_i lie in a cycle of the type

$$\gamma_i = \beta_{i,1} \Rightarrow_G \beta_{i,2} \Rightarrow_G \cdots \Rightarrow_G \beta_{i,m_i+1} = \gamma_i,$$

$i = 1, \ldots, k$, $m_i \geq 1$ (cf. (10.8)). For each γ_i, $i = 1, 2, \ldots, k$, define sets $s_i^{(n)}$ by $s_i^{(0)} = s_i$, $s_i^{(n+1)} = \{s_i^{(n)}\}$, where $s_i \in \Sigma^\gamma$ is the element appearing in (iv) of C2′. Define each $s_i^{(n)}$ with $n \equiv 1 - j \pmod{m_i}$ to be a secondary element resident at $\beta_{i,j}$; also, form all pairs $\{s_i^{(n)}, s_i^{(n+m_i)}\}$, and let all the pairs of this form with $n \equiv -j \pmod{m_i}$ be secondary elements resident at $\beta_{i,j}$ also. Next define singletons $\varrho_i^n = \{\{s_i^{(n)}, s_i^{(n+m_i)}\}\}$, and take each such singleton with $n \equiv -1 - j \pmod{m_i}$ to be resident at $\beta_{i,j}$. Divide the infinite set of these singletons resident at $\beta_{i,j}$ in any convenient way into two disjoint infinite parts; define singletons belonging to one of these parts to be *auxiliary* elements resident at $\beta_{i,j}$, and define the singletons belonging to the other of these parts to be *secondary* elements resident at $\beta_{i,j}$.

Next define further singletons $\varrho_i^{j,n}$ by $\varrho_i^{j,0} = \varrho_i^j$, $\varrho_i^{j,n+1} = \{\varrho_i^{j,n}\}$. Using the fact that $s_i \in^+ s_\ell$ is false for every $i, \ell = 1, \ldots, k$, it is easy to see that we have $\varrho_i^{j,n} \in^+ \varrho_\ell^{p,m}$ if and only if $i = \ell$, $j = p$, and $n < m$. The definitions stated in the preceding paragraph associate infinitely many secondary and auxiliary elements of the form $\varrho_i^{j,0}$ with each place β belonging to any cycle $\alpha_i, \psi(\alpha_i), \psi^2(\alpha_i), \ldots$ with $i \in \{1, 2, \ldots, k\}$, where the map ψ is defined as in Definition 10.8, but we need to extend this association to the remaining cyclic places and to treat the safe places. For this, we exploit much the same construction as in the proof of Lemma 10.4 in the previous section. We use the fact that if α is cyclic (respectively, safe) then $\psi(\alpha)$ is cyclic (respectively, safe or Ω), and that repeated application of the map ψ must eventually bring any place α to one of the places with which auxiliary and secondary places have already been associated. More specifically, regard a cyclic place as having been *treated* if secondary and auxiliary elements $\varrho_i^{j,n}$ have already been associated with it. If any untreated cyclic places remain, choose some α which has already been treated but for which there remain untreated $\beta_1, \ldots, \beta_\ell$ such that $\psi(\beta_1) = \cdots = \psi(\beta_\ell) = \alpha$ (by the observation made just above, such an α must exist). Divide the infinitely many *secondary* elements $\varrho_i^{j,n}$ resident at α into ℓ subsequences, all infinite, and let the elements $\varrho_i^{j,n+1}$ such that $\varrho_i^{j,n}$ belongs to the pth of these subsequences be resident at β_p, $p = 1, \ldots, \ell$. Divide the infinite set of resident items associated in this way with each of the β_ps into two infinite subsequences, and define the elements of one of these subsequences to be auxiliary elements resident at β_p, and the elements of the other subsequence to be secondary elements resident at β_p. Continue in this way as long as there remain any untreated cyclic places.

To handle the safe elements begin with the finite set N of places α such that $\psi(\alpha) = \Omega$. Divide the infinite set of singletons $\{n\}$, where n is an integer and $n \geq H + 1$, into an appropriate number of infinite subsets, and define the elements of each of these subsequences to be resident at a corresponding place α in N. Divide the singletons thereby assigned to α into two infinite subsequences, and define the elements of one of these subsequences to be secondary elements resident at α; the elements of the other subsequence are defined to be auxiliary elements resident at α. Then use the map ψ in the same iterative fashion as in the preceding paragraph, until resident auxiliary and secondary elements have been assigned to all safe places. (Again, we adopt the technical convention of regarding $\{b\}$ as an auxiliary element with residence at Ω whenever b is a singleton secondary element for which $\{b\}$ is not otherwise introduced.)

It is easy to see that the construction just outlined associates infinitely many resident auxiliary elements a and secondary elements b with each non-trapped place α and defines naturally a *good assignment Aux, Sec of auxiliary and secondary elements to the non-trapped nodes of G*, as from Definition 10.9. ∎

In view of the preceding lemma, Lemma 10.17 can now be restated as follows.

Lemma 10.19 (Completeness) *Assume that conditions C1′–C4′ hold and that the Ugraph G may contain trapped nodes. Then P is injectively satisfiable.*
□

From Lemmas 10.2, 10.5, 10.16, and 10.19 and from the discussion in Section 10.1, we have

Theorem 10.1 *The class of formulae MLSU has a solvable satisfiability problem.*
□

In view of Corollary 3.1, Theorem 10.1 yields the following more general decidability result.

Corollary 10.1 (Decidability) *The family of $\in$-left formulae (cf. Definition 3.10) over the language MLSU has a solvable satisfiability problem.*   □

11. THE CHOICE OPERATOR

11.1 Introduction

In this chapter we will consider the satisfiability problem for the unquantified theory MLSSFRC (MultiLevel Syllogistic with Singleton, Finite, Rank comparison and Choice constructs) in the language including $\emptyset$, $\cap$, $\backslash$, $\cup$, $\{\bullet, \ldots, \bullet\}$, $\in$, η, together with *Finite* and $\mathcal{R}$, where *Finite(S)* means that S has finitely many members and $S \, \mathcal{R} \, T$ indicates that $rk \, S \leq rk \, T$.

The choice operator η chooses, in every non-empty set S, the minimum element of S in a well-ordering $<$ of the universe $\mathcal{V}$, such that:

(R_1) $x < y$ whenever $rk \, x \in rk \, y$;

(R_2) $<$ is anti-lexicographic (cf. Section 2.4) on finite sets;

(R_3) $x < y$ whenever $rk \, x = rk \, y$, x is finite and y is infinite.

We think of η as a way of formalizing a certain use of transfinite induction in mathematics, where one's goal is to disprove a statement of the form $s \neq \emptyset$ by considering the minimum element of s with respect to a well-ordering $<$ of the universe that satisfies some property P. Two conditions must be met by the property P, namely:

Consistency: the property P is consistent with the above requirements (R_1), (R_2), (R_3) imposed on $<$;

Completeness: the validity of the mathematical arguments does not depend on how the choice of an ordering $<$ satisfying the extra property P is specifically made.

It is then desirable for a language including η to have an associated non-empty class W of well-orderings such that every formula of the language is either false in every interpretation or satisfiable in every well-ordering that belongs to the class W (cf. the notion of strong satisfiability in Section 3.7).

As much as concerns the theory MLSSFRC considered in the present chapter, the collection of all well-orderings that satisfy conditions (R_1), (R_2) and (R_3) above is certainly non-empty (by Theorem 2.3), but it does not have the completeness property stated in the preceding paragraph. Indeed, the formula

$$y = x \setminus \{\emptyset\} \ \& \ \eta\{x, y\} \neq y$$

of MLSSFRC is unsatisfiable in those well-orderings of $\mathcal{V}$ possessing properties (R_1), (R_2) and (R_3) that arrange the infinite sets to which $\emptyset$ does not belong before all other infinite sets of the same rank. On the other hand, there are well-orderings $<$ of $\mathcal{V}$ with properties (R_1), (R_2) and (R_3) in which $\omega < \omega \setminus \{\emptyset\}$ so that the above formula is satisfied by the assignment $x \mapsto \omega$, $y \mapsto \omega \setminus \{\emptyset\}$.

Thus, we need to impose some extra condition on the well-ordering of $\mathcal{V}$ in order to have the desired completeness property with respect to the theory MLSSFRC.

We will show that a sort of 'randomness' condition on the well-orderings of $\mathcal{V}$ suffices to get the desired result. This condition amounts to requiring that every *ordered syllogistic diagram* of size n is induced by some n-tuple of disjoint non-empty sets. Where an ordered syllogistic diagram is, intuitively, a finite data structure (generalizing the notions of Venn diagram and syllogistic scheme introduced in Chapter 6) which encodes any possible situation of a finite number of pairwise disjoint non-empty sets expressible by the constructs of MLSSFRC.

In order to simplify the presentation, we will initially consider the case devoid of the choice operator, and prove the decidability of this restricted theory, MLSSFR, which is an extension of the theory MLSSF already treated in Chapter 6. We then introduce ordered syllogistic diagrams to deal with the general case in which the choice operator is permitted to occur any number of times.

11.1.1   Scenario of a proof exploiting the properties of η

Before entering into the problem of detecting the validity/unsatisfiability of MLSSFRC formulae, we wish to offer a clue to the interest of the problem. We will illustrate the typical use of η in a formalized argument, by producing a detailed proof of the following important, yet rather elementary, mathematical fact: if x and y are ordinals, and $x \not\subseteq y$, then $y \subseteq x$. This could be proved in a single shot, by resorting to the decision method of Section 7.6—however, we choose to ignore that method here. On the other hand, we will be able to avoid explicit recourse to the transfinite induction principle, thanks to the very powerful semantics of η.

Let us assume by contradiction that x, y are ordinals with $x \not\subseteq y$ and $y \not\subseteq x$. It follows from $y \setminus x \neq \emptyset$ that $\eta(y \setminus x) \in y \setminus x$, whence $\eta(y \setminus x) \subseteq y$ by the transitivity of ordinals (see Section 2.3). Now we observe that $\eta(y \setminus x) \in y \setminus x$ & $\eta(y \setminus x) \subseteq y$ entails $\eta(y \setminus x) \subseteq x$, hence $\eta(y \setminus x) \subseteq x \cap y$, in MLSSFRC.

The reverse inclusion $x \cap y \subseteq \eta(y \setminus x)$ can also be proved. Indeed, if $z \in x \cap y$, then either $\eta(y \setminus x) = z$ or $\eta(y \setminus x) \in z$ or $z \in \eta(y \setminus x)$, because y, which is an ordinal, is totally ordered by $\in$, and because $\eta(y \setminus x)$ and z both belong to y. However one must reject the possibilities $\eta(y \setminus x) = z$, $\eta(y \setminus x) \in z$, because these would imply $\eta(y \setminus x) \in x$. It follows that $z \in \eta(y \setminus x)$, and we readily conclude $x \cap y = \eta(y \setminus x)$.

Symmetrically, one proves that $y \cap x = \eta(x \setminus y)$. Hence $x \setminus y \neq \emptyset$, $y \setminus x \neq \emptyset$, and $\eta(x \setminus y) = \eta(y \setminus x)$ should hold together in MLSSFRC, which is impossible.

11.2   Syllogistic diagrams

We begin by defining the basic data structure used in the present chapter.

Definition 11.1 *Let*

- $n \in \omega$;
- $\beta = (\beta_1, \beta_2, \ldots, \beta_n)$, *with* $\beta_i \subseteq Pow\{1, 2, \ldots, i - 1\}$ *for* $i = 1, 2, \ldots, n$, *and* $\beta_i \cap \beta_j = \emptyset$ *whenever* $i \neq j$;
- $\delta = (\delta_1, \delta_2, \ldots, \delta_n)$, *with* $\delta_i \in \{0, 1, \omega\}$ *for* $i = 1, 2, \ldots, n$, *and* $\delta_i \neq 0$ *whenever* $\beta_i = \emptyset$;
- $\varrho = (\varrho_1, \varrho_2, \ldots, \varrho_n)$, *with* $\varrho_1 = 1$ *and* $\varrho_i \leq \varrho_{i+1} \leq \varrho_i + 1$ *for* $i = 1, 2, \ldots, n - 1$.

ϱ_i *is called the* pseudo-rank *of* i. *Moreover we put* $\beta_0 = \emptyset$, $\delta_0 = \varrho_0 = 0$, *and* $\beta_J = \bigcup_{j \in J} \beta_j$, $\delta_J = max_{j \in J} \delta_j$, $\varrho_J = max_{j \in J} \varrho_j$ *for every* $J \subseteq \{1, \ldots, n\}$, $J \neq \emptyset$. *Again, we call* ϱ_J *the* pseudo-rank *of* J.

Assume that the following properties hold:

(p₁) $J \in \beta_i$ *implies* $\varrho_J < \varrho_i$;

(p₂) $\delta_i = 0$ *implies* $\varrho_i = (max_{J \in \beta_i} \varrho_J) + 1$.

Then the triple (β, δ, ϱ) *is called a* syllogistic diagram *and* n *is called the* size *of* (β, δ, ϱ).   □

The above definition is completely motivated by the following example of a syllogistic diagram.

Let $\sigma_1, \sigma_2, \ldots, \sigma_n$ be pairwise disjoint non-empty sets such that $rk \, \sigma_i \leq rk \, \sigma_{i+1}$, for $i = 1, \ldots, n - 1$.

Let $\Sigma_J = \bigcup_{j \in J} \sigma_j$ for every $J \subseteq \{1, 2, \ldots, n\}$, and let $\partial_i = |\sigma_i \setminus \{\Sigma_J : J \subseteq \{1, \ldots, n\}\}|$ for $i = 1, \ldots, n$. Moreover, let $\sigma_0 = \emptyset$ and $\varrho_0 = 0$.

Put:

(1) $\beta_i = \{J \subseteq \{1, \ldots, n\} : \Sigma_J \in \sigma_i\}$;

(2) $\delta_i = \begin{cases} 0 \text{ if } \partial_i = 0, \\ \omega \text{ if } \partial_i \geq \omega, \\ 1 \text{ otherwise}; \end{cases}$

$$(3) \qquad \varrho_i = \begin{cases} \varrho_{i-1} & \text{if } rk\,\sigma_i = rk\,\sigma_{i-1}, \\ \\ \varrho_{i-1} + 1 & \text{otherwise;} \end{cases}$$

for $i = 1, \ldots, n$. This is—as one easily ascertains—a syllogistic diagram.

Definition 11.2 *The syllogistic diagram satisfying (1)–(3) above is said to be* induced *by the σ_is.*   □

As the reader will easily recognize at this point, the essential aspects of the combinatorics of the Σ_Js are as follows:

Lemma 11.1

$$\begin{aligned} &\Sigma_I \in \Sigma_J &&\textit{iff } I \in \bigcup_{j \in J} \beta_j; \\ &\Sigma_I \subseteq \Sigma_J &&\textit{iff } I \subseteq J; \\ &\Sigma_I = \Sigma_J \cap \Sigma_K &&\textit{iff } I = J \cap K; \\ &\Sigma_I = \Sigma_J \setminus \Sigma_K &&\textit{iff } I = J \setminus K; \\ &\Sigma_I = \Sigma_J \cup \Sigma_K &&\textit{iff } I = J \cup K; \\ &\textit{Finite } \Sigma_I &&\textit{iff } (\forall i \in I)\delta_i < \omega; \\ &\textit{for every } q \textit{ in } \omega, \ \Sigma_J = \{\Sigma_{I_1}, \ldots, \Sigma_{I_q}\} \textit{ iff} \\ &&&\{I_1, \ldots, I_q\} = \bigcup_{j \in J} \beta_j \, \& \forall j \in J \delta_j = 0, \\ &rk\,\Sigma_I \le rk\,\Sigma_J \textit{ iff } \exists j \in J \forall i \in I(j < i \to (\forall k \in \{j+1, \ldots, i\})\varrho_k = \varrho_{k-1}). \end{aligned}$$
   □

Let us assume now that a given syllogistic diagram (β, δ, ϱ) is induced by suitable sets σ_i (we will see in Section 11.2.2 that this is always the case, at least under suitable 'density' conditions concerning the diagram). Indicating by n the size of the diagram, we address the following question:

'Which are the *i*s for which σ_i is hereditarily finite in every tuple $(\sigma_1, \sigma_2, \ldots, \sigma_n)$ that induces (β, δ, ϱ)?'

The set of these *i*s may, in fact, be non-empty. For instance, if

$$\beta_1 = \{\emptyset\}, \ \beta_2 = \{\{1\}\}, \text{ and } \delta_1 = \delta_2 = 0,$$

then σ_1 and σ_2 are forced to be hereditarily finite, as

$$\sigma_1 = \{\Sigma_\emptyset\} = \{\emptyset\}, \quad \sigma_2 = \{\Sigma_{\{1\}}\} = \{\sigma_1\} = \{\{\emptyset\}\}.$$

It will turn out that the *i*s we are seeking to characterize are those with $i \le m$, where m is the number returned by the following procedure:

Algorithm 11.1

 1. *procedure* threshold($\beta, \delta, \varrho, n$);
 2. i ← 1;
 3. (*while* ∃ j = n, n − 1, …,i |

$$\delta_j \; = \; 0 \; \& \; \beta_j \; \subseteq \text{Pow}\{1,\dots,i-1\})$$

4.   claim: $\varrho_i \; = \; \cdots \; = \; \varrho_j \; = \; \varrho_{i-1} \; + \; 1$ and
              $rk \, \sigma_j \; = \; \varrho_j$ in every tuple $(\sigma_1,\dots,\sigma_n)$
              that induces (β,δ,ϱ);

5. $j_* \leftarrow j$;

6.   ($while \; \varrho_{j_*+1} \; = \; \varrho_{j_*}$) $j_* \leftarrow j_* + 1$;   end;

7.   comment: we convene here that $\varrho_{n+1} = \varrho_n + 1$;

8. $i \leftarrow j_* + 1$;

9.   end;

10.   return $i - 1$;

11. end procedure.

□

Definition 11.3 Let (β,δ,ϱ) be a syllogistic diagram of size n. The threshold of (β,δ,ϱ) is the natural number $m \leq n$ returned by the procedure 'threshold' in Algorithm 11.1.

Any i in $\{0,\dots,m\}$, as well as any subset J of $\{1,\dots,m\}$ is said to be trapped with respect to $(\beta,\delta\,\varrho)$.   □

Lemma 11.2 The threshold of a syllogistic diagram is always defined.

Otherwise stated, the procedure 'threshold' always terminates, and the claim-statement contained in it is never violated.
Proof. In order to prove the thesis, one must first 'overload' it with the following additional thesis:

(∗) 'Every value taken by the variable i appearing inside the procedure threshold is such that

- $i \leq n+1$;

- $\varrho_{i-1} + 1 \; = \; \varrho_i$;

- for any j in $\{i,\dots,n\}$ with $\delta_j \; = \; 0$, the first number ℓ such that $\beta_j \subseteq \text{Pow}\{1,\dots,\ell\}$ satisfies the inequality $\varrho_\ell \geq \varrho_i - 1$.'

From the first of these, it readily follows that the procedure terminates (producing a result $m \leq n$), because i increases at each iteration of the outer while-loop.

The thesis (∗) holds initially, because $i = 1 \leq n+1$ and in fact $\varrho_1 = 1 = \varrho_0 + 1$, so that $\varrho_\ell \geq \varrho_i - 1 = 0$ for any ℓ.

Also notice that the claim-statement holds the first time it is encountered —clearly, with $i = 1$. Indeed, since $\beta_j \neq \emptyset$ by the very definition of a syllogistic diagram, one has $\beta_j = \text{Pow} \, \emptyset = \{\emptyset\}$ and therefore $\varrho_j = (max_{J\in\beta_j}\varrho_J) + 1 = \varrho_0 + 1 = 1 = \varrho_1$. Furthermore, $\sigma_j = \{\Sigma_\emptyset\} = \{\emptyset\}$, and therefore $rk \, \sigma_j = 1 = \varrho_j$.

Assuming that the *claim*-statement is satisfied during an iteration of the outer *while*-loop, it is a routine matter to verify that $(*)$ will hold when i is updated, short later, during the same iteration. In particular, let j_* and ℓ be such that $j_* \geq i_{\text{new}}$, $\delta_{j_*} = 0$, and $\beta_{j_*} \subseteq Pow\{1, \ldots, \ell\}$. Since $i_{\text{old}} < i_{\text{new}} \leq j_*$, and since the maximum j for which $\beta_j \subseteq Pow\{1, \ldots, i_{\text{old}} - 1\}$ is smaller than i_{new}, it follows that

$$i_{\text{old}} \leq j < i_{\text{new}} \leq j_* , \quad \beta_{j_*} \not\subseteq Pow\{1, \ldots, i_{\text{old}} - 1\},$$

and therefore

$$i_{\text{old}} - 1 < \ell, \quad \varrho_{i_{\text{new}}} - 1 = \varrho_{i_{\text{old}}} = \varrho_{i_{\text{old}}-1} + 1 \leq \varrho_{\ell}.$$

If the body of the outer *while*-loop is executed again, this presupposes that there is a $j_* \geq i_{\text{new}}$ with $\delta_{j_*} = 0$ and

$$\beta_{j_*} \subseteq Pow\{1, \ldots, i_{\text{new}} - 1\}, \quad \beta_{j_*} \not\subseteq Pow\{1, \ldots, i_{\text{old}} - 1\},$$

so that

$$\varrho_{j_*} = \left(max_{J \in \beta_{j_*}} \varrho_J\right) + 1 = \varrho_{i_{\text{old}}} + 1 = \varrho_{i_{\text{new}}}.$$

Furthermore,

$$\sigma_{j_*} = \{\Sigma_J : J \in \beta_{j_*}\},$$

which implies

$$rk\, \sigma_{j_*} = \left(max_{J \in \beta_{j_*}} rk\, \Sigma_J\right) + 1 = \left(max_{J \in \beta_{j_*}} \varrho_J\right) + 1 = \varrho_{j_*}$$

in every tuple $(\sigma_1, \ldots, \sigma_n)$ that induces (β, δ, ϱ). Hence, the *claim*-statement holds again during the new iteration. ∎

From the validity of the *claim*-statement one easily derives the following:

Corollary 11.1 *If (β, δ, ϱ) has threshold m and is induced by $(\sigma_1, \sigma_2, \ldots, \sigma_n)$, then the sets $\sigma_1, \sigma_2, \ldots, \sigma_m$ are hereditarily finite.* □

11.2.1 Dense syllogistic diagrams

In some pathological cases, a syllogistic diagram (β, δ, ϱ) is *not* induced by any tuple of σ_is. This happens, for instance, when $\delta_i = \omega$ for some i less than or equal to the threshold m of the diagram. More generally, this happens because the von Neumann universe, being devoid of individuals (i.e., member-less objects different from $\emptyset$), offers an insufficent provision of sets of rank ϱ_m or less, where m is the threshold of the diagram. Consider, for instance, the diagram

$$\beta_1 = \{\emptyset\}, \quad \beta_2 = \emptyset, \quad \delta_1 = 0, \quad \delta_2 = 1, \quad \varrho_1 = 1, \quad \varrho_2 = 1,$$

which has threshold 2. Any pair (σ_1, σ_2) inducing this diagram must have $\sigma_1 = \{\emptyset\}$, which implies $rk\,\sigma_2 = rk\,\sigma_1 = 1$. But there exists only one set of rank 1 in the von Neumann universe, whereas σ_2 must be distinct from σ_1.

In fact, it turns out that when a diagram is not induced by sets, then its 'restriction to the threshold', defined as the triple $(\beta', \delta', \varrho')$ with

$$\beta' = (\beta_1, \ldots, \beta_m), \quad \delta' = (\delta_1, \ldots, \delta_m), \quad \varrho' = (\varrho_1, \ldots, \varrho_m),$$

(where m is the threshold) is, in turn, induced by no sets. We can hence get rid of 'embarrassing' diagrams by just imposing suitable constraints on the initial segment of a diagram. To this aim, we introduce below the notion of a *dense* syllogistic diagram.

In what follows, by ε_i we will indicate the i-th singleton set in the anti-lexicographic ordering of all hereditarily finite sets. Therefore, $\varepsilon_1, \varepsilon_2, \ldots, \varepsilon_{|V_\lambda|}$ will be the ordered list of all singletons of rank less than or equal to λ.

Definition 11.4 *A syllogistic diagram (β, δ, ϱ) of size n and threshold m is said to be* dense *if and only if either $m = 0$ or $m = |V_{\varrho m}|$ and the restriction $(\beta', \delta', \varrho')$ of (β, δ, ϱ) to $\{1, 2, \ldots, m\}$ is induced by the singletons $\varepsilon_1, \varepsilon_2, \ldots, \varepsilon_m$.* $\square$

Lemma 11.3 *Let $\sigma_1, \sigma_2, \ldots, \sigma_n$ be pairwise disjoint non-empty sets. Then there exists a dense syllogistic diagram of size $d \leq |V_n| + n$ induced by sets $\gamma_1, \gamma_2, \ldots, \gamma_d$ such that for each i in $\{1, 2, \ldots, n\}$ there exists a $J \subseteq \{1, 2, \ldots, d\}$ such that $\sigma_i = \bigcup_{j \in J} \gamma_j$.*

Proof. Let us put $\sigma_0 = \emptyset$, $\sigma_{n+1} = V_\alpha \setminus (\sigma_1 \cup \cdots \cup \sigma_n)$, where α is the first limit ordinal that exceeds $rk\,\sigma_n$. Let s be the first i in $\{0, \ldots, n\}$ for which $rk\,\sigma_i + 1 < rk\,\sigma_{i+1}$.

Let

$$\begin{aligned}
\{\gamma_1, \ldots, \gamma_m\} &= \{\{x\} : x \in V_{rk\,\sigma_s}\}, \\
\{\gamma_{m+1}, \ldots, \gamma_{m+r}\} &= \{\sigma_\ell \setminus V_{rk\,\sigma_s + 1} : \ell \in \{s+1, \ldots, n\}\},
\end{aligned}$$

with $\gamma_1 = \varepsilon_1 < \gamma_m = \varepsilon_m < \gamma_{m+1} < \cdots < \gamma_{m+r}$. The diagram induced by these γ_ks is clearly dense, with threshold m. Moreover, for each i in $\{1, \ldots, n\}$, the set $\{j \in \{1, \ldots, m\} : \varepsilon_j \subseteq \sigma_i\}$, with a member of the form $m + k$ added if $i > s$, is a $J \subseteq \{1, \ldots, m+r\}$ such that $\sigma_i = \bigcup_{j \in J} \gamma_j$. Finally, one has $m + r \leq m + n = |V_{rk\,\sigma_s}| + n \leq |V_n| + n$, because $rk\,\sigma_s \leq s \leq n$. $\blacksquare$

Corollary 11.2 *Given arbitrary sets $X_1, \ldots, X_q$, there exists a dense syllogistic diagram of size $d < C(q)$, where C is the computable function $C(q) = |V_{2q}| + 2^q$, induced by sets $\gamma_1, \gamma_2, \ldots, \gamma_d$ such that for each k in $\{1, 2, \ldots, q\}$ there exists a $J \subseteq \{1, 2, \ldots, d\}$ such that $X_k = \bigcup_{j \in J} \gamma_j$.*

Proof. In the preceding lemma, let the σ_is be the non-empty regions of the Venn diagram induced by the X_ks. In other words, the σ_is are all non-empty sets of the form

$$\bigcap W \setminus \bigcup (\{X_1,\ldots,X_q\} \setminus W) \,,$$

with $\emptyset \neq W \subseteq \{X_1,\ldots,X_q\}$.

The statement follows immediately, since the Venn regions are fewer than 2^q and since each X_k can be written as a union of σ_is.   ∎

11.2.2   Sets inducing syllogistic diagrams

In this section we want to show that any dense syllogistic diagram is induced by some sets of arbitrarily large rank.

Lemma 11.4 *Let $D = (\beta, \delta, \varrho)$ be a dense syllogistic diagram of size n and threshold m, and let α be a limit ordinal. There are sets $\sigma_{m+1}, \ldots, \sigma_n$ in $V_{\alpha+\omega} \setminus V_\alpha$ such that $\varepsilon_1, \varepsilon_2, \ldots, \varepsilon_m, \sigma_{m+1}, \ldots, \sigma_n$ induce D.*

Proof. In order to achieve this result we need the following:

Lemma 11.5 *There exist sets $S_{j\ell}$, with $j \in \{m+1, \ldots, n\}$, $\ell \in \{0,1,2,\ldots\}$ such that:*

(a) the $S_{j\ell}$s are all infinite;

(b) the $S_{j\ell}$s are pairwise distinct and disjoint;

(c) $rk\, S_{j\ell} = \alpha + \varrho_j$;

(d) for no pair $S_{j\ell}$, $S_{j'\ell'}$ there exists a membership chain of the form $S_{j\ell} \in x_1 \in \cdots \in x_g \in S_{j'\ell'}$, with $g \geq 0$ and $x_1, \ldots, x_g$ arbitrary sets.

Proof. Let $\{s\}^{[\lambda]}$ indicate the set s enclosed in λ pairs of braces. That is:

$$\{s\}^{[0]} = s;$$
$$\{s\}^{[k+1]} = \{\{s\}^{[k]}\}, \text{ for } k \geq 0.$$

Let $M_{j\ell t}$ be pairwise distinct infinite sets of rank α, with $j \in \{m+1, \ldots, n\}$ and $\ell, t \in \{0,1,2,\ldots\}$ (cf. Corollary 2.4).

Put

$$S_{j\ell} = \{\{M_{j\ell t}\}^{[\varrho_j - 1]} : t = 0,1,2,\ldots\} \,.$$

It is immediate to verify that this choice of $S_{j\ell}$ satisfies (a)–(d).   ∎

Next, for $k = 1, 2, \ldots, m$ put $\sigma_k = \varepsilon_k$ and, going up in the ordering of indices k in $\{m+1, \ldots, n\}$, put

$$(*) \qquad \sigma_k = u_k \cup \left\{\bigcup_{j \in J} \sigma_j : J \in \beta_k\right\}, \text{ where}$$

$$u_k = \begin{cases} \emptyset & \text{if } \delta_k = 0, \\ \{S_{k0}\} & \text{if } \delta_k = 1, \\ \{S_{k1}, S_{k2}, \ldots\} & \text{if } \delta_k = \omega. \end{cases}$$

Notice that if $j \in J \in \beta_k$ then, by the property (p_1) of syllogistic diagrams, $\varrho_j < \varrho_k$, $j < k$, so that σ_j is perfectly known when σ_k is about to be defined by $(*)$.

Plainly these sets $\sigma_1 = \varepsilon_1, \ldots, \sigma_m = \varepsilon_m$, $\sigma_{m+1}, \ldots, \sigma_n$ are all nonempty since by Definition 11.1 $\delta_k \neq 0$ whenever $\beta_k = \emptyset$. In order to show that $\sigma_{m+1}, \ldots, \sigma_n$ all belong to $V_{\alpha+\omega} \setminus V_\alpha$ we prove the following

Lemma 11.6 $rk\, \sigma_j = \alpha + \varrho_j + 1$, for $j = m+1, \ldots, n$.

Proof. We proceed by induction on j. If $j = m+1$ then, since $m + 1$ is not trapped, $\delta_{m+1} \neq 0$ and, by (c) of Lemma 11.5, $rk\, u_{m+1} = \alpha + \varrho_{m+1} + 1$. Since $\left\{ \bigcup_{j \in J} \sigma_j : J \in \beta_{m+1} \right\}$ has finite rank then, by $(*)$, $rk\, \sigma_{m+1} = \alpha + \varrho_{m+1} + 1$.

Next, assume the claim true for every j such that $m + 1 \leq j < \lambda \leq n$, and let us show that $rk\, \sigma_\lambda = \alpha + \varrho_\lambda + 1$. Indeed, by induction hypothesis

$$rk \left\{ \bigcup_{j \in J} \sigma_j : J \in \beta_\lambda \right\} \leq \alpha + \varrho_\lambda + 1.$$

We distinguish two cases:

Case $\delta_\lambda \neq 0$. By (c) of Lemma 11.5 we have $rk\, u_\lambda = \alpha + \varrho_\lambda + 1$. It follows, by $(*)$, that $rk\, \sigma_\lambda = \alpha + \varrho_\lambda + 1$.

Case $\delta_\lambda = 0$. By (p_2) of Definition 11.1, $\varrho_\lambda = max_{J \in \beta_\lambda}\, \varrho_J + 1$.

Since λ is not trapped, there must exist some J in β_λ which is not trapped or equivalently $J \not\subseteq \{1, \ldots, m\}$. It follows by the induction hypothesis that

$$
\begin{aligned}
rk\, \sigma_\lambda &= rk \left\{ \bigcup_{j \in J} \sigma_j : J \in \beta_\lambda \right\} = max_{J \in \beta_\lambda} rk\, \bigcup_{j \in J} \sigma_j + 1 \\
&= max_{\substack{J \in \beta_\lambda\, \& \\ J \not\subseteq \{1, \ldots, m\}}} rk\, \bigcup_{j \in J} \sigma_j + 1 \\
&= max_{\substack{J \in \beta_\lambda\, \& \\ J \not\subseteq \{1, \ldots, m\}}} (\alpha + \varrho_J + 1) + 1 \\
&= \alpha + \left(max_{\substack{J \in \beta_\lambda\, \& \\ J \not\subseteq \{1, \ldots, m\}}} \varrho_J + 1 \right) + 1 = \alpha + \varrho_\lambda + 1.
\end{aligned}
$$

This completes the proof of the lemma.   ∎

An immediate consequence of this lemma is the following

Corollary 11.3 $\sigma_j \in V_{\alpha+\omega} \setminus V_\alpha$, for $j = m + 1, \ldots, n$.   □

Moreover, since (β, δ, ϱ) is dense, $rk\, \sigma_i = rk\, \varepsilon_i = \varrho_i$ for $i = 1, 2, \ldots, m$.

Next, we show that the σ_is are pairwise disjoint. We begin by proving the following.

Lemma 11.7 *No element of type $S_{k\ell}$ can equal an element of type $\bigcup_{j \in J} \sigma_j$.*

Proof. Assume by contradiction that $\bigcup_{j \in J} \sigma_j = S_{\kappa \ell}$. Plainly, J cannot be trapped. It follows that there are indices $j_0, j_1, \ldots, j_r$ and sets of indices $J_0, J_1, \ldots, J_r$ such that

- $j_t \in J_t$, $J_t \in \beta_{j_{t+1}}$, $t = 0, 1, \ldots, r-1$;

- $j_r \in J_r = J$;

- $\delta_{j_0} \neq 0$.

Consequently, by $(*)$, for some $\ell' \geq 0$,

$$S_{j_0 \ell'} \in \cdots \in \bigcup_{j \in J} \sigma_j = S_{k\ell},$$

contradicting (d) of Lemma 11.5. ∎

We are now ready to show that $\sigma_1, \ldots, \sigma_m, \sigma_{m+1}, \ldots, \sigma_n$ are pairwise disjoint. Assume $j \neq k$ and let us show that $\sigma_j \cap \sigma_k = \emptyset$. We proceed by induction on j. If $j = 1$ then

$$\sigma_1 = u_1 \cup \begin{cases} \{\emptyset\} & \text{if } \emptyset \in \beta_1 \\ \emptyset & \text{if } \emptyset \notin \beta_1 \end{cases} \quad \text{and } \sigma_k = u_k \cup \Big\{ \bigcup_{j \in J} \sigma_j : J \in \beta_k \Big\}.$$

If $\sigma_1 \cap \sigma_k \neq \emptyset$ then, by (b) of Lemma 11.5 and by Lemma 11.7, it would be $\emptyset \in \beta_1 \cap \beta_k$, which is impossible since (β, δ, ϱ) is a syllogistic diagram. Consequently, $\sigma_1 \cap \sigma_k = \emptyset$.

Next assume $\sigma_{j_0} \cap \sigma_{k_0} = \emptyset$ for every $j_0 < j$, $k_0 \neq j_0$, and let $k \neq j$. If $\sigma_j \cap \sigma_k \neq \emptyset$, then, by (b) of Lemma 11.5 and by Lemma 11.7, it would be $\bigcup_{j_0 \in J_0} \sigma_{j_0} = \bigcup_{k_0 \in K_0} \sigma_{k_0}$ for some $J_0 \in \beta_j$, $K_0 \in \beta_k$. Since $j_0 \in J_0 \in \beta_j$ implies $j_0 < j$ then, by induction hypothesis it must be $J_0 = K_0$. This gives $\beta_j \cap \beta_k \neq \emptyset$, which is impossible since $j \neq k$. Therefore, $\sigma_j \cap \sigma_k = \emptyset$, completing the verification of pairwise disjointness of the sets $\sigma_1, \sigma_2, \ldots, \sigma_n$.

In order to finish the proof of our main Lemma 11.4 it is sufficient to notice that, by $(*)$ and Lemma 11.6, $\sigma_1, \ldots, \sigma_n$ induce D. ∎

11.3 Decidability of the theory MLSSFR

In this section we want to show that the theory MLSSFR, obtained from our main theory MLSSFRC by dropping the choice operator η, is decidable. where x', y', u, v, z are variables not appearing in p.

In view of the normalization process Algorithm 3.1 and of the subsequent Remark 3.3, it is sufficient to show how to decide satisfiability of a finite conjunction q of literals of types:

$$(=) \qquad\qquad x = y \cup z, \quad x = y \setminus z,$$

$$(\{\bullet\}) \qquad\qquad y = \{x\},$$

$$(F, \neg F) \qquad\qquad \text{Finite } x, \quad \neg \text{Finite } x,$$

$$(\mathcal{R}, \neg \mathcal{R}) \qquad\qquad x \mathcal{R} y, \quad \neg x \mathcal{R} y.$$

Let $y_1, y_2, \ldots, y_k$ be all variables in q.

Theorem 11.1 *The conjunction q is satisfiable if and only if there exist:*

- *a dense syllogistic diagram (β, δ, ϱ) of size d with $d < |V_{2^k}| + 2^k$;*
- *a set $\Gamma = \{\pi_1, \pi_2, \ldots, \pi_d\}$ of places of the set of atoms of type $(=)$ in q (cf. Definition 8.1);*

such that the sets $\Delta_i = \{j : \pi_j(y_i) = 1\}$ fulfil the following conditions:

(c_1) if $y_i = \{y_j\}$ is in q, then $\beta_{\Delta_i} = \{\Delta_j\}$ and $\delta_{\Delta_i} = 0$;
(c_2) if Finite y_i appears in q, then $\delta_{\Delta_i} < \omega$;
(c_3) if $\neg$ Finite y_i appears in q, then $\delta_{\Delta_i} = \omega$;
(c_4) if $y_i \mathcal{R} y_j$ appears in q, then $\varrho_{\Delta_i} \le \varrho_{\Delta_j}$;
(c_5) if $\neg y_i \mathcal{R} y_j$ appears in q, then $\varrho_{\Delta_i} > \varrho_{\Delta_j}$.

Proof. Let M be a model of q and let $My_1, \ldots, My_k$ be the sets corresponding to $y_1, \ldots, y_k$ in the model M.

By Corollary 11.2, there exists a dense syllogistic diagram (β, δ, ϱ) of size $d < |V_{2^k}| + 2^k$ induced by sets $\gamma_1, \gamma_2, \ldots, \gamma_d$ such that for each j in $\{1, 2, \ldots, k\}$ there is a $J \subseteq \{1, 2, \ldots, d\}$ with $My_j = \bigcup_{t \in J} \gamma_t$, for $j = 1, \ldots, k$.

We put

$$\pi_i(y_j) = \begin{cases} 1 & \text{if } \gamma_i \subseteq My_j, \\ 0 & \text{otherwise;} \end{cases}$$

where $i \in \{1, 2, \ldots, d\}$ and $j \in \{1, 2, \ldots, k\}$.

It is immediate to verify that conditions (c_1)–(c_5) are all satisfied.

Conversely, let $\Gamma = \{\pi_1, \pi_2, \ldots, \pi_d\}$ and (β, δ, ϱ) satisfy conditions (c_1)–(c_5). By Lemma 11.3 there exist sets $\sigma_1, \sigma_2, \ldots, \sigma_d$ inducing (β, δ, ϱ). Put:

$$(*) \qquad My_i = \bigcup_{j \in \Delta_i} \sigma_j.$$

We claim that this defines a model of q. Indeed if $y_i = y_j \cup y_k$ appears in q then

$$My_i = \bigcup_{r \in \Delta_i} \sigma_r = \bigcup_{\pi_r(y_i)=1} \sigma_r = \bigcup_{\substack{\pi_r(y_j)=1 \lor \\ \pi_r(y_k)=1}} \sigma_r$$

$$= \bigcup_{\pi_r(y_j)=1} \sigma_r \cup \bigcup_{\pi_r(y_k)=1} \sigma_r = My_j \cup My_k .$$

Similarly if $y_i = y_j \setminus y_k$ appears in q, since the σ_rs are pairwise disjoint we have:

$$My_i = \bigcup_{r \in \Delta_i} \sigma_r = \bigcup_{\pi_r(y_i)=1} \sigma_r = \bigcup_{\substack{\pi_r(y_j)=1 \& \\ \pi_r(y_k)=0}} \sigma_r$$

$$= \bigcup_{\pi_r(y_j)=1} \sigma_r \setminus \bigcup_{\pi_r(y_k)=1} \sigma_r = My_j \setminus My_k .$$

If $y_i = \{ y_j \}$ is in q, then $\delta_{\Delta_i} = 0$ by (c$_1$). This gives, again by (c$_1$),

$$My_i = \bigcup_{r \in \Delta_i} \sigma_r = \left\{ \bigcup_{r \in J} \sigma_r : J \in \beta_{\Delta_i} \right\}$$

$$= \left\{ \bigcup_{r \in \Delta_j} \sigma_r \right\} = \{ My_j \}.$$

Furthermore, since $\sigma_1, \ldots, \sigma_d$ induce (β, δ, ϱ) it follows that

- σ_i is finite if and only if $\delta_i < \omega$;

- $rk\, \sigma_i \leq rk\, \sigma_j$ if at only $\varrho_i \leq \varrho_j$.

This together with (c$_2$)–(c$_5$) imply that M correctly models all literals of types $(F, \neg F)$ and $(\mathcal{R}, \neg \mathcal{R})$. We have thus completed the verification of Theorem 11.1. ∎

As an immediate consequence we have the following:

Corollary 11.4 (Decidability) *The theory MLSSFR is decidable.* □

Lemma 3.4 then leads to the following more general decidability result.

Corollary 11.5 *The family of left formulae (cf. Definition 3.10) over the language MLSSFR has a solvable satisfiability problem.* □

11.4   Ordered syllogistic diagrams

Let (β, δ, ϱ) be a syllogistic diagram of size n induced by some sets, and let $<$ be a well-ordering of $\mathcal{V}$ satisfying the properties (R$_1$), (R$_2$), and (R$_3$) mentioned in Section 11.1. Then there exists a total ordering $\lhd$ of $Pow\{ 1, 2, \ldots, n \}$ such that:

(p$_3$) $J \lhd K$ when $\varrho_J < \varrho_K$;

(p$_4$) $\{ j \} \lhd \{ k \}$ when $1 \leq j < k \leq n$;

(p$_5$) $J \lhd K$ when the following conditions hold together: $\delta_J < \omega$, $\delta_K < \omega$, and J precedes K in the anti-lexicographic ordering induced by the standard ordering of natural numbers;

(p₆) $J \lhd K$ when $\delta_J < \omega = \delta_K$ and $\varrho_J = \varrho_K$ hold together;

(p₇) when $\delta_j = \delta_k = 0$, $\varrho_j = \varrho_k$, and $j < k$, then β_j anti-lexicographically precedes β_k with respect to $\lhd$.

Definition 11.5 *A quadruple* $(\beta, \delta, \varrho, \lhd)$ *where* (β, δ, ϱ) *is a dense syllogistic diagram of size* n *and* $\lhd$ *is a total ordering of* $Pow\{1, 2, \ldots, n\}$ *satisfying properties (p₃)–(p₇), is called an* ordered syllogistic diagram *(abbreviated OSD) of size* n.

Let (β, δ, ϱ) *be a dense syllogistic diagram induced by* $\sigma_1, \sigma_2, \ldots, \sigma_n$. *Assume that* $\sigma_1 < \cdots < \sigma_n$ *in some well-ordering* $<$ *of* $\mathcal{V}$ *satisfying* (R_1), (R_2) *and* (R_3). *Put*

$$J \lhd K \text{ if and only if } \bigcup_{j \in J} \sigma_j < \bigcup_{k \in K} \sigma_k.$$

Then we say that the OSD $(\beta, \delta, \varrho, \lhd)$ *is* induced *by the sets* $\sigma_1, \ldots, \sigma_n$ *in the well-ordering* $<$. □

We need a stronger notion for expressing the completeness assumption:

Definition 11.6 *Let* $(\beta, \delta, \varrho, \lhd)$ *be induced by* $\sigma_1, \sigma_2, \ldots, \sigma_n$ *in the well-ordering* $<$ *of* $\mathcal{V}$, *exploited for interpreting the choice operator* η. *We say that* $\sigma_1, \sigma_2, \ldots, \sigma_n$ strongly induce $(\beta, \delta, \varrho, \lhd)$ *in* $<$ *if and only if the following property holds:*

'Let $I \subseteq \{1, 2, \ldots, n\}$, $J \in \beta_I$ be such that $\varrho_J < \varrho_i$ for every i in I. Then $\eta\left(\bigcup_{i \in I} \sigma_i\right) = \bigcup_{t \in T} \sigma_t$ for some T in β_I.' □

11.4.1 The completeness assumption

We are now ready to state the extra semantical constraint we need to impose on the well-orderings of the universe $\mathcal{V}$ in order to achieve the strong completeness (cf. Chapter 6) of the theory MLSSFRC considered in this chapter. We impose that the well-ordering $<$ of $\mathcal{V}$ with respect to which η chooses the minimum in every non-empty set, satisfies the following property, in addition to the properties (R_1), (R_2), and (R_3) mentioned in Section 11.1:

(R₄) Every OSD $(\beta, \delta, \varrho, \lhd)$ is strongly induced, in the well-ordering $<$, by some finite sequence of sets.

In the next sections we will show that orderings of $\mathcal{V}$ satisfying (R_1)–(R_4) do exist. Moreover, every formula of MLSSFRC is either false in every interpretation or satisfiable in *any* well-ordering satisfying (R_1)–(R_4). Indeed, an algorithm will be produced that is able to detect which of the two cases occurs.

11.4.2   Sets inducing ordered syllogistic diagrams

In this section we extend Lemma 11.4 to OSDs. More precisely we prove the following basic proposition:

Lemma 11.8 *Let $D = (\beta, \delta, \varrho, \vartriangleleft)$ be an OSD of size n and threshold m, and let α be a limit ordinal. Assume that V_α has been well-ordered by means of an $<$ satisfying properties (R_1), (R_2), and (R_3). Then it is possible to extend $<$ into a well-ordering $<^*$ of $V_{\alpha+\omega}$ such that:*

- *analogues of the properties (R_1), (R_2), and (R_3), with $<^*$ in place of $<$, hold;*

- *there are sets $\sigma_{m+1}, \ldots, \sigma_n$ in $V_{\alpha+\omega} \setminus V_\alpha$ such that $\varepsilon_1, \varepsilon_2, \ldots, \varepsilon_m,$ $\sigma_{m+1}, \ldots, \sigma_n$ strongly induce D in $<^*$.*

Proof. Let $\sigma_{m+1}, \ldots, \sigma_n$ be the sets, constructed in Lemma 11.4, such that $\sigma_1 = \varepsilon_1, \ldots, \sigma_m = \varepsilon_m, \sigma_{m+1}, \ldots, \sigma_n$ induce (β, δ, ϱ). Define the well-ordering $<^*$ of $V_{\alpha+\omega}$ in the following way:

- take $<^*$ to be $<$ on V_α;

- slightly modify the construction of the well-ordering of $V_{\alpha+\omega} \setminus V_\alpha$ described in the proof of Theorem 2.3, by requiring infinite sets of the same rank β, $\alpha \leq \beta < \alpha + \omega$ to satisfy the extra conditions:

  (a) $\bigcup_{j \in J} \sigma_j <^* \bigcup_{k \in K} \sigma_k$ whenever $J \vartriangleleft K$;

  (b) $S_{j0} <^* S_{k0}$ whenever $j < k$;

  (c) $\bigcup_{j \in J} \sigma_j <^* S_{k\ell}$ for every k, ℓ.

To this end it is sufficient to order infinite sets of the same rank β, by arranging all infinite sets of rank β of type $\bigcup_{j \in J} \sigma_j$ by (a), followed by the sets of rank β of type S_{j0} well-ordered by (b), followed by the remaining infinite sets of rank β well-ordered arbitrarily.

The well-ordering of $V_{\alpha+\omega} \setminus V_\alpha$ given in Theorem 2.3, modified in the above sense, is a well-ordering of $V_{\alpha+\omega} \setminus V_\alpha$. Plainly, the concatenation of the well-ordering $<$ of V_α with this well-ordering of $V_{\alpha+\omega} \setminus V_\alpha$ gives a well-ordering $<^*$ of $V_{\alpha+\omega}$.

We want to show that the sets $\varepsilon_1, \ldots, \varepsilon_m, \sigma_{m+1}, \ldots, \sigma_n$ strongly induce $D = (\beta, \delta, \varrho, \vartriangleleft)$ with respect to $<^*$. Indeed, by (a) we have that $\sigma_1 = \varepsilon_1, \ldots, \sigma_m = \varepsilon_m, \sigma_{m+1}, \ldots, \sigma_n$ induce $D = (\beta, \delta, \varrho, \vartriangleleft)$. Therefore, it remains to be shown that if $J \subseteq \{1, 2, \ldots, n\}$, $J' \in \beta_J$, and $\varrho_{J'} < \varrho_j$ for every j in J, then the minimum of $\bigcup_{j \in J} \sigma_j$ is achieved by some $\bigcup_{t \in T} \sigma_t$ with T in β_J. Indeed, since $J' \in \beta_J$, then by $(*)$ of Section 11.2.2 it follows $\bigcup_{j' \in J'} \sigma_{j'} \in \bigcup_{j \in J} \sigma_j$. Moreover, $\varrho_{J'} < \varrho_j$ for every j in J. It follows by Lemma 11.5 that $rk \left(\bigcup_{j' \in J'} \sigma_{j'} \right) < rk\, \sigma_j \leq \alpha + \varrho_j + 1$ for every $j \in J$.

Hence $rk \left(\bigcup_{j' \in J'} \sigma_{j'} \right) \leq rk\, S_{j\ell}$, for every $j \in J$, $\ell = 0, 1, 2, \ldots$, by (c) of Lemma 11.5. It follows by (c) above that $\bigcup_{j' \in J'} \sigma_{j'} <^* S_{j\ell}$ for every $j \in J$, $\ell = 0, 1, 2, \ldots$. This, by $(*)$ of Section 11.2.2, immediately implies that the minimum of $\bigcup_{j \in J} \sigma_j$ with respect to $<^*$ is achieved by some element of type $\bigcup_{t \in T} \sigma_t$ with T in β_J.

Consequently $\sigma_1 = \varepsilon_1, \ldots, \sigma_m = \varepsilon_m, \sigma_{m+1}, \ldots, \sigma_n$ strongly induce D, completing the proof of our lemma.   ∎

11.4.3 Consistency of the completeness assumption

Let us show how to construct a well-ordering of $\mathcal{V}$ satisfying conditions (R$_1$), (R$_2$), (R$_3$), and (R$_4$). Let $D_1, D_2, \ldots, D_k, \ldots$ be an enumeration of all possible ordered syllogistic diagrams. Denoting by $i \cdot \omega$ the limit ordinal $\omega + \cdots + \omega$, i times, let $<_i$ be inductively defined as follows:

- $<_0$ is the unique anti-lexicographic well-ordering of $V_\omega = \mathcal{H}$;

- for $i \geq 1$, $<_i$ is the well-ordering $<^*$ of $V_{(i+1) \cdot \omega}$ given by Lemma 11.8 when $D = D_i$, $\alpha = i \cdot \omega$, and $V_{i \cdot \omega}$ has been ordered by $<_{i-1}$.

Now, if we take $\bigcup_{i \in \omega} <_i$ and concatenate it with the well-ordering of $\mathcal{V} \setminus V_{\omega \cdot \omega}$ given in Theorem 2.3, we obtain a well-ordering of $\mathcal{V}$ satisfying the properties (R$_1$), (R$_2$), (R$_3$), and (R$_4$).

11.5 Decidability of the theory MLSSFRC

Let us prove the main result of this chapter, namely that every formula of MLSSFRC is either false or satisfiable in every well-ordering satisfying the conditions (R$_1$), (R$_2$), (R$_3$), and (R$_4$). Moreover an algorithm to decide which case occurs will be given.

For this purpose, by the same argument developed in Section 11.3, it is sufficient to restrict ourselves to considering finite conjunctions of literals of the types $(=)$, $(\{\bullet\})$, $(F, \neg F)$ and $(\mathcal{R}, \neg \mathcal{R})$ of Section 11.3, together with atoms of type

(η) $y = \eta\, x.$

Thus, let q be such a conjunction and let $y_1, y_2, \ldots, y_k$ be all variables in q.

Theorem 11.2 (Decidability) *The conjunction q is satisfiable if and only if there exist:*

- *a set $\Gamma = \{\pi_1, \pi_2, \ldots, \pi_d\}$ of places of the set of literals of type $(=)$ in q, with $d < |V_{2^k}| + 2^k$;*

- *an OSD $(\beta, \delta, \varrho, \lhd)$ of size d satisfying (c_1)–(c_5) of Theorem 11.1, together with the following new condition:*

(c_6) *If $y_i = \eta \, y_j$ appears in q then $\Delta_i \in \beta_{\Delta_j}$; moreover for all $J \subseteq \{1, \ldots, d\}$*

    (a) $k \in \Delta_j$ implies $\varrho_{\Delta_i} < \varrho_k$;

    (b) $J \in \beta_{\Delta_j}$ implies $\Delta_i \lhd J$ or $\Delta_i = J$.

Proof. The proof of the necessity of conditions (c_1)–(c_6) is completely analogous to that of Theorem 11.1.

The converse uses property (R_4). Indeed, let Γ and $(\beta, \delta, \varrho, \lhd)$ satisfy condition (c_1)–(c_6). By property (R_4), there are sets $\sigma_1, \sigma_2, \ldots, \sigma_d$ strongly inducing $(\beta, \delta, \varrho, \lhd)$. Putting, as in Theorem 11.1,

$$(*) \qquad M y_i = \bigcup\nolimits_{j \in \Delta_i} \sigma_j,$$

one can prove that M defines a model of q. Indeed, by the very same arguments used in the proof of Theorem 11.1, M is a model of all literals of type $(=)$, $(\{\bullet\})$, $(F, \neg F)$ and $(\mathcal{R}, \neg \mathcal{R})$ in q.

In order to show that M correctly models also literals of type (η), we assume that $y_i = \eta \, y_j$ is in q. Then $\Delta_i \in \beta_{\Delta_j}$ holds, yielding

$$M y_i = \bigcup_{r \in \Delta_i} \sigma_r \in \bigcup_{t \in \Delta_j} \sigma_t = M y_j \, .$$

To show that $M y_i$ is indeed the least element in $M y_j$, we argue as follows. Let $k \in \Delta_j$; then $\varrho_{\Delta_i} < \varrho_k$, by (a) of (c_6). Since $\sigma_1, \ldots, \sigma_d$ strongly induce $(\beta, \delta, \varrho, \lhd)$, the least element of $\bigcup_{t \in \Delta_j} \sigma_t$ is $\bigcup_{r \in R} \sigma_r$ for some R in β_{Δ_j}. On the other hand, either $\Delta_i \lhd R$ or $\Delta_i = R$ holds by (b) of (c_6), so that $M y_i = \bigcup_{t \in \Delta_j} \sigma_t \leq \bigcup_{r \in R} \sigma_r = \eta(M y_j)$.

We can then conclude that $M y_i = \eta(M y_j)$, completing the proof of our decidability result. ∎

Our decidability result concerning MLSSFRC, namely Theorem 11.2, can then be lifted to the quantified case in the same manner Corollary 11.5 ensues from Corollary 11.4.

REFERENCES

[Abi65]   A. Abian. *The theory of sets and transfinite arithmetic.* W.B. Saunders Co., Philadelphia and London, 1965.

[Ack62]   W. Ackermann. *Solvable cases of the decision problem.* North-Holland Publishing Co., Amsterdam, 1954, 1962.

[Acz88]   P. Aczel. *Non-well-founded sets.* Vol 14, Lecture Notes, Center for the Study of Language and Information, Stanford, 1988.

[Ahl66]   L. V. Ahlfors. *Complex analysis.* McGraw-Hill, New York, 2nd edn, 1966.

[AHU76]   A.V. Aho, J.E. Hopcroft, and J.D. Ullman. *The design and analysis of computer algorithms.* Addison-Wesley, 1976.

[Art44]   E. Artin. *Galois theory.* Vol 2, Notre Dame Mathematical Lectures, Notre Dame, Indiana, 1944.

[Bee86a]   M. Beeson. Normalization of terms in logic and computer science. In *Logic and computer science: new trends and applications*, Turin, Italy, 1986.

[Bee86b]   M. Beeson. *Towards a computation system based on set theory.* Unpublished manuscript, 1986.

[Beh22]   H. Behmann. Beiträge zur Algebra der Logik insbesondere zum Entscheidungsproblem. *Math. Annalen*, 86,163–220, 1922.

[Bet69]   E.W. Beth. Semantic entailment and formal derivability. In J. Hintikka, editor, *The philosophy of mathematics*, Oxford University Press, London, 1969.

[BF84]   M. Breban and A. Ferro. Decision procedures for elementary sublanguages of set theory. III. Restricted classes of formulas involving the powerset operator and the general set union operator. *Advances in Applied Mathematics*, 5,147–215, 1984.

[BFOS81] M. Breban, A. Ferro, E.G. Omodeo, and J.T. Schwartz. Decision procedures for elementary sublanguages of set theory. II. Formulas involving restricted quantifiers, together with ordinal, integer, map, and domain notions. *Communications on Pure and Applied Mathematics*, 34,177–195, 1981.

[Bib82] W. Bibel. *Automated theorem proving.* Vieweg Verlag. Wiensbaden, 1982.

[BLM*86] R. Boyer, E. Lusk, W. McCune, R. Overbeek, M. Stickel, and L. Wos. Set theory in first-order logic: Clauses for Gödel's axioms. *Journal of Automated Reasoning*, 2,287–327, 1986.

[BM67] G. V. Birkhoff and S. McLane. *Algebra.* The MacMillan Co., 1965-1967.

[Boc72a] J. M. Bocheński. *La logica formale dai presocratici a Leibniz.* Giulio Einaudi Editore, Torino, 1972.

[Boc72b] J. M. Bocheński. *La logica matematica.* Giulio Einaudi Editore, Torino, 1972.

[Bre82] M. Breban. *Decision algorithms for a class of set-theoretic formulae involving one occurrence of the union set operator.* Ph.D thesis, New York University, 1982.

[Bro78] F.M. Brown. Towards the automation of set theory and its logic. *Artificial Intelligence*, 10(3),281–316, 1978.

[Can87] D. Cantone. *A decision procedure for a class of unquantified formulae of set theory involving the powerset and singleton operators.* Ph.D thesis, New York University, 1987.

[Can88] D. Cantone. *Decision procedures for elementary sublanguages of set theory. X. Multilevel syllogistic extended by the singleton and powerset operators.* Technical Report 373, New York University, 1988.

[Car70] R. Carnap. *Meaning and necessity.* The University of Chicago Press, Chicago and London, 1970. Sixth impression.

[Car77] L. Carrol. *Symbolic logic.* The Harvester Press Limited, Hassocks, Sussex, 1977.

[CC88] D. Cantone and V. Cutello. *Some decidability results on quantified sublanguages of set theory.* Technical Report 378, New York University, 1988.

[CC89a]   D. Cantone and V. Cutello. *A decision procedure for set-theoretic formulae involving rank and cardinality comparison.* Technical Report 440, New York University, 1989.

[CC89b]   D. Cantone and V. Cutello. *Decision procedures for elementary sublanguages of set theory. XVI. Multilevel syllogistic extended by singleton, rank comparison and unary intersection.* Technical Report 439, New York University, 1989.

[CC89c]   D. Cantone and V. Cutello. Decision procedures for elementary sublanguages of set theory. XV. Multilevel syllogistic extended by the predicate *Finite* and the operators singleton and $pred_<$. *Journal of Automated Reasoning*, 1989.

[CCF88]   D. Cantone, V. Cutello, and A. Ferro. Decision procedures for elementary sublanguages of set theory. XIV. Three languages involving rank related constructs. In P. Gianni, editor, *International Symposium on Symbolic and Algebraic Computation*, Vol 358, Lecture Notes in Computer Science, Springer-Verlag, 1988.

[CCP89]   D. Cantone, V. Cutello, and A. Policriti *Set-theoretic reductions of Hilbert's tenth problem.* Technical Report 449, New York University, 1989.

[CF86]   D. Cantone and A. Ferro. Some recent decidability results in set theory. In *X incontro di logica matematica - La logica nell'informatica*, Siena, 1986.

[CFMS87]   D. Cantone, A. Ferro, B. Micale, and G. Sorace. Decision procedures for elementary sublanguages of set theory. IV. Formulae involving a rank operator or one occurrence of $\Sigma(x) = \{\{y\}|y \in x\}$. *Communications on Pure and Applied Mathematics*, 40,37–77, 1987.

[CFO88]   D. Cantone, A. Ferro, and E.G. Omodeo. Decision procedures for elementary sublanguages of set theory. VIII. A semidecision procedure for finite satisfiability of unquantified set-theoretic formulae. *Communications on Pure and Applied Mathematics*, 41,105–120, 1988.

[CFOP88]   D. Cantone, A. Ferro, E.G. Omodeo, and A. Policriti. *Sillogistica con operatore di scelta.* Research Report 34, University of Udine, 1988.

[CFOP89]   D. Cantone, A. Ferro, E. G. Omodeo, and A. Policriti. Scomposizione sillogistica disgiuntiva. In P. Mello, editor, *GULP - 89*,

Atti del quarto convegno nazionale sulla programmazione logica, p.199–209, 1989.

[CFOS87]   D. Cantone, A. Ferro, E.G. Omodeo, and J.T. Schwartz. Decision algorithms for some fragments of analysis and related areas. *Communications on Pure and Applied Mathematics,* 40,281–300, 1987.

[CFS85]   D. Cantone, A. Ferro, and J.T. Schwartz. Decision procedures for elementary sublanguages of set theory. VI. Multi-level syllogistic extended by the powerset operator. *Communications on Pure and Applied Mathematics,* 38,549–571, 1985.

[CFS87]   D. Cantone, A. Ferro, and J.T. Schwartz. Decision procedures for elementary sublanguages of set theory. V. Multilevel syllogistic extended by the general union operator. *Journal of Computer and System Sciences,* 34(1),1–18, 1987.

[CGO88]   D. Cantone, S. Ghelfo, and E.G. Omodeo. The automation of syllogistic. I. Syllogistic normal forms. *Journal of Symbolic Computation,* 6(1),83–98, 1988.

[CL73]   C. Chang and C. Lee. *Symbolic logic and mechanical theorem-proving.* Academic Press, 1973.

[CO88]   D. Cantone and E.G. Omodeo. *Topological syllogistic with continuous and closed functions.* Research Report 35, University of Udine, 1988.

[CO89a]   D. Cantone and E.G. Omodeo. On the decidability of formulae involving continuous and closed functions. In *Proceedings of the 11th International Joint Conference of Artificial Intelligence,* p.425–430, 1989.

[CO89b]   D. Cantone and E.G. Omodeo. Topological syllogistic with continuous and closed functions. *Communications on Pure and Applied Mathematics,* 1989.

[COP89]   D. Cantone, E.G. Omodeo, and A. Policriti. *The automation of syllogistic. II. Optimization and complexity issues. Journal of Automated Reasoning,* 1989.

[CS88]   D. Cantone and J.T. Schwartz. *Decision procedures for elementary sublanguages of set theory. XI. Multilevel syllogistic extended by some elementary map constructs.* Technical Report 374, New York University, 1988.

[DG79]   B. Dreben and W. D. Goldfarb. *The decision problem: solvable classes of quantificational formulas.* Addison-Wesley, 1979.

[DM79]   N. Dershowitz and Z. Manna. Proving termination with multiset ordering. *Communications of the ACM*, 22(8),465–476, 1979.

[DW83]   M. D. Davis and Elaine J. Weyuker. *Computability, complexity and languages. Computer Science and Applied Mathematics*, Academic Press, 1983.

[End72]   H. B. Enderton. *A mathematical introduction to logic.* Academic Press, New York and London, 1972.

[Fer81]   A. Ferro. *Decision procedures for some classes of unquantified set theoretic formulae.* Ph.D thesis, New York University, 1981.

[Fer85]   A. Ferro. A note on the decidability of MLS extended by the powerset operator. *Communications on Pure and Applied Mathematics*, 38,367–374, 1985.

[Fer88a]   A. Ferro. *Multilevel syllogistic extended with rank comparison, the Finite predicate, singleton and choice operators.* Unpublished manuscript, 1988.

[Fer88b]   A. Ferro. Decision procedures for elementary sublanguages of set theory. XII. Multilevel syllogistic extended with singleton and choice operators. Submitted to *Journal of Automated Reasoning*, 1988.

[FO78]   A. Ferro and E.G. Omodeo. An efficient validity test for formulae in extensional two-level syllogistic. *Le Matematiche*, 33,130–137, 1978.

[FO87]   A. Ferro and E.G. Omodeo. Decision procedures for elementary sublanguages of set theory. VII. Validity in set theory when a choice operator is present. *Communications on Pure and Applied Mathematics*, 40,265–280, 1987.

[FOS80a]   A. Ferro, E.G. Omodeo, and J.T. Schwartz. Decision procedures for elementary sublanguages of set theory. I. Multi-level syllogistic and some extensions. *Communications on Pure and Applied Mathematics*, 33,599–608, 1980.

[FOS80b]   A. Ferro, E.G. Omodeo, and J.T. Schwartz. Decision procedures for some fragments of Set Theory. In *Proceedings of the 5th conference on Automated Deduction*, Vol 87, Lecture Notes in Computer Science, p.88–96, Springer-Verlag, 1980.

[Gar68]   M. Gardner. *Logic machines, diagrams, and Boolean algebra.* Dover Publications, New York, 1968.

[GO85]   S. Ghelfo and E.G. Omodeo.   Towards practical implementations of syllogistic. In B.F. Caviness, editor, *EUROCAL '85, European Conference on Computer Algebra - Proceedings Vol.2*, Vol 204, Lecture notes in Computer Science, p.40–48, Springer-Verlag, 1985.

[Gog78]   D. Gogol. The $\forall_n \exists$-completeness of Zermelo-Fraenkel set theory. *Zeitschr. f. math. Logik und Grundlagen d. Math.*, 1978.

[Gog79]   D. Gogol. Sentences with three quantifiers are decidable in set theory. *Fundamenta Mathematicae*, 102, 1979.

[HA49]   D. Hilbert and W. Ackermann.   *Grundsuge der theoretischen Logik.* Springer, Berlin, 3rd edn, 1949.

[HA50]   D. Hilbert and W. Ackermann. *Mathematical logic.* Chelsea Publishing Co., 1950.

[Hal83]   L. Hällnas. *On normalization of proofs in Set theory.* Ph.D thesis, Filosofiska istitutionen, Stockholms universitet, 1983.

[Jec78]   T.J. Jech. *Set theory.* Academic Press, 1978.

[Kel55]   J.L. Kelley. *General topology.* Van Nostrand, New York, 1955.

[Knu75]   D. E. Knuth. *Fundamental algorithms.* Vol 1, Addison-Wesley, 2nd edn, 1975.

[Lei69]   A.C. Leisenring. *Mathematical Logic and Hilbert's ε-symbol.* MacDonald & Co., 1969.

[Lew79]   H. L. Lewis.   *Unsolvable classes of quantificational formulas.* Addison-Wesley, 1979.

[LJ52]   J. Łukasiewicz. *Aristotle's syllogistic from the standpoint of modern formal logic.* Clarendon Press, Oxford, 2nd edn, 1952.

[Lov78]   D. W. Loveland.   *Automated theorem proving: a logical basis.* North-Holland, Inc., New York, 1978.

[LS80a]   D.W. Loveland and R.E. Shostak. Simplifying interpreted formulas. In *Proceedings of the 5th conference on Automated Deduction*, Vol 87, Lecture Notes in Computer Science, p.97–109, Springer-Verlag, 1980.

[LS80b]   D.W. Loveland and R.E. Shostak. *Simplifying interpreted formulas*. Technical Report CSL-117, SRI, 1980. Project 8752.

[Man77]   Yu.I. Manin. *A course in mathematical logic*. Springer-Verlag, New York, 1977.

[McC63]   S. McCall. *Aristotle's modal syllogisims*. North-Holland Publishing Co., Amsterdam, 1963.

[Men64]   E. Mendelson. *Introduction to mathematical logic*. Van Nostrand - Reinhold, Princeton, New Jersey, 1964.

[Mor65]   A. Morse. *A Theory of sets*. Academic Press, New York, 1965.

[Mos69]   A. Mostowski. *Constructible sets with applications*. North Holland, 1969.

[NO78]   G. Nelson and D.C. Oppen. A simplifier based on efficient decision algorithms. In *Conference Record of the Fifth Annual ACM Symposium on Principles of Programming Languages*, p.141–150, 1978.

[NO79]   G.D. Nelson and D. Oppen. Simplification by cooperating decision procedures. *ACM Transactions on Programming Languages and Systems*, 1(2), 1979.

[NO80]   G. Nelson and D.C. Oppen. Fast decision procedures based on congruence closure. *J.ACM*, 27(2),356–354, 1980.

[Omo82]   E.G. Omodeo. Decidibilità e validità, in presenza di operatori di scelta. 1982. *Atti degli incontri di Logica Matematica, Siena*.

[Omo84]   E.G. Omodeo. *Decidability and Proof Procedures for Set Theory with a Choice Operator*. Ph.D thesis, New York University, 1984.

[Omo86]   E.G. Omodeo. L'automazione della sillogistica. *Le Scienze*, 37(218),120–128, 1986.

[Omo88]   E. Omodeo. Hints for the design of a set calculus oriented to automated deduction. In M. Boscarol, L. Carlucci Aiello, G. Levi, eds, *Foundations of Logic and Functional Programming*, Vol 306, Lecture Notes in Computer Science, Springer-Verlag, 1988.

[ON77]   D.C. Oppen and G. Nelson. Fast decision algorithms based on union and find. In *Proceedings of 18th IEEE Symposium on Foundations of Computer Science, Princeton, N.J.*, p.114–119, 1977.

[Opp79]   D.C. Oppen. Complexity of combinations of quantifier-free theories. In *4th Conference on Automated Deduction, Austin, Texas*, 1979.

[Pai86]   R. Paige. Programming with invariants. *IEEE Software*, 3(1),56–69, 1986.

[Pas78]   D. Pastre. Automatic theorem proving in set theory. *Artificial Intelligence*, 10, 1978.

[Pel86]   F.J. Pelletier. Seventy-five problems for testing automatic theorem provers. *Journal of Automated Reasoning*, 2,191–216, 1986.

[Per88a] D. Perlis. Commonsense Set Theory. In P. Maes and D. Nardi, eds, *Meta-Level Architectures and Reflection*, p.87–98, Elsevier Science Publishers - North Holland, 1988.

[Per88b] D. Perlis. Meta in Logic. In P. Maes and D. Nardi, eds, *Meta-Level Architectures and Reflection*, p.37–49, Elsevier - North Holland, 1988.

[PH85]   R. Paige and F. Henglein. *Mechanical translation of set-theoretic problem specifications into efficient RAM code - a case study*. In B.F. Caviness, editor, *EUROCAL '85, European Conference on Computer Algebra - Proceedings Vol.2*, Vol 204, Lecture notes in Computer Science, p.554–567, Springer-Verlag, 1985.

[Pol87]   A. Policriti. Completeness and decidability of the deducibility problem of some classes of formulas of set theory. *Le Matematiche*, 42(1), 1987.

[PP88a]   F. Parlamento and A. Policriti. Decision procedures for elementary sublanguages of set theory. IX. Unsolvability of the decision problem for a restricted subclass of the Δ_0-formulas in set theory. *Communications on Pure and Applied Mathematics*, 41,221–251, 1988.

[PP88b]   F. Parlamento and A. Policriti. *Decision procedures for elementary sublanguages of set theory. XIII. Model graphs, reflection and decidability*. Research Report 2, University of Udine, 1988.

[PP88c]   F. Parlamento and A. Policriti. The logically simplest form of the infinity axiom. *Proceedings of the AMS*, 103(1),274–276, 1988.

[PP89]   F. Parlamento and A. Policriti. Note on "The logically simplest form of the infinity axiom". *Proceedings of the AMS*, 1989.

[Pre29]   M. Presburger. Über die Vollständigkeit eines gewissen Systems
            der Arithmetic ganzer Zahlen, in welchem die addition als einzige
            Operation hervortritt. In *Comptes-rendus du Premier Congrès
            des Mathematiciens des Pays Slaves*, p.192–201,395, Warsaw,
            1929.

[Qui61]   W. V. Quine. *Methods of logic*. Holt, Rinehart and Winston, New
            York, 1961.

[Sch78]   J.T. Schwartz. *Instantiation and decision procedures for certain
            classes of quantified set-theoretic formulae*. Technical Report 78–
            10, Institute for Computer Applications in Science and Engineer-
            ing, NASA Langley Research Center, Hampton, Virginia, 1978.

[SDDS86]   J.T. Schwartz, R.K.B. Dewar, E. Dubinsky, and E. Schonberg.
            Programming with sets. Texts and Monographs in Computer Sci-
            ence, Springer-Verlag, 1986.

[She56]   J.C. Shepherdson. On the interpretation of Aristotelian syllogis-
            tic. *The Journal of Symbolic Logic*, 21,137–147, 1956.

[Sho78]   R.E. Shostak.   An algorithm for reasoning about equality.
            C.ACM, 21(7),583–585, 1978.

[Sho82]   R.E. Shostak. Deciding combination of theories. In *6th Confer-
            ence on Automated Deduction*, New York, 1982.

[Sik64]   R. Sikorski. *Boolean algebras*. Springer-Verlag, 2nd edn, 1964.

[Ven34]   J. Venn. *Symbolic logic*. Chelsea, 2nd edn, 1934.

[Vil71]   F. Ville. Décidabilité des formules existentielles en théorie des
            ensembles. *C. R. Acad. Sc.*, t.272, Serie A,513–516, 1971.

[vN77a]   J. von Neumann. An axiomatization of set theory. In J. van Hei-
            jenoort, editor, *From Frege to Gödel*, p.393–413, Harvard Univer-
            sity Press, Cambridge, Massachusetts, 3rd printing, 1977.

[vN77b]   J. von Neumann. On the introduction of transfinite numbers. In
            J. van Heijenoort, editor, *From Frege to Gödel*, p.346–354, Har-
            vard University Press, Cambridge, Massachusetts, 3rd printing,
            1977.

[Wey77]   R.W. Weyhrauch.   *FOL: A proof checker for first-order logic*.
            Technical Report Memo AIM-235.1, Stanford Artificial Intelli-
            gence Laboratory, 1977.

[WOLB84] L. Wos, R. Overbeek, E. Lusk, and J. Boyle. *Automated reasoning. Introduction and applications.* Prentice Hall, Englewood Cliff, New Jersey, 1984.

[Wos88]   L. Wos. *Automated reasoning. 33 basic research problems.* Prentice Hall, Englewood Cliff, New Jersey, 1988.

INDEX OF REFERENCED STATEMENTS

Algorithm 3.1 p.97

Algorithm 5.1 p.145

Algorithm 5.2 p.152

Algorithm 7.1 p.198

Algorithm 7.2 p.202

Algorithm 7.3 p.205

Algorithm 7.4 p.209

Algorithm 11.1 p.320

Association Algorithm p.220

Claim 6.1 p.186

Condition C1 p.261, 290

Condition C1′ p.301

Condition C2 p.261, 291

Condition C2′ p.305

Condition C3 p.262, 293

Condition C3′ p.306

Condition C4 p.262, 293

Condition C4′ p.306

Condition C5 p.263

Condition C6 p.264

Condition C7 p.264

Condition C8 p.264

Condition C9 p.264

Condition C10 p.265

Condition C11 p.265

Construction Process p.267

Corollary 2.1 p.50

Corollary 2.4 p.61

Corollary 2.6 p.73

Corollary 2.7 p.73

Corollary 3.1 p.101

Corollary 4.2 p.106

Corollary 4.3 p.107

Corollary 4.4 p.108

Corollary 4.5 p.109

Corollary 4.7 p.116

Corollary 4.8 p.118

Corollary 4.9 p.119

Corollary 4.10 p.121

Corollary 4.11 p.121

Corollary 4.12 p.128

Corollary 5.1 p.136

Corollary 5.2 p.137

Corollary 5.3 p.144

Corollary 5.4 p.150

Corollary 5.6 p.159

Corollary 5.7 p.159

Corollary 6.1 p.173

Corollary 6.2 p.173

Corollary 6.4 p.179

Corollary 6.5 p.181

Corollary 6.6 p.189

Corollary 6.7 p.189

Corollary 7.2 p.203

Corollary 8.1 p.216

Corollary 8.3 p.235

Corollary 9.1 p.282

Corollary 11.2 p.323

Corollary 11.4 p.328

Corollary 11.5 p.328

Definition 1.3 p.27

Definition 1.5 p.29

Definition 1.6 p.29

Definition 1.7 p.31

Definition 2.4 p.62

Definition 2.5 p.67

Definition 2.6 p.70

Definition 3.1 p.78

Definition 3.2 p.79

Definition 3.3 p.81

Definition 3.9 p.96

Definition 3.10 p.100

Definition 4.1 p.102

Definition 4.2 p.106

Definition 4.5 p.109

Definition 4.9 p.124

Definition 4.10 p.126

Definition 5.1 p.141

Definition 6.1 p.166

Definition 6.2 p.175

Definition 6.3 p.176

Definition 6.4 p.178

Definition 8.1 p.217

Definition 8.2 p.217

Definition 8.4 p.223

Definition 8.6 p.225

Definition 9.1 p.258

Definition 9.2 p.258

Definition 9.3 p.259

Definition 9.4 p.261

Definition 9.5 p.263

Definition 9.6 p.264

Definition 9.7 p.266

Definition 10.1 p.287

Definition 10.4 p.293

Definition 10.6 p.302

Definition 10.8 p.307

Definition 10.9 p.307

Definition 11.1 p.319

Instantiated Variant of the Association Algorithm p.227

Example **2.2** p.64

Example **2.4** p.66

Example **2.6** p.66

Example **2.7** p.67

König's lemma p.107

Lemma **1.1** p.10

Lemma **1.5** pp.27, 28

Lemma **1.7** p.31

Lemma **2.1** p.45

Lemma **2.2** p.49

Lemma **2.4** p.53

Lemma **2.8** p.66

Lemma **2.9** p.68

Lemma **2.10** p.69

Lemma **2.11** p.71

Lemma **3.3** p.96

Lemma **3.4** p.99

Lemma **4.1** p.104

Lemma **4.2** p.105

Lemma **4.3** p.107

Lemma **4.4** p.110

Lemma **4.5** p.110

Lemma **4.6** p.111

Lemma **4.7** p.113

Lemma **4.8** p.114

Lemma **4.9** p.115

Lemma **4.10** p.117

Lemma **4.11** p.124

Lemma **4.12** p.127

Lemma **4.13** p.127

Lemma **5.1** p.133

Lemma **5.3** p.143

Lemma **5.7** p.150

Lemma **5.8** p.151

Lemma **5.9** p.151

Lemma **6.1** p.167

Lemma **6.2** p.167

Lemma **6.3** p.168

Lemma **6.4** p.171

Lemma **6.5** p.175

Lemma **6.7** p.176

Lemma **6.10** p.180

Lemma **6.11** p.188

Lemma **7.1** p.199

Lemma **7.2** p.199

Lemma **7.3** p.203

Lemma **7.5** p.206

Lemma **7.6** p.206

Lemma **8.1** p.216

Lemma **8.2** p.217

Lemma 8.3 p.218

Lemma 8.4 p.221

Lemma 8.5 p.222

Lemma 8.6 p.223

Lemma 8.7 p.224

Lemma 8.8 p.225

Lemma 8.10 p.226

Lemma 8.11 p.232

Lemma 8.12 p.233, 234

Lemma 8.13 p.233, 247

Lemma 8.14 p.233

Lemma 8.15 p.237

Lemma 8.17 p.249

Lemma 9.1 p.258

Lemma 9.2 p.260

Lemma 9.3 p.263

Lemma 9.4 p.265

Lemma 9.5 p.274

Lemma 9.8 p.278

Lemma 9.9 p.279

Lemma 9.10 p.279

Lemma 9.11 p.281

Lemma 9.12 p.281

Lemma 9.13 p.282

Lemma 9.14 p.284

Lemma 10.1 p.288

Lemma 10.2 p.293

Lemma 10.3 p.294

Lemma 10.4 p.299

Lemma 10.5 p.300

Lemma 10.6 p.301

Lemma 10.7 p.302

Lemma 10.13 p.303

Lemma 10.16 p.307

Lemma 10.17 p.309

Lemma 10.19 p.315

Lemma 11.3 p.323

Lemma 11.4 p.324

Lemma 11.5 p.324

Lemma 11.6 p.325

Lemma 11.7 p.325

Lemma 11.8 p.330

Main Inductive Lemma p.233, 234

Normalization Process p.97

Transfinite Induction Principle p.43, 44

Transfinite Recursion Principle p.65

Problem 6.1 p.183

Procedure 4.1 p.122

Procedure 4.2 p.130

Remarks 3.2 p.83

Remark **3.3** p.99

Remark **4.1** p.108

Remarks **5.1** p.149

Remark **9.1** p.260

Remark **10.4** p.288

Remark **10.5** p.291

Remark **10.6** p.292

Termination Lemma p.233, 247

Theorem **1.1** p.11

Theorem **2.1** p.44

Theorem **2.3** p.57

Theorem **3.1** p.83

Theorem **6.2** p.169

Theorem **6.3** p.171

Theorem **10.1** p.315

Theorem **11.1** p.327

Theorem **11.2** p.331